Current Perspectives in Primate Biology

CURRENT PERSPECTIVES IN PRIMATE BIOLOGY

Edited by

David M. Taub

Department of Psychiatry and Behavioral Science
Medical University of South Carolina
Charleston, South Carolina
and
Yemassee Primate Center
Yemassee, South Carolina

and

Frederick A. King

Director, Yerkes Regional Primate Research Center
Emory University
Atlanta, Georgia

VNR VAN NOSTRAND REINHOLD COMPANY
New York

Library of Congress Catalog Card Number: 85-22778
ISBN: 0-442-28311-3

Manufactured in the United States of America.

Published by Van Nostrand Reinhold Company Inc.
115 Fifth Avenue
New York, New York 10003

Van Nostrand Reinhold Company Limited
Molly Millars Lane
Wokingham, Berkshire RG11 2PY, England

Van Nostrand Reinhold
480 Latrobe Street
Melbourne, Victoria 3000, Australia

Macmillan of Canada
Division of Gage Publishing Limited
164 Commander Boulevard
Agincourt, Ontario MIS 3C7, Canada

15 14 13 12 11 10 9 8 7 6 5 4 3 2 1

Library of Congress Cataloging in Publication Data
Main entry under title:
Current perspectives in primate biology.
Papers from the IXth Congress of the International Primatological Society, held Aug. 8-13, 1982, in Atlanta, Ga.
Includes bibliographies and index.
1. Primates—Congresses. 2. Physical anthropology—Congresses. I. Taub, David Milton. II. King, Frederick A. III. International Primatological Society. Congress (9th : 1982 : Atlanta, Ga.)
QL737.P9C86 1986 599.8 85-22778
ISBN 0-442-28311-3

CONTENTS

PART II: Physical Anthropology

*This article has been presented in collaboration with B. Senut at the 1982 Physical Anthropology and Anatomy meeting in Atlanta, Georgia.

PART III: Biomedicine

PREFACE

The IXth Congress of the International Primatological Society was held August 8 through 13, 1982, in Atlanta, Georgia. Hosted by the Yerkes Regional Primate Research Center of Emory University, it was the first I.P.S. Congress held in the United States in ten years. In terms of attendance, scientific presentations, countries represented, and scope of primatological themes covered, it was the largest meeting ever held devoted exclusively to the exchange of scientific information on the Order Primata. During the five-day meeting, there were 270 paper presentations in 32 scientific paper sessions, 229 presentations in 33 symposia and workshops, 9 videotapes and films, 24 poster presentations, and 2 keynote addresses for a total of 534 individual scientific presentations. Over 720 persons registered and attended the congress.

The international flavor of the congress was reflected in the fact that in 32% of all presentations, the senior author was from a country other than the United States. Scientists, students, and guests came from Argentina, Australia, Belgium, Brazil, Canada, The People's Republic of China, Colombia, Denmark, France, Federal Republic of Gabon, Germany, Grenada, India, Indonesia, Israel, Italy, Japan, Kenya, Malaysia, Mexico, The Netherlands, Pakistan, Puerto Rico, Rwanda, Singapore, Republic of South Africa, Spain, Sri Lanka, St. Kitts, Switzerland, the United Kingdom, and the United States of America.

While some authors of papers presented at the congress have elected to publish their contributions elsewhere, we believe that a collection of papers selected for quality and interest should appropriately emerge from the meeting. Therefore, congress participants were invited to submit manuscripts of papers that they had presented. Each submitted manuscript was reviewed by scientists of the Editorial

Review Board. Several other scientists who also served as reviewers are listed as Invited Editorial Consultants. We wish to express our heartfelt appreciation to these colleagues who consented to undertake the arduous task of reviewing the manuscripts. Sixty-four papers were finally accepted for inclusion in two books.

The papers have been grouped in two topical divisions. *Current Perspectives in Primate Social Dynamics* is concerned with primates as social animals. *Current Perspectives in Primate Biology* is concerned with primates as biological entities. Within each book, papers have been arranged according to subject matter. *Current Perspectives in Primate Social Dynamics* contains three parts: social organization and behavioral development, ecology, and communication. In *Current Perspectives in Primate Biology*, three topical parts cover reproduction, physical anthropology, and biomedical research.

Finally, we would like to acknowledge the efforts of individuals and organizations who were essential to the completion of these two books. The Yerkes Center made available personnel, resources, and facilities. Cathy Yarbrough has been indispensible in coordinating the many activities involved in developing these volumes and we wish to express our warm gratitude to her for a job well done. We thank Michele Newman-Higgins for her tireless, thorough, and professional job of entering the manuscripts into the word processor and proofreading each paper. The editorial and production staff at Van Nostrand Reinhold, especially Eric Rosen, Susan Munger, and Mimi Mikels, have provided constant support and expert guidance through all phases of this production, and we gratefully acknowledge their assistance.

DAVID M. TAUB
FREDERICK A. KING

CONTRIBUTORS

E. Alberts
H. A. Grove Animal Research Center
Private Bag X169
Pretoria 0001, South Africa

J. E. Beeber
Department of Anatomy
Department of Anthropology
Wayne State University School of Medicine
540 E. Canfield Avenue
Detroit, Michigan 48201

K. Ben
Kunming Institute of Zoology
Academia Sinica
Kunming, Yunnan
China

Ch. Berge
Laboratoire d'Anatomie Comparee
Museum National d'Histoire Naturelle
55 Rue Buffon
75005 Paris, France

B. Burnett
Primate Research Institute
New Mexico State University
Holloman AFB, New Mexico 88330

R. Cambre
Denver Zoological Gardens
City Park
Denver, Colorado 80205

P. J. Chan
Endocrine Research Unit
Michigan State University
East Lansing, Michigan 48824

O. J. Colillas
Centro Argentino de Primates (CAPRIM), CONICET
Serrano 665
1414 Buenos Aires, Argentina

J. Czelusniak
Department of Anatomy
Department of Biological Sciences
Wayne State University School of Medicine
540 E. Canfield Avenue
Detroit, Michigan 48201

W. Dai
Kunming Institute of Zoology
Academia Sinica
Kunming, Yunnan
China

G. Darai
Universitat Heidelberg
Institut fur Medizinische Virologie
Im Neuenheimer Feld 324
D-6900 Heidelberg, West Germany

F. J. DeMayo
Endocrine Research Unit
Michigan State University
East Lansing, Michigan 48824

W. R. Dukelow
Endocrine Research Unit
Michigan State University
East Lansing, Michigan 48824

J. G. Else
Institute of Primate Research
National Museums of Kenya
Nairobi, Kenya

H. C. Engele
Department of Psychology
University of Tubingen
Aussenstelle Weissenau
Rasthalde 3
7980 Ravensburg, West Germany

G. J. Fox
Department of Sociology, Anthropology, and Economics
East Carolina University
Greenville, North Carolina 27834

J. W. Froehlich
Department of Anthropology
University of New Mexico
Albuquerque, New Mexico 87131

P. H. Froehlich
Department of Anthropology
University of New Mexico
Albuquerque, New Mexico 87131

G. B. Fuller
Primate Research Institute
New Mexico State University
Holloman AFB, New Mexico 88330

T. Furusaka
Department of Psychology
Faculty of General Education
Tokyo University of Agriculture and Technology
Fuchu 183, Japan

G. P. Goo
Hazelton Research Animals, Inc.
Texas Primate Center
P.O. Box 549
Alice, Texas 78333-0549

M. Goodman
Department of Anthropology
Department of Anatomy
Department of Biological Sciences
Wayne State University School of Medicine
540 E. Canfield Avenue
Detroit, Michigan 48201

D. J. Goosen
H. A. Grove Animal Research Center
Private Bag X169
Pretoria 0001, South Africa

M. Gunther
Fachbereich Biologie
Freie Universitat Berlin
Berlin, East Germany

W. C. Hobson
Primate Research Institute
New Mexico State University
Holloman AFB, New Mexico 88330

H. O. Hofer
Department of Zoology and Comparative Anatomy
University of Kassel
D-3500 Kassel, West Germany
and
Professor Emeritus
Tulane University
New Orleans, Lousiana

R. J. Hutz
Endocrine Research Unit
Michigan State University
East Lansing, Michigan 48824

F. K. Jouffroy
Laboratoire d'Anatomie Comparee
Museum National d'Histoire Naturelle
55 Rue Buffon
75005 Paris, France

F. A. King
Yerkes Regional Primate Research Center
Emory University
Atlanta, Georgia 30322

T. Kito
Department of Psychology
Faculty of Letters
Kyushu University
Fukuoka 812, Japan

T. Kubo
Department of Otolaryngology
Osaka University
Yamadaoka 1-2, Suita
Osaka 565, Japan

P. P. LeFloch-Prigent
Laboratoire d'Anatomie
Faculte de Medecine Paris-Ouest
U. E. R. Biomedicale des Saints-Peres
75270, Paris Cedex 06, France

S. Matano
Department of Biological Anthropology
Osaka University
Yamadaoka 1-2, Suita
Osaka 565, Japan

S. Matsumoto
Primate Research Institute
Kyoto University
Kanrin, Inuyama
Aichi 484, Japan

T. Matsunaga
Department of Otolaryngology
Osaka University
Yamadaoka 1-2, Suita
Osaka 565, Japan

L. Miller
Department of Anthropology
University of Colorado
1100 14th Street
Denver, Colorado 80202

P. Moller
Universitat Heidelberg
Pathologisches Institut
Im Neuenheimer Feld
D-6900 Heidelberg, West Germany

M. Natori
Primate Research Institute
Kyoto University
Kanrin, Inuyama
Aichi 484, Japan

C. Niemitz
Fachbereich Biologie

Freie Universitat Berlin
Berlin, East Germany

M. W. Odendaal
H. A. Grove Animal Research Center
Private Bag X169
Pretoria 0001, South Africa

T. Oyama
Department of Psychology
Faculty of Letters
University of Tokyo
Hongo
Tokyo 113, Japan

B. Preilowski
Department of Psychology
Aussenstelle Weissenau
University of Tubingen
Rasthalde 3
7980 Ravensburg, West Germany

D. Quiatt
Department of Anthropology
University of Colorado
1100 14th Street
Denver, Colorado 80202

M. Reger
Department of Psychology
Aussenstelle Weissenau
University of Tubingen
Rasthalde 3
7980 Ravensburg, West Germany

E. L. Segura
Instituto Nacional de Diagnostico e Investigacion de la Enfermedad de Chagas (INDIECH)
Paseo Colon 568
1063 Buenos Aires, Argentina

A. Seidel
Kernforschungszentrum Karlsruhe
Institut fur Genetik und fur Toxikologie von Spaltstoffen
Postfach 3640
D-7500 Karlsruhe 1, West Germany

B. Senut
Laboratoire d'Anthropologie du Museum National d'Histoire Naturelle
Musee de l'Homme
Place du Trocadero
75116 Paris, France

T. Setoguchi
Primate Research Institute
Kyoto University
Inuyama City
Aichi 484, Japan

D. A. Shafer
Genetics Research Laboratory
Georgia Mental Health Institute and Department of Psychiatry
Emory University School of Medicine
Atlanta, Georgia 30306

D. G. Smith
Department of Anthropology and the California Primate Research Center
University of California
Davis, California 95616

B. Song
Kunming Institute of Zoology
Academia Sinica
Kunming, Yunnan
China

W. Sontag
Kernforschungszentrum Karlsruhe
Institut fur Genetik und fur Toxikologie von Spaltstoffen
Postfach 3640
D-7500 Karlsruhe 1, West Germany

G. G. Stott
Department of Veterinary Anatomy
College of Veterinary Medicine
Texas A & M University
College Station, Texas 77843

C. Tardieu
C. N. R. S. Research Group "Locomotion Animale"
Laboratoire d'Anatomie Comparee
Museum National d'Histoire Naturelle
55 Rue de Buffon
75005 Paris, France

D. M. Taylor
Kernforschungszentrum Karlsruhe
Institut fur Genetik und fur Toxikologie von Spaltstoffen
Postfach 3640
D-7500 Karlsruhe 1, West Germany

W. G. Thiess
Kernforschungszentrum Karlsruhe
Institut fur Genetik und fur Toxikologie von Spaltstoffen
Postfach 3640
D-7500 Karlsruhe 1, West Germany

B. L. Travi
Centro Argentino de Primates (CAPRIM), CONICET
Casilla de Correo 145
3400 Corrientes, Argentina

H. Van Der Laarse
H. A. Grove Animal Research Center
Private Bag X169
Pretoria 0001, South Africa

R. Zhou
Kunming Institute of Zoology
Academia Sinica
Kunming, Yunnan
China

T. E. Ziegler
Department of Psychology
University of Wisconsin
W. J. Brogden Psychology Building
1202 West Johnson Street
Madison, Wisconsin 53706

EDITORIAL REVIEW BOARD

INVITED EDITORIAL CONSULTANTS

Dr. William J. Hamilton, III
Department of Environmental Studies
University of California
Davis, CA 95616

Dr. Donald Lindburg
San Diego Zoological Society
San Diego, CA 92112

Dr. Terry Maple
Department of Psychology
Georgia Institute of Technology
Atlanta, GA 30332

Dr. Henry McHenry
Department of Anthropology
University of California
Davis, CA 95616

Dr. Frank Mendel
Department of Anatomy
State University of New York
Buffalo, NY 14214

Dr. Donald Meyer
Laboratory of Comparative Physiology/Psychology
Ohio State University
Columbus, OH 43212

Dr. Paul I. Musey
Veterans Administration Medical Center Research Services
Decatur, GA 30033

Dr. Jane Phillips-Conroy
Division of Biology and Medicine
Brown University
Providence, RI 02912

Dr. Duane Rumbaugh
Department of Psychology
Georgia State University
Atlanta, GA 30303

Dr. David G. Smith
Anthropology Department
University of California
Davis, CA 95616

Dr. William Stone
Department of Genetics
Trinity University
San Antonio, TX 78284

Dr. Richard W. Thorington, Jr.
Smithsonian Institution
Washington, D.C. 20560

Dr. Laura Greer Vick
Duke Primate Center
Duke University
Durham, NC 27706

Dr. Patricia L. Whitten
Department of Obstetrics and Gynecology
Yale University
New Haven, CT 06510

Welcoming Address: Primatology Today

Frederick A. King

I am delighted to welcome you to the Congress on behalf of the Yerkes Regional Primate Research Center, host for this ninth meeting of the International Primatological Society. For the first time, the IPS is joined by the American Society of Primatologists and the International Society of Human Ethology in a joint venture that brings together more primatologists than have ever before been associated in a single meeting. We had hoped for at least 500 participants, and I am told that as of this morning we have now registered over 700.

Behavioral, biological and biomedical research, evolution and ecology, and conservation are all well represented at this meeting. Part of the work to be reported is fundamental and essential to our understanding of the very nature of primate species, and a portion is focused upon the practical matters of animal and human disease, breeding, and concern for threatened and endangered populations in the wild.

The relevance of much of the work for human affairs and some of the world's urgent problems is clear. As for example in: (1) the relationship between primate social studies and human aggression and violence that we see about us daily in our streets and across the world; (2) the relationship of research in reproductive biology to the ever-growing problem of the earth's expanding human population . . . a crisis which underlies many of man's current and future woes; (3) studies of primate ecology and conservation which are critical to our understanding of the immense problem of the accelerating depletion of our natural resources and environments; (4) research on the physical health of primates in which we are constantly discovering new knowledge with significance for human health and disease.

There are many credits to be given for whatever success this meeting will enjoy. Our appreciation particularly goes to: (1) the Animal Resources Branch, Division of Research Resources, National Institutes of Health, for financial support that filled an essential need; (2) the corporate contributors, advertisers and exhibitors who helped financially and added an important dimension to the Congress; (3) the Program Committee and the many participants of all three societies for their guidance, advice and time contributed; and (4) Cathy Yarbrough of the Yerkes Center, who organized and conducted the logistics of the meeting and has worked extraordinarily hard for many months to insure its success.

The formation of the International Primatological Society can be traced to a meeting held in Giesen, Germany, 20 years ago. Approximately 100 people interested in primatology attended, and among them was Professor Helmut Hofer, who will give our opening address this morning.

Born in Austria, Professor Hofer attended the University of Vienna where he shifted from his early interest in medicine to zoology and paleontology. There he was influenced by Professors J. Versluys, W. Marinelli, F. Hochstetter and O. L. Abel. He was in Vienna when Konrad Lorenz began his career, and he witnessed the early experiments on jackdaws by Lorenz. Dr. Hofer has wide interests in vertebrate ethology, morphology and paleobiology. His concern with brain structure brought him to the Max Planck Institutes in Giesen and Frankfurt. His name will forever be associated with the circumventricular organs of the brain, a term he coined in referring to an array of structures sharing in common their size, position, and degree of vascularization.

At the Max Planck, Professor Hofer found the time and resources to develop his ideas on primatology. Together with anthropologist A. H. Schultz of Switzerland, and anatomist, D. Starck of Frankfurt, he founded the _Folia Primatologica_, the _Bibliotheca Primatologica_ and the standard reference, the _Handbook on Primates_.

Dr. Hofer, a founding father of the International Primatological Society, is presently at the University of Kassel, West Germany, and is Professor Emeritus at Tulane University in this country.

And now, as I wish you an intellectually stimulating and satisfying meeting, it is my great pleasure to introduce our invited speaker, Professor H. O. Hofer.

Invited Address: The Origins of Modern European Primatology and the IPS

Helmut O. Hofer

The central problem of primatology is the interrelationship between the nonhuman primates and man. The similarities are sometimes striking, which is a source of surprise to all who have dealt with monkeys or apes. Despite some occasionally grotesque anatomical peculiarities, one is immediately impressed by the external and internal anatomical resemblances among the primates. But this is only one point; the nonhuman primates exhibit behavioral patterns which we can relate to in an instant, even if this behavior may be strange. I recall a delightful tale about a mother who was showing her little boy the monkeys and apes in a zoo. When the child saw the primates playing around, chasing each other and grimacing at the people he burst out in laughter. Suddenly he stopped laughing and began meditating, evidently facing a problem his young brain could not solve. Then he asked: "Mom, are those animals?" Broadly speaking, primatology is beginning to deal with this question; with reference to man, it is the ultimate question we have to answer.

The roots of modern European primatology can be traced back to the late 17th century, when the Englishman Edward Tyson (1650-1708) dissected the corpse of a chimpanzee, probably the first of this species to come to England. I now quote Vernon Reynolds (1967, p. 47):

> "Tyson compared this specimen, point by point, at each stage of the anatomical dissection, with monkeys on the one hand, and with man on the other. In this way he drew up two lists. The first consisted of thirty-four characters in which his 'Pygmie' resembled a monkey more closely than it resembled man."

With this statement Tyson foreshadowed T. H. Huxley's affirmation that the pongids share with man more anatomical characters than they share with monkeys. Apart from his own experience, Huxley very likely referred to Tyson's statement. However, Huxley moved one step farther in 1863 by cautiously expressing the idea that these resemblances can also be found in behavior. Recall the question of the little boy in the zoo: "Mom, are those animals?" Tyson, Huxley, and the little boy identified essentially the same problem: What are the similarities between man and pongids and, by extension, how do we explain them? I do not hesitate to call the perplexing resemblances between man and primates, especially the Great African apes, one of the most challenging facts biology has brought to light. This is no less true today than it was in Tyson's time. Unfortunately, the problem of how to explain the astonishing similarity between the nonhuman primates and man has been the source of considerable controversy.

In 1699, when Tyson published the results of his studies, he stated literally: "In this chain of the creation, as an intermediate link between an Ape (that is, monkey) and a Man, I would place our Pygmie"; (from Reynolds , l.c.p. 47). Reynolds is correct in stating that Tyson's classification was not an evolutionary one, but a "gradational" classification. The concept of an "intermediate link" in a "chain of the creation" does not mean any natural relation in a phylogenetic sense. It refers to a relation of the forms in the creative ideas or thoughts of God.

However, Tyson can be considered the founder of scientific primatology since he not only demonstrated the problem (the similarity with man), but he also offered an explanation for it (the Creation). In retrospect, Tyson's explanation seems scientifically naive. However, a science is not the sum of certified knowledge; facts must be connected by concepts and theories so that ultimately a comprehensive understanding is achieved. The explanation for a problem must not necessarily be the correct explanation. Very often the first explanation is incorrect; it is more or less an attempt to understand, which has to be amended as new discoveries come to light. Nevertheless, every explanation, right or wrong, is an endeavor to achieve an understanding, and this attempt is an essential aspect of science. In this sense, Tyson founded primatology as a science.

Nothing of importance happened in the following years in the field of primatology until 1859, when Darwin published his revolutionary theory of evolution. Four years later, in 1863, Thomas Henry Huxley in England and Ernst Haeckel in Germany used Darwin's concept to explain the origin of man. By this they determined "Man's place in Nature." This was actually the year in which post-Darwinian primatology was begun; Darwin himself did not follow until 1871. Darwin's theory rapidly assumed the position of the only logically acceptable explanation for the origin of species, their variety and diversification, their appearance in geological space and time, and most importantly, their historical characterization. This includes man, since there is not the slightest reason to exclude man from the general causative principles of the origin of new forms.

The main purpose of primatology in the time after Darwin was to demonstrate the origin and ascendance of man by collecting and comparatively evaluating the fossil record. In those decades, fundamental knowledge was obtained on almost all systems of organs of different groups of primates. This was carried out with the intention of better understanding the burgeoning fossil material, and of tentatively reconstructing the phylogenetic process of hominization from a comparative perspective. One specific complex of problems which still commands particular interest may be mentioned in the context of the problem of hominization: What are the anatomical alterations which our ancestors underwent when they acquired exclusive bipedal locomotion? I mention this problem because it is still controversial, and because it demonstrates the complexities we face when we try to analyze one of the basic problems of primatology. The relations between causes and effects remain obscure. Therefore, it is better to use the neutral term "anatomical relations" to describe a condition; thus we avoid the terms "cause" and "effect," which indicate that the former is prior in phylogenetic time to the latter. Actually, all of the following anatomical modifications involved in bipedalism are mutually related: The prebasic kyphosis of the skull which brings the face into the direction of upright locomotion, the kyphoses and lordoses of the backbone, the shape of the vertebrae and the intervertebral discs, the orientation of the posterior extremities and the morphology of the joints, the anatomy of the foot including the shape of the terminal phalanx of the big toe!

In the decades between the turn of the century and the outbreak of World War II, the first monographic descriptions of the phylogeny of primates and of the self-interpretation of man appeared. However, there was an important problem which we could not overcome until a few years ago. The problem was as follows: At first fossils were unearthed which represented specialized endforms of the phylogenetic radiation, or at least they occurred very late in geological times. These were <u>Homo neanderthalensis</u> and its relatives, and <u>Homo (Pithecanthropus) erectus</u> and the relatives of this species.

It is rather simple to outline the diagram of an adaptive phylogenetic radiation if representative material from different geological times is available. We have only to arrange the fossils according to their anatomical similarities along a time continuum, and we can "read" -- sit venia verbo -- the course of evolution. However, the worst that can happen to a paleontologist has happened to the paleoanthropologists: Because the endforms were found first, it was necessary to think counter-current to the course of human evolution. This explains many erroneous concepts and conclusions in paleoanthropology, which are unavoidable under the circumstances even to highly distinguished scientists.

Dart described in 1925 the first _Australopithecus africanus_ from South Africa, and Broom added many more fossil records from the same area after World War II. Louis S. B. Leakey, who was in my opinion one of the most ingenious primatologists, found _Australopithecus_ in East Africa. It eventually became clear that there was a long-lasting, widely distributed and varying population of bipedal primates which shared pongid and hominid characters; the latter were probably in the majority. The place of the Australopithecines in the hominine radiation was controversially discussed. It was von Koenigswald, to my best knowledge, who stated that according to our experience with other phylogenetic radiations we must assume that the basic evolutionary radiation must have taken place in pre-pleistocene times. In the last decade or so, hominine fossils were unearthed, of course by the Leakey family, demonstrating that the evolution of man took at least the same time as the evolution of other primate species. In my opinion it took even more time, because the ultimate development of the brain occurred in pleistocene times within the genus _Homo_.

It is not my intention to deal with these problems, despite their importance. I would just like to mention that the process of hominization took place in Africa, where an enormous number of well-preserved fossils have been collected, and whence many more can be expected. They have been described in short articles, or incompletely in short notes. Only the skull of _Zinjanthropus_ has been analyzed in a classical manner, by Philip Tobias. The recent discovery of "Lucy" (_Australopithecus afarensis_) has also contributed significantly to our understanding of human evolution. It is now urgent that other fossils are studied as carefully so that we may come to an understanding of the alterations and specializations in anatomy which occurred during the process of hominization.

The behavior of primates was studied much later than anatomy and phylogeny. The reasons for this surprising fact are evident: Modern psychology, especially experimental psychology, has only relatively recently emerged from the old psychology. Secondly, a prerequisite for the study of many kinds of behavior is the ability to maintain and breed primates in captivity for a long time. This was not accomplished before the years after World War II. Thirdly, controlled behavioral studies can be carried out effectively only in special institutions which have adequate facilities. These were not available until recent times. However, some early behavioral studies with apes were quite successful. Before the outbreak of World War I, Wolfgang Kohler, a Gestalt psychologist supported by the Prussian Academy of Sciences, established the research station on Tenerife to study the intelligence of chimpanzees. He reported his results in the book The Mentality of Apes, which became a classic.

Almost contemporary with Kohler, Nadine Kohts in Moscow studied the behavior of a baby chimpanzee maintained in a human family. At approximately the same time, Robert M. Yerkes and his wife Ada began to study primates in the U.S. His book "Almost Human" appeared in 1925. Although Kohler was the first to carry out psychological studies on chimpanzees, the real founder of primate psychology was Yerkes, who established primate behavioral research as a field of scholarly pursuit. In short, the publications which mark the beginning of systematic studies in primate psychology as a new science appeared within the short span of five years.

For the psychologists of these early times, the major problem was not a problem of human evolution. The primary problem was to study the psychology of a nonhuman primate, closely related to man, which could be used as an experimental model. The "almost human" organism could be exposed to a variety of experimental conditions, and the results would help us to understand basic human psychology. The early primate psychologists were mainly interested in the proximal causes of human behavior. However, the comparative approach necessarily raised questions about the problem of human evolution, because it is self-explanatory that the next step is a phylogenetic attempt to understand the psychological development of man. The very same question faces the paleoanthropologists who are trying to analyze the factors determining the unique evolutionary course of man; one of those determinants is behavior. It is therefore just a consequence of his paleontological work that the late Louis S. B. Leakey initiated the field studies on pongids in the wild by, among others, Jane Goodall and Dian Fossey. The study of primate behavior became the predominant field of primatology after World War II, as scientists began to appreciate that no other group of animals exhibits such an enormous variety of behavioral patterns intragenerically as well as intraspecifically, and from population to population.

The historic development of primatology in the last years can be seen largely as an attempt to explain why the primates exhibit such a wide spectrum of behaviors. The primates are characterized by anatomical specializations which do not limit the variety of behavioral patterns. In fact, the opposite is true, since some of the specializations bring about an increase in the variety of behavioral patterns. All primates possess an opposable big toe which enables them to become tree dwellers, at least temporarily; the opposability of the big toe does not, however, preclude a terrestrial life. This enables the primates to live in two different habitats, which widens their behavioral potential. All primates have hands which are able to manipulate; although not as precise as in man, the variety of manipulations which they perform is surprising. Finally primates possess a brain which has developed considerably in the course of evolution, making possible many new and specific functions. The greater complexity of the central nervous system provides the neuroanatomical substrate for the ever-evolving behavior which characterizes many primates. Given the capacity for the development and evolution of behavior, including cultural behavior, in the order of primates, we may be tempted to state that the primates are the only order of mammals from which a being like man could emerge.

A few years after the last World War, primatology began to develop surprisingly fast in every subdiscipline. An important influence on the development of European primatology at this time was the initiation of two series of volumes on primates. In Great Britain W. C. Osman Hill began his series of monographs entitled _Primates_ in 1953; in 1956, Hofer, Schultz, and Starck in Germany and Switzerland published the first of the edited volumes entitled _Primatologia: Handbook of Primatology_. From these series, the breadth and depth of the growing field of primatology was made evident, and scientists in the field were provided with a firm substrate from which to launch new investigations. However, primatology was also beginning to exhibit what I once called the "delta phenomenon": This is the splitting off into several disciplines, something like the delta of a river. Because of the fragmentation, the science becomes cumbersome. To avoid a Babylonian disorder of the scientific language which invariably occurs in the "delta situation" of a science, symposia are organized, reviews and books are published, and societies are founded. Very often, however, the new sciences are not considered "classical" sciences, and are therefore not academically represented in the universities.

In the early 1960's it became obvious that an organization was urgently needed to unify the field of primatology, which was at the time deeply in the "delta situation" at an international level. The spectrum of different kinds of research on primates became increasingly unsurveyable for a single scientist. The paleontology of primates got a tremendous impulse in Great Britain, South Africa and Kenya. The behavioral sciences in Great Britain, Japan and particularly in the U.S.A., comparative anatomy in Germany and Great Britain, and cytogenetics, immunobiology and serology in several countries were progressing apace. Finally, it had become obvious that primates are suitable experimental animals for many types of medical research.

In 1962 three symposia on primatology were held almost simultaneously in New York, London and in Giessen, Germany. This unfortunate coincidence was possible only because the committees organizing the symposia had no knowledge of the plans of the others. Bernhard Rensch, who was participating with us at the symposium in Giessen suggested the formation of a group of scientists representing primatology in the I.U.B.S. This group was established by mail ballots. I think the names of the scientists who met in Frankfurt at the Max Planck Institute for Brain Research should be mentioned. The president was F. Bourliere (Paris). The others are listed in alphabetical order: Cynthia Booth, Tom and Barbara Harrisson, Helmut O. Hofer, G.H.R. von Koenigswald, Adriaan Kortlandt, Louis S. B. Leakey, W. B. Spatz and Johannes Tigges. All of the participants of this brief but very successful meeting agreed that an international primatological society should be formed as a legal entity protected by the laws of the Federal Republic of Germany. The legal seat of this society should be Frankfurt. In preparing the articles and bylaws of the society, I enjoyed the help of Dietrich Starck a great deal. In deepest appreciation and gratitude I have to mention my research associates of that time: Dr. Sprankel, Dr. Spatz and Dr. Tigges, who were of immeasurable help in both their advice and their activities. We had to organize the society without a single penny, but driven by our enthusiasm we were successful.

One year later, in the Fall of 1964, on the occasion of the opening celebration of the Delta Regional Primate Research Center, a representative group of primatologists was gathered. Dr. Arthur Riopelle, at that time the director of the center, had organized a small symposium. He was most enthusiastic about the formation of an International Primatological Society, and he deserves all of the credit for helping us take the crucial final step. At an informal meeting during this symposium the articles were read, approved, and signed, and subsequently the first Board of the Society was elected by written ballots. The first president of the society was Dr. Leonard Carmichael, a distinguished scientist who was with Robert Yerkes for many years. I, myself, was the first Secretary General. Thus, the baby was born in Covington, Louisiana! We had achieved what we wanted, but we had no money! "There is no music without any money" as a German adage says. We organized international congresses to promote communication -- this would be the music -- and received money through the membership fees, which were at that time two dollars per year, and later three dollars! All of our meetings were sponsored by funding agencies such as the National Science Foundation, the National Institutes of Health, the Wenner Gren Foundation, the German Research Association, and corresponding Swiss organizations. Without this help it would have been impossible to organize the congresses. Since the year the society was founded, congresses have been held in Germany, the U.S.A., Switzerland, Japan, India and Italy. This congress is the second meeting at the Yerkes Center. In addition, several national societies have become affiliated with our society, and we now have our own journal.

The function of an academic society is to promote the particular discipline by permanently maintaining communication among the scientists working in different fields. The open discussion of diverse topics from varying perspectives is a major stimulus for innovative scientific work. Such communication also serves to stimulate young students to begin their scientific work in our field. Our society has served this purpose: Primatology has become a part of the academic curriculum in anthropology at least in Great Britain and the U.S.A. The primatological literature appearing every year is almost unsurveyable. The proceedings of all of our meetings are published as books or as a series of volumes. The foundation of our society was an important landmark for primatology, and the activities of the society were a major impulse for the development of primatology in all of its subdisciplines. This will last for the time to come.

ACKNOWLEDGMENTS

I am happy to express my appreciation and gratitude to Dr. Frederick A. King, Director of the Yerkes Regional Primate Research Center, Emory University, for the privilege of presenting the introductory address at the IXth Meeting of the International Primatological Society. I am also grateful to Dr. Margaret Long Walker and Dr. Lary C. Walker for correcting the drafts and for their advice. If I had not enjoyed their help, the timely completion of this paper would not have been possible.

REFERENCES

Hill, W. C. O.: Primates, Vol. 1. The Edinburgh University Press, Edinburgh. (1953).

Hofer, H. O., Schultz, A. H., and Starck, D. eds: Primatologia, Vol. 1. Karger, Basel, New York. (1956).

Kohler, W.: The Mentality of Apes. English translation from 2nd revised edition. Harcourt, Brace, New York. (1925).

Reynolds, V.: The Apes. Dutton, New York. (1967).

Yerkes, R. M.: Almost Human. Century, New York, London. (1925).

PART I

Reproductive Behavior, Biology, and Endocrinology

1

Evolution of Sex Pheromones in Higher Primates

G. J. Fox

"Copulins" were first discovered in *Macaca mulatta* vaginal secretions. These volatile fatty acids (acetic, propionic, isobutyric, butyric, isovaleric, and isocaproic) appeared to release mounting and ejaculation and were proposed as sex pheromones. Copulins have also been detected in vaginal secretions of *M. fascicularis*, *M. nemestrina*, *M. arctoides*, *Papio cynocephalus doguera*, *Erythrocebus patas*, *Pan troglodytes*, *Homo sapiens*, and *Saimiri sciureus*. These aliphatic acids may occur generally among higher primates. Attraction and/or sexual behavior is released by sex pheromones acting via olfactory pathways. It is unlikely that copulins would operate mainly as immediate releasers of sexual behavior since our analysis, based on gas liquid chromatography of 94 and 52 vaginal lavages from *M. fascicularis* and *P. troglodytes*, respectively, indicates that concentrations of these acids do not provide distinctive or reliable cues of fertility. Other biochemical and behavioral studies of *M. mulatta* support our interpretation. Copulins may elicit attraction by enhancing species' uniqueness since there are differences between species in presence and quantities of these acids. Attraction may or may not result in copulation, depending upon a variety of biological and social factors. Immediate release of sexual behavior may have been the initial role of primate sex pheromones, but as selection for behavioral plasticity in higher primates occurred, attraction may have become the principal function of copulins.

INTRODUCTION

Pheromones were probably the first kind of signals used for communication (Wilson, 1968). Since higher primates are relatively microsmatic (Michael, 1969), the possible existance of pheromones acting via olfactory pathways received little study in comparison with investigations of communication through the visual or vocal-auditory channels. However, experiments by Michael et al., led to the proposition that certain volatile fatty acids in vaginal secretions of rhesus monkeys, Macaca mulatta, were constituents of a sex pheromone or pheromones (Michael et al., 1971; Curtis et al., 1971). These short-chain fatty acids, acetic, propionic, isobutyric, butyric, isovaleric, and isocaproic, appeared to increase the frequency of male mounting attempts and ejaculation (Keverne and Michael, 1971; Michael et al., 1972, 1976). The term, "copulins" was coined for these acids (Michael and Keverne, 1970). Goldfoot et al. (1976) found that these acids were not effective in evoking ejaculatory behavior. In a further study, high levels of copulatory activity continued when males were unable to detect the olfactory stimuli which might be provided by these acids (Goldfoot et al., 1978). Keverne (1974, 1976b, 1978) and Michael and Zumpe (1979) acknowledged variability in male responses to copulins. Thus, debate ensued over the role of copulins as sex pheromones (e.g., Goldfoot et al., 1976; Keverne, 1976a).

Substances functioning as sex pheromones play either one or both of two roles: release of attraction and/or sexual behavior (Wilson and Bossert, 1963). Our analyses, based on examination of aliphatic acid contents of vaginal lavages from long-tailed macaques (M. fascicularis) and chimpanzees (Pan troglodytes) indicate that attraction is the potential role of copulins as sex pheromones. This interpretation of copulins as attractants is congruent with the seemingly contradictory observations cited for male rhesus since attraction might not necessarily result in sexual behavior. A coherent perspective on the evolution of sex pheromones and behavior in higher primates is also furnished by this interpretation of C_2 - C_6 vaginal fatty acids as serving principally to elicit attraction.

METHODS AND MATERIALS

Fifty-two samples of vaginal secretions from 33 chimpanzees of various age classes and reproductive states were analyzed for aliphatic acids by gas liquid chromatography. These vaginal lavages were supplied by Yerkes Regional Primate Research Center of Emory University. Thirty-three long-tailed macaques received vaginal lavages on days 5, 12, and 20 from the onset of menses. In addition, daily vaginal washings were collected throughout an entire menstrual cycle for each of two females (#524 and #781). All these monkeys were housed at the Research Farm associated with the Department of Comparative Medicine at Bowman Gray Medical School, Wake Forest University. Volatile fatty acid contents of these samples were also measured by means of gas liquid chromatography.

One milliliter of distilled water was used for vaginal washing. Immediately after collection, each lavage received 0.25 ml of 50% sulfuric acid to stop further acid production since aliphatic acids are evidently fermentation by-products of vaginal flora (Bonsall and Michael, 1971, 1978; Michael et al., 1972, 1975; Keverne, 1974, 1976b, 1978, Goldfoot et al., 1976). Samples were prepared for gas liquid chromatography by adding 0.5 gm sodium chloride to each. The samples were then extracted into 1 ml of ether. This ether layer was dried over half the volume of anhydrous magnesium sulfate as ether for 10 minutes and 1 microliter of this ether layer was injected into a Varian Model 1700 chromatograph. A glass column packed with SP 1000 plus 3% orthophosphoric acid on 100/120 mesh with acid washed Chromosorb W support (Supelco, Inc.) was used in the chromatograph. Settings for the detector, injector, and oven were 200, 170, and 150°C, respectively; 10^{-11} for electrometer; and 16 for attenuator. Detection was carried out in the flame ionization mode. As a control, a standard series of volatile fatty acids concentrated to 100 micrograms/ml was run through the chromatograph. We chose octanoic acid as a concentration marker for each sample. Although others have relied upon n-pentanol for this purpose, we could not separate this alcohol from our solvent peak. Octanoic acid is a fatty acid and, therefore, we screened every sample for octanoic acid before inserting our standard concentration marker. Peak areas were calculated manually and expressed as ratios of the area to octanoic acid. Aliphatic acids occurring in any quantity under 0.1 micrograms were noted as a "trace" since accurate measurement below this amount was not practicable.

RESULTS

Our data suggest that copulins are components of signals of the identity of a species. Acetic, propionic, isobutyric, butyric, isovaleric, and isocaproic acids have been discerned in vaginal secretions of M. mulatta, M. fascicularis, M. nemestrina, M. arctoides, Papio cynocephalus doguera, Erythrocebus patas, Saimiri sciureus, Pan troglodytes, and Homo sapiens (Michael et al., 1971, 1972, 1974, 1975, 1976; Curtis et al., 1971; Fox, 1982). Some of these volatile fatty acids probably occur in every anthropoid taxon. However, only three of the copulins were usually present in the 52 vaginal washings from 33 chimpanzees housed at the Yerkes Regional Primate Research Center. Gas liquid chromatography revealed acetic, isobutyric, and isocaproic acids in 100%, 77%, and 67%, respectively, of these lavages. Other aliphatic acids were detected in less than 18% of all samples. Thus, only three copulins - acetic, isobutyric and isocaproic - appear to be typical of chimpanzee vaginal secretions. Of these three acids, isocaproic exhibited the greatest mean concentration: 90.2 micrograms/ml. The mean concentration of isobutyric acid was 8.8 micrograms/ml (Fox, 1982). Isocaproic and isobutyric acids have relatively low detection thresholds (0.81 and 8.5 ppm, respectively) among humans (Amoore, 1970). Assuming that chimpanzee olfactory capacities are at least as sensitive as those of humans, isocaproic acid, representing 63% of the total aliphatic acid content of the chimpanzee lavages, is a major contributor to chimpanzee vaginal odor. Isobutyric acid also seems to be a significant constituent of copulin-produced scent emanating from the chimpanzee vaginal tract.

Very different results were obtained when 94 vaginal lavages from 33 long-tailed macaques living at the Research Farm associated with the Department of Comparative Medicine at Bowman Gray Medical School were analyzed. Acetic, propionic, and butyric acids were revealed in 100%, 76%, and 74%, respectively, of all samples. All other copulins (including isobutyric and isocaproic acids, which seem to be characteristic of chimpanzee vaginal secretions) were manifested in 35% or fewer of these long-tailed macaque specimens. As in the case for chimpanzees, acetic acid should probably be discounted as a very important source of vaginal odor since it has a relatively high detection threshold for humans (30.7 ppm) (Amoore, 1970). The mean concentration of butyric acid in long-tailed macaque vaginal lavages was 2.93 micrograms/ml, well above the 1.11 ppm detection threshold for humans (Amoore, 1970). Chimpanzee and long-tailed macaque vaginal odor therefore seem to be distinctive from each other.

Since there is substantial variability of copulins in anthropoid vaginal secretions, the interesting hypothesis that these volatile fatty acids might distinguish individual females within a species was explored. Fourteen of the chimpanzees investigated had received two or three vaginal lavages, but no regular pattern in presence of certain acids, relative concentrations of particular acids, or total acid volume could be discerned for any of these females (Fox, 1982).

The copulin composition of vaginal secretions from two long-tailed macaques, each of whom was lavaged daily throughout a menstrual cycle, appeared to differentiate these females from each other. Female 524's vaginal secretions contained virtually no propionic acid. On day 13 of her cycle, 0.1 micrograms was recorded. Only a trace (less than 0.1 micrograms) was found on nine days; otherwise, this acid was completely absent. Washings from 781 evidenced propionic acid on 15 days, in amounts ranging as high as 8.6 micrograms (Figure 1), well above the human detection threshold of 4.03 ppm (Amoore, 1970). Isobutyric acid was almost lacking in samples from 781. A trace of this acid was discerned on six days and only 0.7

micrograms on the other day (day 22) that it was present. This same acid was discriminated on 11 days and in quantities up to 3.3 micrograms for 524 (Figure 1). Butyric acid was more frequent in specimens from 524 than in those from 781 (Figure 1). The mean concentration of butyric acid for 524 (2.66 micrograms/ml) was far beyond the human detection threshold of 1.11 ppm (Amore, 1970). However, the mean amount of butyric acid for 781 (1.12 micrograms/ml) was barely above that threshold. The range and largest concentration of butyric acid were also greater for 524 than for 781 (Figure 1). Isovaleric acid showed a pattern of contrast between these two females similar to that of butyric acid (Figure 1).

Considering that acetic acid is probably not an important contributor to vaginal scent resulting from copulins, that propionic and butyric acids are characteristic in long-tailed macaque vaginal secretions, and that the greatest contrast between these two females seemed to be due to propionic, butyric, and isovaleric acids, a two-way analysis of variance test using amounts of propionic, butyric, and isovaleric acids as repeated measures was performed. The F ratio obtained, 0.03 ($df = 1,4$, $p > 0.95$), was not statistically significant, indicating that differences in copulin content of vaginal secretions do not distinguish individual females from each other within a species.

A mid-cycle rise in the concentration of copulins has been reported for rhesus macaques, olive baboons, and

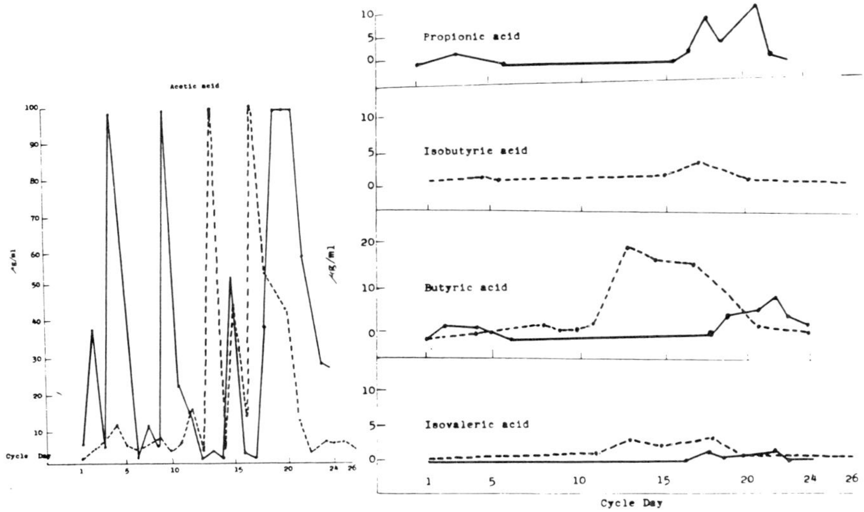

Figure 1. Amounts of copulins (acetic, propionic, isobutyric, butyric, and isovaleric acids) in daily vaginal lavages throughout a menstrual cycle for each of two long-tailed macaques, *Macaca fascicularis*, females 781 (solid line) and 524 (broken line).

humans (Michael et al., 1972, 1975). This increase could be associated with the approach and onset of ovulation, thus providing an olfactory signal for males. Although a mid-cycle rise in total copulin content of vaginal lavages collected on days 5, 12, and 20 from the beginning of menses for each of 33 long-tailed macaques studied in our laboratories is evident when day 12 samples are compared with those from day 5, Figure 2 also shows that this concentration was greater on day 20 than on day 12. (A menstrual cycle of about 24-28 days is usual for Macaca fascicularis.)

Two of the long-tailed macaques in our investigation received daily lavages during an entire menstrual cycle. Female 524 displayed larger amounts of copulins on days 13-18 of her 26-day cycle but these acids peaked on days 18-22 of the shorter 24-day cycle of 781 (Figure 1). Goldfoot, et al. (1976) found that these acids reached maximum concentration 4-5 days after ovulation by the rhesus macaques in their colony.

Most of the copulin content of the 99 lavages from the 33 long-tailed macaques was acetic acid (66%, 82%, and 78% on days 5, 12, and 20, respectively). As noted

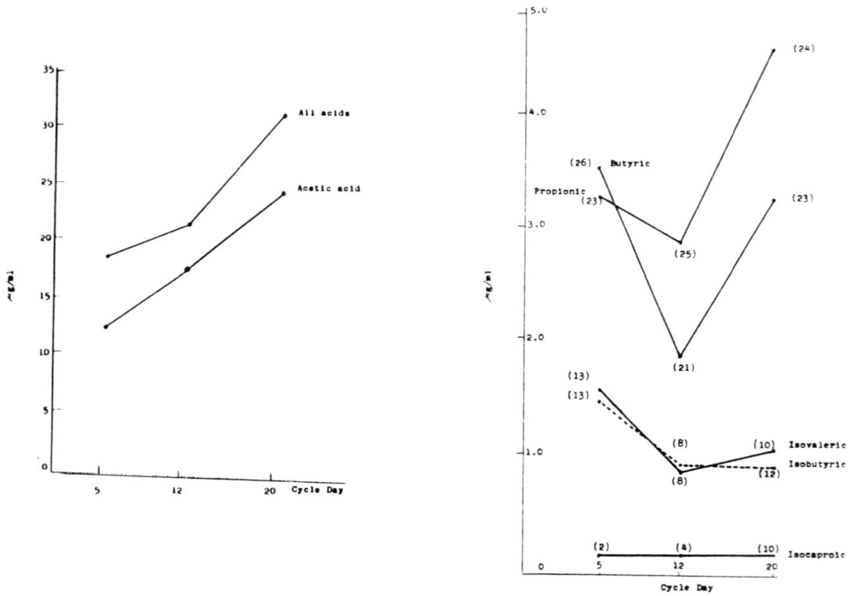

Figure 2. Mean concentration of copulins (acetic, propionic, isobutyric, butyric, isovaleric, and isocaproic acids) in vaginal lavages collected on days 5, 12, and 20 from the onset of menses for each of 33 long-tailed macaques, Macaca fascicularis. Numbers in parentheses indicate actual number of samples in which the particular acid was detected; acetic acid was present in all samples.

above, acetic acid is unlikely to be a major source of vaginal odor. Propionic and butyric acids appear to be characteristic in *M. fascicularis* vaginal secretions. Both these acids are present at elevated levels when postovulatory samples are contrasted with those at mid-cycle (Figure 2). Isobutyric acid is the only copulin to exhibit a decrease in mean quantity from day 12 to day 20 and that drop is very small (0.02 micrograms/ml). Since isocaproic acid was rare in these lavages (occurring in 16 of 99) and more than a trace (less than 0.1 micrograms/ml) was not detected in any of them, its fluctuation during the cycle is probably inconsequential. An analysis of variance using the six copulins as repeated measures yielded an F ratio of 0.13 ($df = 2,15$, $p > 0.95$). The null hypothesis cannot be rejected on the basis of this statistical test; therefore, differences in fatty acid content of long-tailed macaque vaginal secretions do not seem to be significant for discriminating among preovulatory (day 5), ovulatory (day 12), or postovulatory (day 20) conditions.

Vaginal lavages received from chimpanzees at the Yerkes Regional Primate Research Center were accompanied by information on the degree of swelling in the anogenital region. We divided these into two groups. One group included 13 samples from chimpanzees exhibiting one-half, three-fourths, or maximum swelling of the anogenital area. This grouping would correspond roughly to the follicular phase of the cycle. Twenty-eight lavages from individuals in the luteal stage comprised the other group. The mean amount, 149.8 micrograms/ml, of total aliphatic acids was higher for females who were one-half to fully turgescent than for those in the luteal stage, 140.2 micrograms/ml. Nevertheless, an analysis of variance test with the six copulins used as repeated measures did not attain statistical significance since the F ratio was 0.14 ($df = 1,10$, $p > 0.95$). Inability to reject the null hypothesis suggests that copulin composition of vaginal secretions does not distinguish the most fertile (follicular) from the least fertile (luteal) phase of the estrous cycle. This interpretation is further confirmed by comparing the concentrations of the characteristic chimpanzee copulins (acetic, isobutyric, and isocaproic acids) relative to each other at different periods of the cycle. Relative concentrations of these acids are most similar at maximum swelling (ovulation likely) and in luteal stage (ovulation very unlikely; Fox, 1982). Thus, copulins do not appear to furnish definite or reliable cues of female fertility.

DISCUSSION

If the volatile fatty acids termed "copulins" are considered to constitute a sex pheromone or pheromones in higher primates, behavioral studies of male rhesus macaques indicate clearly that these acids do not effect immediate release of sexual behavior, one of the primary functions of sex pheromones. Responses of males exposed to these acids are variable and do not inevitably result in copulatory behavior (Keverne, 1974, 1976b, 1978; Goldfoot et al., 1976; Michael and Zumpe, 1979). Furthermore, anosmic males may continue to copulate as often with their previous partners in the absence of such olfactory cues (Goldfoot et al., 1978). Our biochemical analyses of copulins throughout the estrous cycle of long-tailed macaques and chimpanzees strongly suggest that these acids would be inefficient as immediate releasers of sexual behavior since they do not provide a definitive signal of fertility nor do they manifest consistent association with stages in the cycle.

The other principal function of sex pheromones is attraction. Judging by numerous reports of anthropoid males sniffing the genital region of females and/or inserting a finger into the vagina and then smelling the finger after it is withdrawn (e.g., Moynihan, 1966 for *Callicebus moloch*; Baldwin, 1968 for *Saimiri sciureus*; Altmann, 1962 for *Macaca mulatta*; Jay, 1965 for *M. sinica*; Rahaman and Parthasarathy, 1969 for *M radiata*; Blurton-Jones and Trollope, 1968 for *M. arctoides*; van Hooff, 1969 for *M. nemestrina*: Nishida, 1970 for *Pan troglodytes*; Hess, 1973 for *Gorilla g. gorilla*), males of many anthropoid species actively seek vaginal odors. Copulins may be the source of this attraction.

Data obtained in our laboratories show that the copulin content of chimpanzee and long-tailed macaque vaginal secretions and, as far as can be ascertained, vaginal scent due to copulins sharply distinguish these species from each other. Michael et al. (1975, 1976) have noted that there are obvious species differences when copulins of *Homo sapiens* and *Macaca mulatta* are compared. Humans display almost three times as much propionic acid as rhesus while rhesus exhibit about twice as much butyric acid as humans. If one reviews figures on copulins that have been published for higher primate taxa (Michael et al., 1972, 1976), interspecific variability in presence and amount of these volatile fatty acids is evident. An illustrative case is furnished by *M. arctoides*. Isocaproic acid was not recorded for 26 specimens from stump-tailed macaques (Michael et al., 1976). The species to which an anthropoid belongs may be discriminated by an olfactory signal generated by a particular pattern of copulins. By identifying species, the odor(s) produced by copulins may facilitate intraspecific attraction.

Attraction between anthropoids may or may not result in mating. Higher primates are complex animals whose behavior is flexible and heavily influenced by experience and learning. Social as well as physical factors of present and past environments can greatly affect their behavior. Immediate release of sexual behavior may have been the primary evolutionary function of sex pheromones.

During the course of selection for behavioral plasticity in higher primates, attraction, with it probabilistic instead of deterministic impact on sexual behavior, may have become the principal role of copulin-composed messages. Some extremely interesting data reported by Keverne (1976b) from experimental exposure of male rhesus macaques to copulins suggest a futher adaptation of the role of these acids as attractants. Several males did not respond to copulins by engaging in sexual activity with their partners but did decrease aggression toward them and/or prolong their grooming of these females. As components of anthropoid pheromones, the role of copulins may have evolved to elicit general social, rather than strictly sexual, attraction among the behaviorally flexible higher primates.

ACKNOWLEDGMENTS

Dr. David Taub generously provided the _M. fascicularis_ vaginal lavages and designed the sampling framework for collection of the _M. fascicularis_ materials. I also appreciate sincerely the gracious hospitality he and his wife, Pam, extended on my visits to Winston-Salem. In addition to giving liberally of his time, knowledge, and technical advice, Dr. Robert Fulghum afforded full access to the Varian chromatograph and all the resources of his laboratory in the Department of Microbiology of the East Carolina University School of Medicine. Mr. John Worthington, his research assistant, was also very helpful. I thank Mr. Walter Shepherd, Assistant to the Dean of the Medical School, for being instrumental in making many arrangements. Mr. Robert Matthews and Ms. Susan Whalen Simpson were my diligent and enthusiastic research assistants who ran many of the macaque and chimpanzee samples, respectively. A grant from the East Carolina University Research Committee partially funded the chimpanzee investigation. Further financial support for the chimpanzee and macaque projects was furnished by the Department of Sociology, Anthropology and Economics.

REFERENCES

Altmann, S. A.: A field study of the sociobiology of rhesus monkeys. Ann. N. Y. Acad. Sci. 102:338-435. (1962).

Amoore, J. E.: Molecular Basis of Odor. Thomas, Springfield. (1970).

Baldwin, J. D.: The social behavior of adult male squirrel monkeys (_Saimiri sciureus_) in a seminatural environment. Folia Primat. 9:281-314. (1968).

Blurton-Jones, N. G. and Trollope, J.: Social behaviour of stump-tailed macaques in captivity. Primates 9:365-394. (1968).

Bonsall, R. W., and Michael, R. P.: Volatile constituents of primate vaginal secretions. J. Reprod. Fertil. 27:478-479. (1971).

Bonsall, R. W., and Michael, R. P.: Volatile odoriferous acids in vaginal fluid. In Hafez and Evans. The Human Vagina. North-Holland, Amsterdam. pp. 167-177. (1978).

Curtis, R. F., Ballantine, J. A., Keverne, E. B., Bonsall, R. W., and Michael, R. P.: Identification of primate sexual pheromones and the properties of synthetic attractants. Nature, London. 232:396-398. (1971).

Fox, G. J.: Potentials for pheromones in chimpanzee vaginal fatty acids. Folia Primat. 37:255-266. (1982).

Goldfoot, D. A., Kravetz, M. A., Goy, R. W., and Freeman, S. K.: Lack of effect of vaginal lavages and aliphatic acids on ejaculatory responses in rhesus monkeys: behavioral and chemical analyses. Horm. Behav. 7:1-27. (1976).

Goldfoot, D. A., Essock-Vitale, S. M., Asa, C. S., Thornton, J. E., and Leshner, A. I.: Anosmia in male rhesus monkeys does not alter copulatory activity with cycling females. Science 199:1095-1096. (1978).

Hess, J. P.: Some observations on the sexual behaviour of captive lowland gorillas, *Gorilla g. gorilla* (Savage and Wyman). In Michael and Crook. Comparative Ecology and Behaviour of Primates. pp. 507-581. Academic Press, London. (1973).

Jay, P.: Field studies. In Schrier, Harlow and Stollnitz. Behavior of Nonhuman Primates. Vol. 2. pp. 525-591. Academic Press, New York. (1965).

Keverne, E. B.: Sex-attractants in primates. New Scient. 63:22-24. (1974).

Keverne, E. B.: Reply to Goldfoot et al. Horm. Behav. 7:369-372. (1967a).

Keverne, E. B.: Sex attractants in primates. J. Soc. Cosmet. Chem. 27:257-269. (1976b).

Keverne, E. B.: Olfactory cues in mammalian sexual behaviour. In Hutchison. Biological Determinants of Sexual Behavior. pp. 727-763. Wiley, New York. (1978).

Keverne, E. B., and Michael, R. P.: Sex-attractant properties of ether extracts of vaginal secretions from rhesus monkeys. J. Endocr. 51:313-322. (1971).

Michael, R. P.: The role of pheromones in the communication of primate behaviour. Proc. 2nd Int. Congr. Primat., Vol. 1. pp. 101-107. (1969).

Michael, R. P., and Keverne, E. B.: Primate sex pheromones of vaginal origin. Nature, London. 225:84-85. (1970).

Michael, R. P., and Zumpe, D.: Biological factors in the organization and expression of sexual behaviour. In Rosen. Sexual Deviation. pp. 441-480. Oxford University Press, Oxford. (1979).

Michael, R. P., Bonsall, R. W., and Kutner, M.: Volatile fatty acids, 'copulins', in human vaginal secretions. Psychoneuroendocrinology 1:153-163. (1975).

Michael, R. P., Bonsall, R. W., and Warner, P.: Human vaginal secretions: volatile fatty acid contents. Science 186:1217-1219. (1974).

Michael, R. P., Bonsall, R. W., and Zumpe, D.: Evidence for chemical communication in primates. Vitams Horm. 34:137-186. (1976).

Michael, R. P., Keverne, E. B., and Bonsall, R. W.: Pheromones: Isolation of male sex attractants from a female primate. Science 172:964-966. (1971).

Michael, R. P., Zumpe, D., Keverne, E. B., and Bonsall, R. W.: Neuroendocrine factors in the control of primate behavior. Recent Prog. Horm. Res. 28:665-706. (1972).

Moynihan, M.: Communication in the titi monkey, Callicebus. J. Zool., London. 150:77-127. (1966).

Nishida, T.: Social behavior and relationship among wild chimpanzees of the Mahali Mountains. Primates 11:46-87. (1970).

Rahaman, H., and Parthasarathy, D.: Studies on the social behaviour of bonnet monkeys, Primates 10:149-162. (1969).

van Hooff, J. A. R. A. M.: The facial displays of the catarrhine monkeys and apes. In Morris. Primate Ethology. pp. 9-88. Doubleday, Garden City. (1969).

Wilson, E. O.: Chemical systems. In Sebeok. Animal Communication. pp. 75-102. Indiana University Press, Bloomington. (1968).

Wilson, E. O., and Bossert, W. H.: Chemical communication among animals. Recent Prog. Horm. Res. 19:673-716. (1963).

2

A Primate Model for Assessing Estrogenicity—The Castrate Female Rhesus Monkey

G. B. Fuller, B. Burnett, and W. C. Hobson

Rodents have traditionally been used for bioassay of estrogenic activity in food and drugs intended for human use. The differences between rodents and humans have raised the concern that rodents may not be valid predictors of oral estrogenic activity. In this study, a primate model was developed and used for testing the comparative oral estrogenicity of several estrogens. The endpoint used was suppression of serum gonadotropin levels in castrated rhesus monkeys. Six castrated females and two castrated males were treated once per month. Either vehicle or an estrogen was administered orally at 0800 hours on days 2 and 3. Blood samples (2-3 ml) were obtained hourly for 6 hours (1200 to 1700) on days 1 and 3. Follicle stimulating hormone (FSH) and leutinizing hormone (LH) were determined by radioimmunoassay. There was no change in FSH or LH levels following dosing with the vehicle. The minimal effective oral dose of each estrogen which significantly suppressed serum concentrations of FSH or LH ($\underline{p} < 0.05$) was as follows: estradiol-17 beta, 10 micrograms/kg; estradiol-17 alpha, 100 micrograms/kg; diethylstilbestrol-glucuronide, 100 micrograms/kg; estrone 1 micrograms/kg; zearalenol, 10 mg/kg; and coumestrol, > 10 mg/kg. In general, the relative estrogenicity was similar to the potency reported for rodent bioassay; however, estradiol-17 beta and estrone were several-fold more active orally in primates than in rodents, and surprisingly, estrone appeared to be more potent than estradiol-17 beta. Male castrates were less sensitive than females. These data indicate that the castrate female rhesus monkey is a potential model for establishing the "no effect" levels for estrogenic substances in oral drugs and food substances for human consumption.

INTRODUCTION

The possible health risks of hormonally-active contaminants in the human food chain have become a major concern. The estrogenic activity of contraceptive steroids, animal feed additives, mycotoxins, and phytoestrogens has stimulated much research on their teratological, carcinogenic, hepatotoxic, and pathophysiological effects (Rodricks et al., 1977; Norvell and Shellenberger, 1977). However, to assess the potential health hazards a determination of the compound's estrogenicity must be made in a system which can reliably predict its hormonal activity in humans.

Historically, bioassays for estrogenic activity are based upon an increase in uterine weight in rats or mice injected with the estrogen in question (Astwood, 1938; Rubin, et al., 1951). The validity of a rodent test for prediction of hormonal effects in humans is questionable. The differences between reproductive and endocrine systems of rodents and humans, the potential for errors relating to different routes of administration and amounts of absorption from the gastrointestinal tract, and differences in the levels at which rodents respond to stimulation by hormonal compounds detract from the reliability of rodent-based predictions of human safety.

Consequently, this communication describes our efforts to develop a sensitive nonhuman primate model for assessing estrogenic activity of compounds having possible human health significance. Suppression of gonadotropin levels in castrate rhesus monkeys was the endpoint utilized for testing the biological potency of several orally administered estrogens. Castrate monkeys were used to avoid the difficulties of interpreting the results of estrogen exposure superimposed on the changing endocrine patterns of the menstrual cycle. The castrated monkey has the additional advantage that its gonadotropins are several-fold higher than in intact animals (Atkinson, et al., 1970), thus requiring smaller samples and less sensitive gonadotropin radioimmunoassays to detect estrogen-induced differences.

METHODS AND MATERIALS

Animals

Six ovariectomized female and two castrate male rhesus monkeys were housed individually in air-conditioned quarters. These colony-reared animals were between 4 and 10 years of age when gonadectomized. Males were included to provide a comparison with the response of castrate females. Purina Monkey Chow, (Ralston-Purina Company, St. Louis, MO) was provided daily and fresh fruit was given twice weekly. Water was available ad libitum.

Body weights were recorded monthly for dose level determination. In addition, daily observations were made to assess the general condition and well being of each monkey. Leg tattoos provided permanent identification.

Experimental

Estrogens evaluated were estradiol-17 beta (E_2-17 beta); estradiol-17 alpha, (E_2-17 alpha); estrone (E_1); diethylstilbestrol-glucuronide (DES-G), zearalanol, and coumestrol. E_2-17 beta, E_2-17 alpha, and E_1 were purchased from Sigma Chemical Company, St. Louis, Missouri. DES-G was purchased from Aldrich Chemical Company, Milwaukee, Wisconsin, and coumestrol from Pfaltz and Bauer, Inc., Stamford, Connecticut. Zearalanol was kindly supplied by International Minerals and Chemical Corp., Terra Haute, Indiana. The compounds were dissolved in ethanol, dried on a sugar cube, and administered at 0800 hours on two consecutive days (days 2 and 3). If any animal refused the sugar cube, the dosage was provided on a slice of apple or orange. Each monkey was observed to assure that consumption of the dosage was complete.

The only exception to this treatment procedure was for the highest dosage of coumestrol (10 mg/kg). This plant estrogen was administered via capsule due to the large amount of compound and its low solubility. Each animal was lightly anesthetized with ketamine hydrochloride at 0800 hours on days 2 and 3. A capsule containing the appropriate amount of coumestrol was placed in the distal end of a stomach tube. Animals were restrained, intubated, and the capsule flushed into the stomach with warm water.

Blood samples (2-3 ml) were collected by venipuncture of the brachial vein of unanesthetized animals. The monkeys were trained to voluntarily submit to blood sampling procedures in order to minimize the possible effects of restraint. Blood samples were obtained at 1200, 1300, 1400, 1500, 1600, and 1700 hours on days 1 and 3. Samples were centrifuged and the serum stored frozen until radioimmunoassay. Compounds or controls were evaluated at monthly intervals. Vehicle controls were interposed every 3-4 months. Treatment with each estrogen was initiated with a low dose estimated to be near the "no effect" level and then proceeding to a lower or higher dose as appropriate to establish a no effect level. The low initial dose prevented confounding the results by giving "supraphysiologic" doses which could have carryover effects on the next experiment a month later.

Radioimmunoassay

FSH and LH were assayed by previously described heterologous radioimmunoassay methods (Faiman, et al.,

1975) using a crude monkey gonadotropin standard LER-1909-2 (donated by Dr. L. E. Reichert, Albany Medical College, Albany, NY). This standard contained 0.05 NIH-FSH-S1 u/mg and 0.003 NIH-LH-S1 u/mg by bioassay.

Statistical analyses

The experimental design of this study called for estimation of the minimal effective dose for each of the orally administered estrogens. If the first dose of a particular estrogen administered caused a statistically significant decrease in FSH or LH, the next dose of the same estrogen was given at a 10-fold lower level. If the first dose had no effect, the second dose was 10-fold higher. This procedure would continue until a "no-effect" level was determined. The pre-treatment hormone levels (day 1) were used for the statistical comparisons.

The design of the statistical analysis was as follows: (1) monkeys in each experiment, whether receiving an estrogen or vehicle, were classed as responders and non-responders by the following criterion: A monkey was designated as a responder if the mean plus or minus the standard error of its six pre-treatment hormone values was higher and did not overlap with the mean plus or minus the standard error of its six post-treatment hormone values; (2) using Fishers Exact Test, a procedure for small samples based on the chi-square Statistic (Dixon and Massey, 1964), the number of responders in the control and treatment groups were compared and a probability that the effects observed in the treatment groups were not due to chance variation was determined; and (3) if either the LH or FSH decrease was significant ($p < 0.05$), the dose level was classed as having a significant effect.

No statistical analyses were conducted on the responses of castrate males due to the small number.

RESULTS

The relative acute estrogenicity of estradiol-17 alpha, estrone, DES-glucuronide, zearalanol, and coumestrol for inhibiting FSH or LH release in castrate females are compared with the vehicle control and estradiol-17 (positive control) in Table 1. There was no effect of the vehicle on either FSH or LH release in four repetitions over a 15-month period. The no-effect level for each compound as determined by the lack of a significant effect on either FSH or LH (Table 1) was as follows: 1 micrograms/kg E_2-17beta; 10 micrograms/kg E_2-17alpha; 0.1 micrograms/kg estrone; 10 micrograms/kg DES-G; 1 mg/kg zearalanol; and possibly 10 mg/kg coumestrol. The dose of coumestrol used to test the monkeys was not increased to 100 mg/kg due to the prohibitive cost.

TABLE 1. The estimated minimum dose of estrogens which will depress serum gonadotropin concentrations in castrate female rhesus monkeys

Compound	Dose Level (mg/kg)	Number Responding* FSH	LH
Vehicle control		0/6	2/6
Vehicle control		2/6	1/6
Vehicle control		0/6	1/6
Vehicle control		1/6	1/6
17beta-estradiol	0.001	0/6	2/6
	0.01	5/6#	5/6#
17alpha-estradiol	0.001	2/6	3/6
	0.01	1/6	3/6
	0.1	5/6#	0/6
Estrone	0.0001	1/6	2/6
	0.001	6/6#	1/6
	0.01	6/6#	4/6#
	0.1	4/6#	4/6#
DES-glucuronide	0.001	2/6	1/6
	0.01	2/6	1/6
	0.1	4/6#	6/6#
Zearalanol	0.01	2/6	1/6
	0.1	2/6	1/6
	1.0	0/6	1/6
	10.0	5/6#	5/6#
Coumestrol	1.0	2/6	2/6
	10.0	3/6	2/6

* Number responding/number treated. A positive response was designated if the mean plus or minus the SE of its six pre-treated hormone values did overlap the mean plus or minus the SE of its six post-treatment hormone values.

\# $\underline{p} < 0.05$, represents 4 or more of the 6 treated females with a significant decrease in FSH or LH as analyzed using Fishers Exact Test (1964).

Surprisingly, estrone was 10 times more potent than estradiol-17beta. Inhibition of FSH with estrone required a dose level of 1 micrograms/kg, whereas E_2-17beta required 10 micrograms/kg body weight. Otherwise, relative potency of the other estrogens as assessed by gonadotropin inhibition were (in decreasing activity) estradiol-17alpha, DES-glucuronide, zearalanol, and coumestrol. Male castrates were generally less sensitive to the inhibiting effect of estrogens than were the females (Table 2).

The average of 6 pre-treatment FSH levels in the castrate females ranged from a low of approximately 15 micrograms/ml in rhesus #1377 to 96 micrograms/ml in rhesus #4006. Average pre-treatment FSH levels in the male castrates ranged from 15 to 45 micrograms/ml across all experiments. Circulating LH levels were less variable

TABLE 2. Effect of various estrogens on serum gonadotropin concentrations in male castrate rhesus monkeys

Compound	Dose Level (mg/kg)	Number Responding* FSH	LH
Vehicle control (4)#		2/8	1/8
17beta-estradiol	0.001	1/2	0/2
	0.01	1/2	2/2
17alpha-estradiol	0.001	0/2	0/2
	0.01	0/2	0/2
	0.1	0/2	0/2
Estrone	0.0001	0/2	0/2
	0.001	1/2	1/2
	0.01	2/2	0/2
	0.1	2/2	2/2
DES-glucuronide	0.001	0/2	0/2
	0.01	0/2	0/2
	0.1	1/2	0/2
Zearalanol	0.01	0/2	1/2
	0.1	1/2	0/2
	1.0	1/2	1/2
	10.0	2/2	0/2
Coumestrol	1.0	1/2	0/2
	10.0	1/2	0/2

* Number responding/number treated. A positive response was designated if the mean plus or minus the SE of its six pre-treatment hormone values did overlap the mean plus or minus the SE of its six post-treatment hormone values.

\# All vehicle control experiments were combined.

averaging from 2 to 10 micrograms/ml in females and from 2 to 6 micrograms/ml in males. Basal gonadotropin levels were quite variable between animals, but the within-animal variation was relatively small.

DISCUSSION

The experimental design utilized in this study was based on earlier experiments (Hobson, et al., 1977) in which LH and FSH concentration were measured at various times and frequencies after estrogen exposure. The six hourly blood samples on days 1 and 3 and the 2-day oral exposure on days 2 and 3 provided the optimum combination of sensitivity with minimal numbers of blood collections and radioimmunoassays required. Single oral exposures yielded more variable results, whereas three or more exposures did not appear to increase the assay sensitivity significantly. At least 6 control and 6 post-treatment blood samples were required to reduce hour-to-hour variability in gonadotropin values (unpublished observations). More than a total of 12 blood samples increased both the stress on the animals due to blood loss and the expense of the bioassay.

The results of this study have confirmed the utility of the castrate rhesus monkey bioassay for studying the endocrine impact of the natural estrogens. Utilizing this model to assess gonadotropin depression following 2 oral doses of an estrogen has provided us with an oral potency estimate for each specific estrogen tested. In view of traditional bioassay results from rodents (Rubin, et al., 1951), we were surprised to find that estrone at a dose level of 1 micrograms/kg caused a significant reduction in circulating FSH levels, and in contrast that estradiol-17 beta was only active at the 10 micrograms/kg level. Thus, this model may lead to a change in our predictions of the bioavailability of estrone and estradiol-17 beta when consumed by humans and the relative estrogenicity of these drugs.

No substantive explanation can be offered for the difference between estradiol-17 beta and estrone. However, the relative estrogenic activities of the other compounds were similar to those reported for other bioassay systems. When compared with the oral activities of the parent compounds reported earlier (Hobson, et al., 1977), we found that DES-glucuronide was 400 times less active than DES itself and that zearalanol was approximately 100 times less potent than zearalenone. Although we were unable to find a level of coumestrol required to reduce gonadotropin levels, we can conclude that the effective dosage would be greater than 10 mg/kg. In addition, the relative potency of coumestrol in this study was in agreement with the estimates reported for forage plant estrogens by Livingston (1978).

The apparent difference in sensitivity of FSH and LH to suppression by 100 micrograms/kg of estradiol 17-alpha is difficult to resolve. Although one could assume negative feedback threshold differences between LH and FSH, the lack of a comparable effect of several other estrogens examined does not justify such speculation at this time.

Although it is difficult to draw conclusions from small numbers, male castrate rhesus monkey seemed less sensitive to the estrogens than the females as the differences between pre- and post-treatment gonadotropin levels were generally smaller. Interestingly, Plant, et al. (1978) reported that physiological concentrations of testosterone failed to depress gonadotropin secretions in chronically (4+ years) castrate male rhesus monkeys. However, the males used in our project were castrated just prior to initiation of the study. The females had been ovariectomized 2-3 years previous to this investigation but remained sensitive to the negative feedback effects of the estrogens. Thus, it appears that males would not replace females in this bioassay system.

In conclusion, we have investigated the utility of

the castrate rhesus monkey bioassay model for predicting safe levels of acute oral estrogen exposure for humans. Because of the physiologic similarities of the post-castration gonadotropin response between rhesus monkeys and humans, we maintain that this test system more accurately estimates the "no-effect" level for human exposure than other non-primate models. In addition, the rhesus monkey model is more sensitive than rodent models for acute oral exposure estrogen bioassays.

ACKNOWLEDGMENTS

This work was supported in part by FDA contract #223-78-2230. The authors thank Mr. Scott Murray and Mr. Larry Moe for their skilled technical assistance. We also wish to thank the National Hormone and Pituitary Program for their generous gift of materials used in the gonadotropin assays.

REFERENCES

Astwood, E. B.: A six hour assay for the quantitative determination of estrogen. Endocrinology. 23:25-31. (1938).

Atkinson, L. E., Bhattacharya, A. N., Monroe, S. E., Dierschke, D. J., and Knobil, E.: Effects of gonadectomy on plasma LH concentration in the rhesus monkey. Endocrinology 87:847-849. (1970).

Dixon, W. J., and Massey, F. J.: Introduction to Statistical Analysis. McGraw-Hill, New York. (1964).

Fairman, C., Stearns, E. L., Winter, J. S. D., Reyes, F. I., and Hobson, W. C.: Radioimmunoassay for rhesus monkey gonadotropins. Proc. Soc. Exp. Biol. Med. 149:670-676. (1975).

Hobson, W., Bailey, J., and Fuller, G. B.: Hormone effects of zearalenone in nonhuman primates. J. Toxicol. Environ. Hlth. 3:43-57. (1977).

Livingston, A. L.: Forage plant estrogens. J. Toxicol. Environ. Hlth., 4:301-324. (1978).

Norvell, M., and Shellenberger, T., Eds.: Hormone Research III. Hemisphere, Washington, D. C. (1977).

Plant, T. M., Hess, D. L., Hotchkess, J., and Knobil, E.: Testosterone and the control of gonadotropin secretion in the male rhesus monkey (<u>Macaca mulatta</u>). Endrocrinology 103:535-541. (1978).

Rodricks, J. V., Hesseltine, C. W., and Mehlman, M. A., Eds.: Mycotoxins. Pathotox Publishers Inc., Park Forest South, Illinois. (1977).

Rubin, B. L., Dorfman, A. S., Black, L. and Dorfman, R. I.: Bioassay of estrogens using the mouse uterine response. Endocrinology 49:429-439. (1951).

3

Overt Behavior Correlates of Menstruation and Ovulation in a Lowland Gorilla

D. Quiatt, L. Miller, and R. Cambre

We report the results of behavior studies of gorillas in the Denver Zoo conducted in the summers of 1981 and 1982. The specific question addressed was whether observations of social behavior might help pinpoint ovulation sufficiently to increase chances of successful artificial insemination. Behavior was studied both in conjunction with and blind with respect to daily measurement of blood/hormone levels in urine (blood only in the case of the second study). In the first series of observations, but not in the second, correlations were noted between overt behavior and rises in blood/hormone levels indicating menstruation and estrus. A comparison of the two series affords insights into problems of predicting future behavior of a single individual on the basis of past performance of that same individual.

INTRODUCTION

In 1980 it was decided to make a concerted effort to breed the Denver Zoo's two female gorillas, Bibi and Maguba. Laparoscopic examination of Maguba in April of that year revealed numerous luteal scars on the ovaries, including one fresh scar which afforded an estimate of date of last ovulation. Subsequent analysis of urine samples confirmed that Maguba was cycling with fair regularity. Laparoscopy of Bibi showed no such scarring; however, that she had ovulated at least once was evidenced by a short-lived infant born in 1974. Analysis of semen of the Zoo's one male gorilla, Thomas, showed morphologically normal motile sperm in low but acceptable proportions.

All three animals were about the same age, Thomas and Bibi about 14 years, Maguba a year or two younger. Thomas and Bibi were close companions, and to test whether association with other males might prove more productive, Bibi was shipped to the Cincinnati Zoo. It was hoped that Thomas and Maguba would mate in her absence; failing that, Thomas might serve as donor for artificial insemination.

The behavior studies, ancillary to the breeding program, were conducted on a shoestring and piecemeal. The first was initiated just before Bibi's removal, to see how in her absence social interaction between Maguba and Thomas might be affected by ovulation and menstruation. Observers included graduate and undergraduate students and zoo volunteers, many with little or no prior experience and with minimal training beforehand. Such constraints and others associated with the gorillas' zoo habitat (relatively barren), social environment (impoverished), and sexual experience (highly limited) protected us against unwarranted optimism concerning the prediction of ovulation from overt social behavior.

METHODS AND RESULTS

That many of our observers were inexperienced, apart from the training which we provided, presented a procedural problem. Effective use of such observers depends in large part on the character of observational units of behavior. These should be simply defined, sharply bounded, and non-redundant. A previous study (Chiszar and Caldwell, 1980) indicated an increase, possibly linked with ovulation, in the rate at which Maguba solicited social interactions with both Thomas and Bibi. However, we found it difficult to define such soliciting behavior according to the guidelines just mentioned. In two months of preliminary observations, we were unable to isolate reliably and consistently solicitation of general social interaction, and we observed no instances of specifically sexual solicitation as described by Hess (1973), Harcourt and Stewart (1978), and Keiter and Pichette (1979). The behaviors we settled on as likely indicators of change in internal state were socio-spatial adjustments, monitoring actions, and displays indicative of arousal, as in Figure 1.

Observational units which we used in recording these behaviors were:

Displacement: One individual "causes" another to move away from the location he/she has been occupying. The first may or may not replace the second in space. Cause may be assumed from apparent intent of the displacer (inferred from gaze or eye aversion) or attentiveness of the displaced individual. But rather than make inferences about cause, and to insure consistency, we recorded as displacement any whole body movement of one individual which was one foot or more away from the other, when the other is in movement and has approached within six feet.

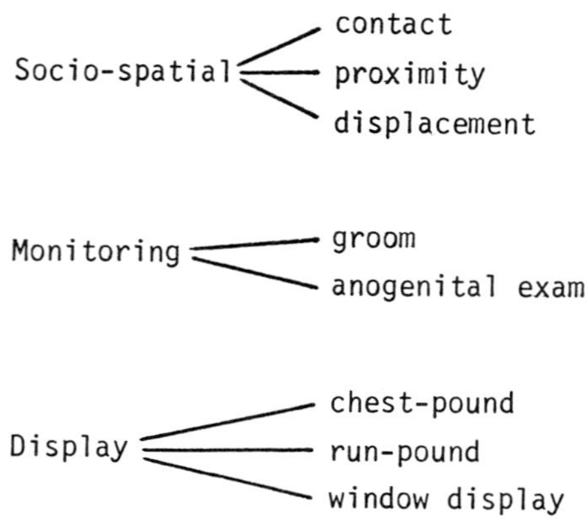

Figure 1. Summary of index behaviors.

Proximity: One individual moves to within arm's reach of another and remains without other interaction for at least 5 seconds. If the approached individual moved off before 5 seconds was up, we scored the behavior as "Displacement." (Note the qualifier, "without other interaction." This is what distinguished "Proximity" as an observational unit.)

Contact: Any tactile communication other than "Examination" e.g., touching, brushing, hitting, pushing, sitting, lying in contact, etc.

Groom: No grooming of either animal by the other occurred, so we eliminated this unit.

Examination: An inspection of one animal by another or a self-inspection. Maguba rarely examined Thomas and Thomas rarely examined himself. Examinations involved looking, touching, smelling, mouthing, and combinations thereof. Examinations of primary interest were those by Thomas or by Maguba herself of Maguba's genital area.

Display: A stereotypical hammering of the chest with the knuckles. We noted two levels of intensity: (1) a low level, without accompanying locomotion or other signs of agitation - 1 or 2 taps or a series of light taps (the displaying individual might be sitting or lying), and (2) a rapid series of forceful chest-pounding while the displaying individual was standing and perhaps locomoting. Many such displays were directed toward spectators. If a display involved contact with the window, we noted that contact.

Figure 2 contains histograms for all behavior categories except Groom and Window Display. As mentioned, no allogrooming was observed. Window displays, since they were directed at human beings including observers, presented problems of interpretation and analysis which we discuss in the next section.

Note that what Figure 2 shows is all events in each behavior category per day, not rates per invariant time unit. Most days (over 90%) involved 1 hour of observation

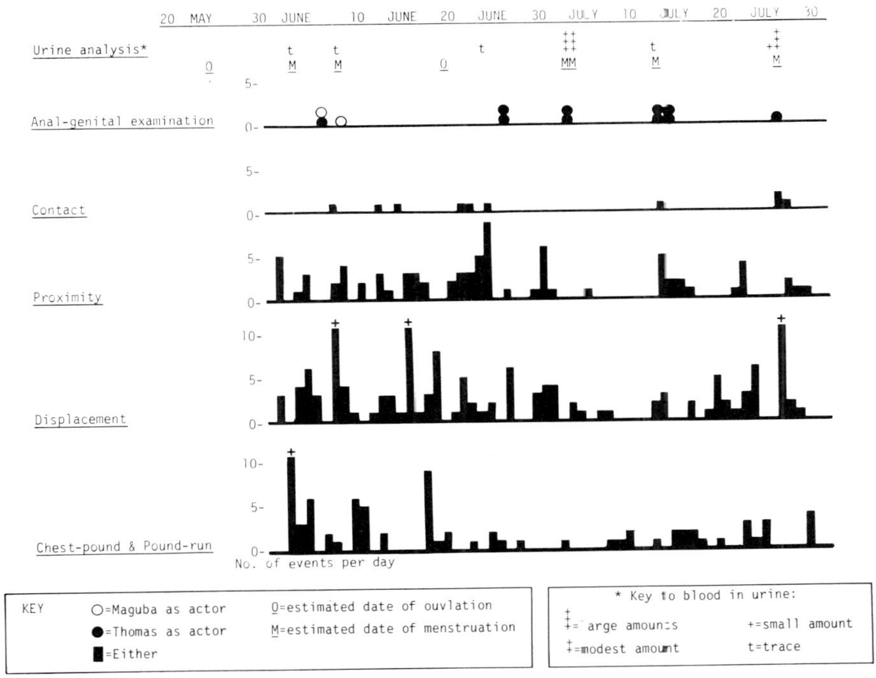

Figure 2. Summer 1981. Summary of behavior on a daily basis (8 weeks).

in the morning and 1 in the afternoon. On a few days observations were made only in the morning or in the afternoon. No attempt was made to sample behavior throughout the day; observation sessions were scheduled for periods when the animals were most active, during the 2 hours following morning feeding (coincident with the first influx of zoo visitors) and the 2 hours preceding afternoon feeding. No statistical tests have been applied to these data; they are presented in full primarily to give an idea of the variability of behavior during this series of observations. On June 2 Bibi was removed to a holding area for transshipment to Cincinnati, which probably accounts for the high frequency of displays on that day as well as the absence of other social interactions - there were no displacements of Maguba by Thomas and no proximities or contacts recorded on June 2. Similarly the trace of blood in Maguba's urine on this date may be due to stress. The next highest frequency of chest-pounding displays falls on June 17, with ovulation a day or two later.

The main item of interest in Figure 2 is a marked association of genital examinations with menstruation as reckoned from urinalysis. Maguba's urine was collected daily and tested for blood with reagent strips (Ames Bili-Labstix). Hormone analysis of urine shipped to the research department of the Zoological Society of San Diego afforded precise retrodiction of ovulation (for their procedures in constructing hormone profiles see Lasley et al., 1982). Menstruation dates were calculated on the basis of positive hemostix test and/or hormone analysis. As noted, the behavior study was conducted blind throughout with respect to urine analysis.

The omission from record of a July ovulation cannot be taken to mean that none occurred. Maguba was cycling regularly before and after July; however, records for the period in question have been lost - a loss which seems particularly unfortunate given the suggestive occurrence of genital examinations on July 13 and 14.

We concluded from this series of observations that monitoring overt ordinary behavior might enable us to confirm, if not predict, ovulation. Meanwhile, it was decided to attempt to inseminate Maguba artificially, using fresh sperm drawn from Thomas. Since timing is of critical importance in artificial insemination, we asked ourselves what other kinds of observations or readily obtainable information could contribute to on-the-spot procedural decisions. It seemed to us that a conjunctive approach was called for. Urine analysis represented Maguba as cycling regularly, but the practical problem remained of how to time insemination precisely and effectively since no program of behavior observation was likely to be of much predictive value far in advance of the event. What was needed was day-by-day assessment of behavior changes during critical periods. We proposed that this might be provided through a conjunctive approach involving: (1) Prolonged close observation and recording of behavior in qualitative detail - beyond frequency counts of social/spatial interactions and stereotypical displays; (2) Closer attention to monitoring and self-monitoring (e.g., genital examination) and to self-stimulation and subtle seductive behavior directed toward familiar males (Hess, 1973) including human males (Patterson, 1981); (3) Strengthening mutual familiarity between at least one regular observer and Maguba, to increase observer responsiveness to subtle behavior cues and to afford regular daily measurement of anogenital cleft (Nadler, 1975); (4) Input from keepers and others with long-established relations with Maguba.

A first attempt at artificial insemination was made in November 1981. It was not successful, possibly because the trauma associated with sedation produced a last-minute suppression of ovulation (Albrecht, 1982). One way to avoid such trauma would be to inseminate Maguba without sedating either her or Thomas, using frozen sperm. Manual insemination presents obvious practical difficulties; however, working toward it fit in with our goal of obtaining more detailed information on physical changes concomitant with estrus (Nadler, 1975).

In late winter 1982, one of us (Lois Miller) began working with Maguba, making rough physical measurements and accustoming her to tactile contact and digital intromission.

The following summer, to conjoin with this work a study of Maguba's own and Thomas's monitoring of her reproductive state, we conducted another 8-week series of behavior observations, the results of which are summarized in Figure 3. Once again behavior observations were conducted blind with respect to urinalysis. No hormone data are available for this series, but timing of presumed menstrual periods (ca. May 12, June 5, July 3) does not suggest any obvious correlation between ovulation cycle events and neatly isolated overt activity. What we find interesting in these data are two features relating to their general distribution. The first has to do with activity rates for all categories of behavior (represented in Figure 3 as number of events per observation hour) except Displays. These are fairly flat, and they perhaps would be flatter if we had more hours. This surprised us, and we were equally surprised - though

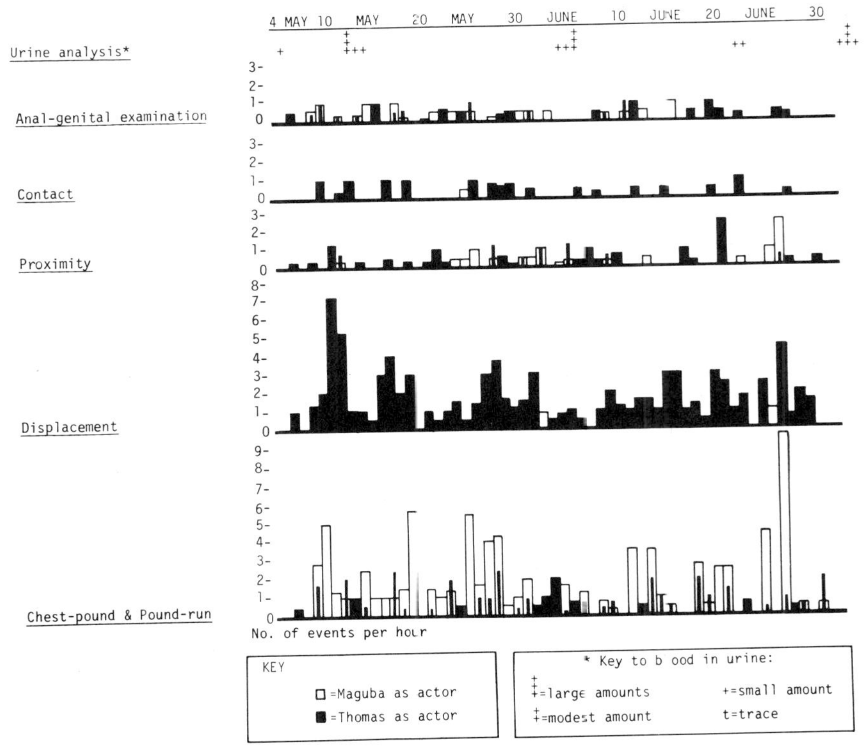

Figure 3. Summer 1982. Summary of behavior on a rate-per-hour basis (8 weeks).

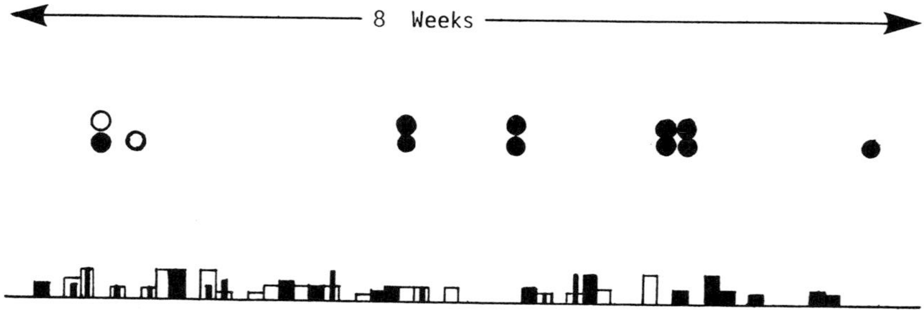

Figure 4. Comparison of anogenital examinations, 1981 and 1982.

probably we should not have been - by the second feature, the marked increase in monitoring of Maguba's reproductive state by both Maguba and Thomas. Figure 4 affords a comparison with the previous summer's 8-week observation series. In 1982, anogenital examinations were almost a daily activity.

DISCUSSION

What conclusions can be drawn from these observations? They seem to us to present important implications for tracking ovulation in gorillas at a distance, not least because of the light they shed on methodological issues. Apart from our own work, we have referred to a study conducted in the summer of 1980 (Chiszar and Caldwell, 1980) which revealed an interesting fluctuation in rate not only of solicitation of social interaction but of chest-pound and pound-run displays. In three successive years, then, observations have been made toward the same goal, relating overt behavior to internal processes. The last two series were designed specifically to facilitate prediction of ovulation. These are distinguished from one another by procedural differences that do not allow pooling of data, and partly for that reason we have taken care not to attribute too much importance to patterns revealed in a particular study. There is, however, a more basic reason for hesitating before grouping data from two or more series of observations where learning may be a factor in the relation between observer and observed. We can best approach it via an outline of methodological conclusions, three of them doubtless familiar, the last a summary inference which is perhaps not so obvious.

The first familiar conclusion, that caution must be observed when drawing inferences from a restricted set of observations, is well illustrated here. The soliciting behavior observed in 1980, which prompted our own work, was suggestive but we saw very little of it in succeeding years. In our 1981 study, anogenital examinations showed a nice correlation with events of the menstrual cycle, but no such neatly patterned correlation was revealed in 1982.

A second conclusion, that changes in context produce changes in behavior, bears a close relation to the first: for while it is well to regard with caution the results of short-term studies, one would not want automatically to discount distinctive differences between them as small sample effects in the statistical sense, i.e., as behaviorally meaningless statistical errors. The 1980 study was of three gorillas - Maguba, Thomas and Bibi. Given the social dynamics of this triad, in which Bibi and Thomas and Bibi and Maguba were the most interactive dyads, soliciting behavior directed by Maguba toward Bibi may have been not uncommon; and it is difficult to say at this point, given the one short study to go by, what the endogenous determinants of that soliciting behavior may have been. Similarly, the 1981 study was of Maguba and Thomas only, but it was begun with the removal of Bibi and was terminated 8 weeks later, so it covered a period of extraordinary disruption of existing social relations. The peaks and valleys of spatial relations between Thomas and Maguba may reflect a readjustment to one another simply as social individuals, while the timing of Thomas's examinations of Maguba suggests heightened interest in her at critical times in the menstrual cycle. A year later such examinations have become frequent, almost a daily feature (as would seem reasonable for effective monitoring of reproductive state), and perhaps more perfunctory.

But Maguba is also examining herself more, and this leads to our third conclusion: the determining context of animal behavior may include human observers. We know that observation affects behavior. Usually, since we cannot assign anything like a precise weight to observer interference in non-experimental settings, we resign ourselves to ignoring it as a methodological problem, apart from whatever care we may take to be discreet. That observation can significantly influence results obtained is evident when we look at chest-pound displays, for example, as measures of general or specifically sexual arousal. The problem is that both Maguba's and Thomas's displays were often occasioned by and/or directed toward familiar people. Some of our observers regularly noted more displays than did others, and two or three male observers became the objects of increasingly frequent window displays by Maguba in the course of the 1982 study, making for obvious difficulties in pooling and interpreting data even within a series of observations.

Observer effect becomes a particular problem where one is seeking to understand an animal's behavior while simultaneously trying to modify it. We referred earlier to our practice, beginning in the spring of 1982, of habituating Maguba to manual exploration of the genital area in order to obtain anogenital cleft measurements and to prepare her for artificial insemination without sedation. It would be naive to assume that this sort of activity had no bearing on subsequent increase in what we have termed Maguba's self-monitoring of reproductive state - or on similar activities by Thomas, who has been present and attentive at all of Lois's sessions with Maguba.

We observed at the outset that the behavior observations were ancillary to practical breeding goals. It may be that what began as an attempt to track ovulation by observations of overt ordinary behavior could be converted into a kind of instructional or sexual therapy program, whereby Thomas's interest in Maguba was quickened and Maguba grew more receptive to Thomas, all as a result of behavior by human beings analogous to the monitoring of reproductive readiness and receptivity which appears to be part of the gorilla repertoire. However, the more general conclusion to which we are led by considerations just noted is that there is a problem inherent in any attempt to combine prediction of infrequent internal events (even those that occur with the regularity of ovulation), on the basis of past and ongoing performance of the individual in question, with analysis of the character of external behavioral predictors. The problem lies in the fact that efforts to sharpen accuracy of prediction by intensifying observation or moving to a closer focus (e.g., from a checklist of social activity to precise measurement of changes in external form) are likely to affect behavior in not easily forseeable ways.

To note that intensified observations are more likely to elicit learned reactions is not to suggest that a principle of uncertainty appropriate for subatomic phenomena be applied to complex organisms. But at each level of behavior organization there may be system limits, characteristic of that level, to which observations of a given individual can be used to predict changes in that same individual's behavior. Establishing such limits is a primary research concern in areas of science and industry to which the man-machine relationship and detection and diagnosis of human or mechanical system failure are critical (e.g., in medicine, aerospace, nuclear power, computer technology; Rasmussen and Rouse, 1980). What is clear, for we do not want to close on such a general and speculative note, is that we have reached a point in our own work where the line between observation and manipulation is drawn exceedingly fine. For some captive gorillas ovulatory periods may be unmistakably indicated, at least in a general way, by overt ordinary behavior; but Maguba, under present circumstances, is not one of these. If we were to continue to think in terms of a behavior study ancillary to but somehow objectively detached from a practical breeding program, the goals of the former almost surely would be confounded by those of the latter.

REFERENCES

Albrecht, B. H.: Update on management of great ape fertility. Internat. J. Primat. 3:241. (1982).

Chiszar, D., and Caldwell, R.: Behavioral changes during Maguba's estrus cycle. Unpublished report prepared for the Denver Zoological Gardens. (1980).

Harcourt, A. H., and Stewart, K. J.: Sexual behaviour of wild mountain gorillas. In D. J. Chivers and J. Herbert. Recent Advances in Primatology. Vol. 6. pp. 611-612. Academic Press, New York. (1978).

Hess, J. P.: Some observations on the sexual behavior of captive lowland gorillas. In R. P. Michael, and J. H. Crook. Comparative Ecology and Behaviour of Primates. pp. 425-473. Academic Press, New York. (1973).

Keiter, M. D. and Pichette, L. P.: Reproductive behavior of captive subadult lowland gorillas (Gorilla g. gorilla). Zool. Garten, N. F., Jena. 49:215-237. (1979).

Lasley, B. L., Czekala, N. M., and Presley, S.: A practical approach to evaluation of fertility in the female gorilla. Amer. J. Primat. Suppl. 1:45-50. (1982).

Nadler, R. D.: Cyclicity in tumescence of the perineal labia of female lowland gorillas. Anat. Rec. 181:791-798. (1975).

Patterson, F. G.: Personal communication to Lois Miller. (1981).

Rasmussen, J., and Rouse, W. B.: Human Detection and Diagnosis of System Failures. Plenum Press, New York. (1980).

4

Determination of Estrogen Concentrations and Ovulation Detection in the Common Marmoset (*Callithrix jacchus*) by an Enzymatic Technique

T. E. Ziegler and G. G. Stott

First morning void urine samples were collected daily for 4 1/2 months from 8 female common marmosets (Callithrix jacchus). These samples were assayed for estrogren (estradiol plus estrone) concentration by an estradiol enzymatic technique, and results were calculated in micrograms/liter. From these data the marmoset's ovarian cycle was charted. Laparoscopic examinations were performed on 7 of the 8 animals to correlate morphological changes of the ovary with estrogen concentrations observed in the assays. Of the 7 females laparoscoped, 5 revealed cyclicity in estrogen excretion, 1 displayed one cycle and thereafter became irregular, and 1 was acyclic. The average cycle length for 19 cycles in 6 cycling marmosets was found to be 18 ± 4 days. The cyclic marmosets had estrogen concentrations ranging from a low of 2 micrograms/liter to a high of 3,326 micrograms/liter. Both intra- and inter-animal variability were high for estrogen concentrations during the cycle of the marmosets. In 4 of the 6 laparoscoped marmosets there was conclusive evidence to validate the estrogen cycle. In one marmoset a preovulatory follicle was seen the day following the estrogen peak. Laparoscopic examination revealed that ovulation in the marmoset may be unilateral or bilateral. The estrogen peak as determined by this estradiol enzymatic assay appears to be useful in determining the time of ovulation in the marmoset.

INTRODUCTION

The high fecundity of marmoset and tamarin species in captivity and the short interval to puberty and maturation have stimulated interest in the marmoset as a primate model for human related studies. The demand for small primates in human related research has increased with the decrease of available primates, and the marmoset species have become a valuable laboratory primate.

Female marmosets, unlike most primates, do not display visual signs during their ovarian cycle, such as menstrual bleeding or sexual skin swelling. Therefore, a female's ovarian cycle cannot be detected by external observation.

Initial studies on the common marmoset (<u>Callithrix jacchus</u>) have revealed the endocrinology of the ovarian cycle, pregnancy, and puberty to be similar to that of humans (Hearn and Lunn, 1975; Hearn, 1977; Hearn et al., 1978; Abbott and Hearn, 1978).

Radioimmunoassay (RIA) of plasma has been utilized to determine Callitrichid ovarian cycles (Preslock et al., 1973; Hearn and Lunn, 1975; Harding et al., 1982). While these assays yield accurate results, the stress of handling the monkeys and the repeated venipunctures may affect their reproductive cycles. The small size and weight (200-700 gm) of the Callitrichids further complicates the problem of repeated blood sampling. For these reasons, urinary estrogen assays are currently being used to monitor the ovarian cycle of marmosets and tamarins (Lunn, 1978; Katz and Epple, 1980; Hodges et al., 1981; Brand, 1981; Hodges et al., 1983; Hodges and Eastman, 1984; French et al., 1983).

This paper describes the use of a urinary enzymatic technique to monitor the estrogen concentrations throughout the reproductive cycle of the common marmoset, and the peak concentration of estrogen is correlated with ovulation as determined through laparoscopy.

The estradiol enzymatic assay measures the concentrations of estrone plus estradiol in the urine by using the transhydrogenase function of the enzyme estradiol-17beta dehydrogenase (Descomps et al., 1980). Estradiol-17beta dehydrogenase catalyzes the reversible interconversion of estradiol-17beta and estrone utilizing either nicotinamide adenine dinucleotide (NAD) or nicotinamide adenine dinucleotide phosphate (NADP) as a cofactor. A linear relationship is established between estrogen (estradiol and/or estrone) concentrations and the transhydrogenase activity measured by the appearance rate of reduced NAD (NADH) in the enzyme system (Nicolas et al., 1979). Large amounts of NADH can be accumulated through numerous oxidative cycles leading to NADH concentrations that can be determined by spectrophotometry (Nicolas et al., 1979).

METHODS AND MATERIALS

The experimental procedure consisted of collecting daily urine samples from 8 female common marmosets (<u>Callithrix jacchus</u>) and assaying the urine for estrogen concentrations by the estradiol enzymatic technique. From these data the marmoset's ovarian cycle was charted.

After the cycle was established, laparoscopic examinations were performed at regular intervals on 6 of the animals to correlate morphologic characteristics of the ovary with estrogen concentrations observed in the assays. Midway through the urine collection period, 2 males were introduced into the room in an attempt to ensure that the females would continue regular cyclicity. One male was place in the cage with 79A035 female who was suspected of being pregnant and the other male was placed with 79B001 female who appeared to be cycling normally.

Experimental animals

Eight adult common marmosets ranging in age from 2 to 4 1/2 years were used for the study (Table 1). The females were caged individually in a specially designed room for laboratory animals with controlled temperature (26°C) and a 12-hour day/night cycle (lights on between 0600 and 1800 hours). Cages (83.2 cm by 36.4 cm by 46.8 cm) were spaced to allow as much room as possible between

TABLE 1. Information on adult female common marmosets

Marmoset Number*	Age of Marmoset at time of Study (in yrs.)	Wgt. (in gm.)	Reproduction History	Remarks
77D003	4.50	289	4 Births 3 sets of twins 1 single birth	
78A011	3.75	232	Mated but no pregnancies	
79A017	3.00	310	Mated but no pregnancies	
79B001	2.75	251	1 birth - twins	Male introduced 5-7-82
79A035	2.50	310	Mated but no pregnancies	Suspected pregnant-male introduced 5-7-82
79B005	2.25	301	Mated but no pregnancies	
80D003	2.00	376	None	Died 5-10-82 during laparoscopic examination
80D001	2.00	239	None	

* The marmoset number indicates: year marmoset was born (first 2 numbers), generation (letter), and if the marmoset is male or female (last 2 numbers: odd=female, even=male)

each one, with four cages placed along each wall. Sleeping buckets, 2 wooden perches, and a water bottle were added to each cage. Fecal collection trays were cleaned daily and the entire cages were cleaned weekly.

Urine collection

Overnight urine was collected daily for 4 1/2 months. The marmoset cages allowed for the insertion of metabolic funnel pans to collect urine at night. The metabolic funnel pans sloped down to a hole in the middle and were covered by screen wire to prevent feces and food from contaminating the urine. A collection beaker (50 ml) was inserted under the pan to collect the urine. A metabolic funnel was placed under each cage at approximately 1800 hours when the monkeys were settling down for sleep. Urine was collected the following morning at approximately 0800 hours. The pans were thoroughly cleaned with hot water spray and left to dry until the next collection. Urine samples were either immediately frozen at -20°C until testing by enzymatic assay or refrigerated at 4°C if to be assayed that day.

Assay and calculation

The estradiol enzymatic reagents were purchased in kit form (BioMerieux, Lyons, France) and contained estradiol-17beta; Tris-HCl buffer; substrates, NAD, NADP, glucose-6-phosphate (G6P), glucose 6-phospate dehydrogenase (G6PDH), and the enzyme estradiol-beta 17beta-dehydrogenase. Estrogen conjugates were hydrolyzed by beta-glucuronidase, Type VII, from Escherichia Coli (Sigma Chemical Co.). Polypropylene test tubes, 12 by 75 mm (VWR Scientific Inc.) were used for all assays and urine preparation. The reagents were prepared for the assay in the following manner: The internal standard, 17-beta-estradiol (160 micrograms/liter), was diluted 1:99 with Tris-HCl buffer pH 7.4 (50 mmol/liter). The regenerating reagent solution was made by reconstituting one vial of lyophilized substrate containing NAD (1 mmol/liter), NADP (1 micromol/liter), and G6P (1 mmol/liter) with 20 ml of Tris-HCl buffer, pH 7.4 (50 mmol/liter). A 50-microliter aliquot of G6PDH (50 U/liter) was added to the solution. To prepare the working solution for the test samples, the regenerating reagent solution was mixed 40:1 with the enzyme, 17beta-estradiol-dehydrogenase (20 U/liter). To prepare the working solution for control samples, the regenerating reagent solution was mixed 40:1 with distilled water.

To prepare the urine for assay, thawed or fresh urine was mixed well and a 1 ml aliquot taken for sampling. The urine samples were hydrolyzed by adding 40 microliters of acetate buffer 0.1M, pH 6.5, and 1 Pasteur drop of beta-glucuronidase (reconstituted with 1 ml of phosphate buffer 0.1M, pH 6.5). Urine samples were mixed well and incubated at 37°C in a waterbath for 30 minutes. The hydrolyzed urine was precipitated with 1 ml of absolute ethanol. Samples were incubated for at least 30 minutes at 2-8°C and centrifuged (Beckman J-21C, 12 in. rotor) for five minutes at approximately 4,000 RPM. Then 400 microliters of the urine supernatant was diluted with 600 microliters of Tris-HCl.

The assay was micromodified to conserve reagents and increase the number of tests possible per test kit. For each urine sample there was a control tube, test tube, and standard tube. The control tubes contained 250 microliters Tris-HCl buffer, 50 microliters urine, and 250 microliters control working solution. The test tubes contained 250 microliters Tris-HCl buffer, 50 microliters urine, and 250 microliters test working solution. The standard tubes contained 250 microliters internal standard, 50 microliters urine, and 250 microliters test working solution. Contents of tubes were mixed and placed in the dark for 20 hours at room temperature. After incubation the contents of each tube were mixed well and the absorbance of each measured on a spectrophotometer (Beckman, model 125) at 340 nm. Zero adjustment was made with distilled water. For each sample the following calculation was performed: Test OD minus Control OD/Standard OD minus Test OD times 40 = $\underline{X}$ micrograms/liter.

Laparoscopy

Laparoscopic examination was performed 24 times on 7 marmosets. The marmosets were anesthetized with 15 mg/animal of ketamine-HCl (Vetalar, Parke-Davis) and a local anesthetic (0.2% Lidocaine-HCl with epinephrine, Elkins-Sinn Inc.) was injected at the incision site. Laparoscopic procedure for the common marmoset was essentially the same as described by Harrison (1980) and a 5 mm pediatric laparoscope was used for the examinations. An Olympus OM-1 camera with a specialized laparoscopic lens adapter was used to photograph the ovaries. ASA 400 Ektachrome, push processed at ASA 800, at 1/8 and 1/15 second shutter speed yielded the highest quality photographs.

RESULTS

Normal cyclicity was observed in 5 of the 7 females studied; one of the others displayed one cycle and became irregular and one was acyclic. A composite pattern of estrogen excretion in 19 cycles from the 6 cycling marmosets is shown in Figure 1. The one cycle from the irregularly cycling marmoset was added to the cycles of the 5 normally cycling marmosets. Estrogen concentration,

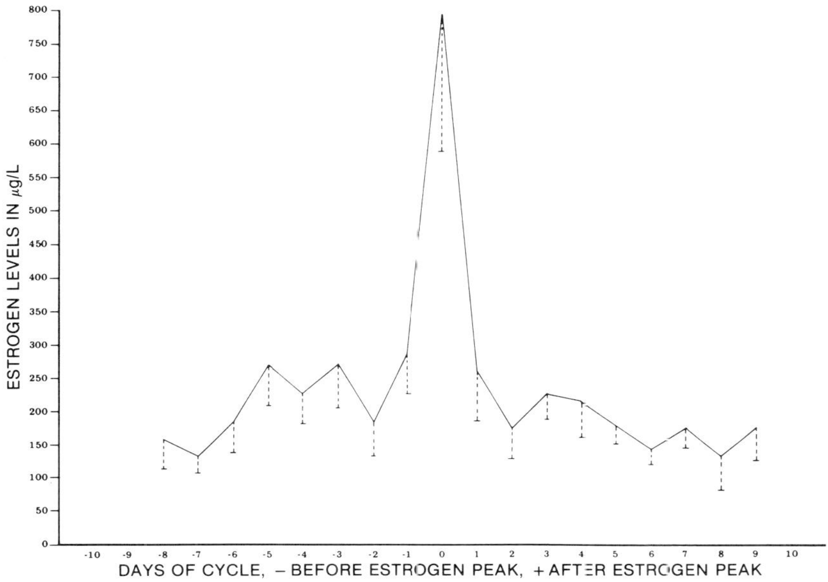

Figure 1. Composite estrogen cycle (mean-SE) for 6 marmosets through 19 cycles.

in micrograms/liter was measured against the days of the cycle. The average cycle was characterized by increasing fluctuations of estrogen concentrations from 9 to 3 days prior to the estrogen peak. A sharp rise in concentration began 2 days before the estrogen peak and culminated in a four-fold increase in estrogen concentration at the peak levels. This was calculated by averaging all levels of estrogen concentration except the peak level and calculating the increase that is associated with the ovulatory peak of estrogen. The estrogen concentration declined rapidly one day after the peak and then fluctuated through the remaining 9 days of the cycle. The average estrogen peak was 779 $\pm$ 185 micrograms/liter. Estrogen levels for one or more cycles per marmoset for the 6 marmosets ranged from a low of 2 micrograms/liter to a high of 3,326 micrograms/liter. The composite cycle revealed an 8-day follicular phase with the estrogen peak on day 9, and a 9-day luteal phase. The average cycle length was 18 $\pm$ 4 days.

Individual marmoset cycles were selected by designating the lowest estrogen concentrations as the beginning of the follicular phase, the highest concentration as the estrogen peak, and the following lowest estrogen concentration as the end of the luteal phase. Marmosets 79A017 and 78A011 were selected to illustrate the estrogen cycle and laparoscopic validation. The estrogen concentrations for marmoset

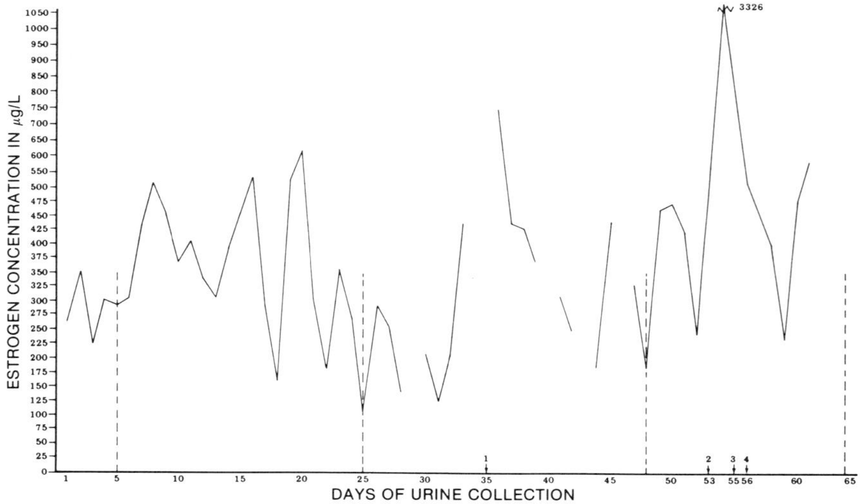

Figure 2. Estrogen concentrations for marmoset 79A017 during 66 days of urine collection. 1 = day male marmosets introduced into room, 2 = first laparoscopic examination, 3 = second laparoscopic examination, 4 = third laparoscopic examination.

79A017 are shown in Figure 2. Analysis of this female's urine revealed 3 cycle lengths extending over 61 days of urine collection. The average cycle length was 20 $\pm$ 3.5 days. The cycles varied considerably from each other with respect to estrogen concentrations and length of follicular and luteal phases. Values of estrogen ranged from 118 micrograms/liter to 3,326 micrograms/liter throughout the collection period. The first cycle (Figure 2) did not display a single definite estrogen peak, but several peaks in estrogen concentration. In the second cycle the open spaces indicate days where insufficient urine was found in the collection beaker for analysis. Even with samples missing it appeared that an estrogen peak occurred. The third cycle revealed a high estrogen peak reaching 3,326 micrograms/liter in concentration.

Laparoscopic examination was performed 3 times during the third cycle of marmoset 79A017. The examinations occurred on days 53, 55, and 56 of urine collection. The first examination, on day 53, occurred during the follicular phase when the estrogen level was 567 micrograms/liter. Examination revealed the left ovary was larger and contained a large corpus luteum on its cranial aspect. The right ovary contained a corpus albicans. A urine sample was not obtained on day 55, resulting in an unknown estrogen concentration; however, on day 54 the estrogen concentration was 3,326 micrograms/liter. This was the highest level of estrogen

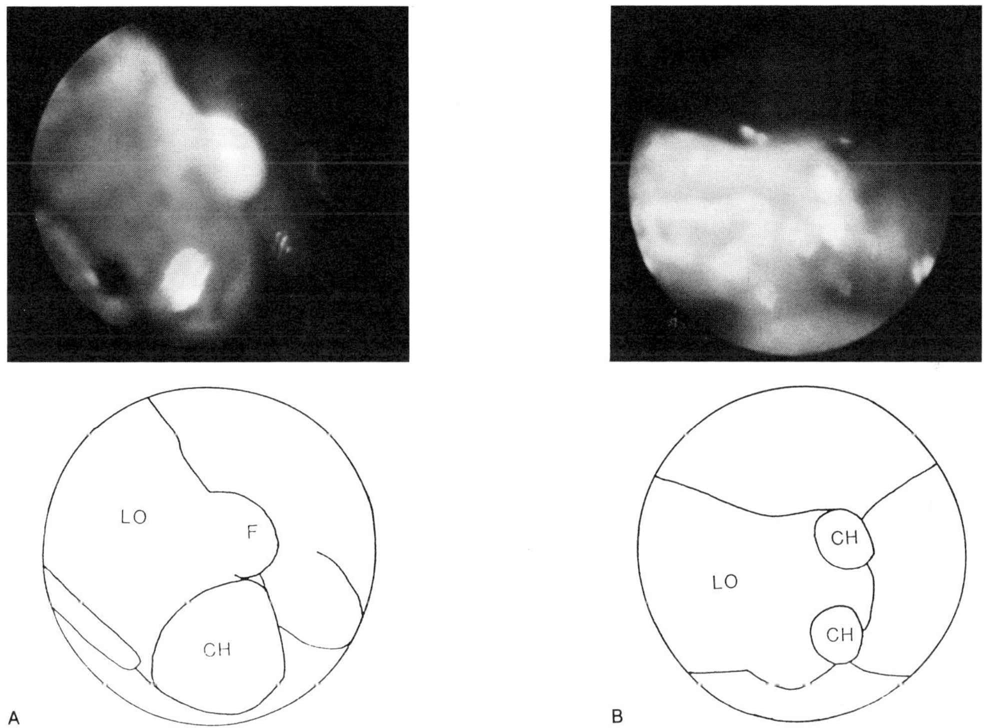

A B

Figure 3. Laparoscopic photographs and descriptive drawings of marmoset 79A017. A: Left ovary (LO) on day 55 demonstrating a preovulatory follicle (F) and a corpus hemorrhagicum (CH). B: Left ovary (LO) on day 56 demonstrating two corpra hemorrhagica (CH).

recorded in any of the marmosets. Examination of the ovary on day 55 revealed an estremely elongated left ovary with a large corpus luteum on the cranial portion of the ovary. A corpus hemorrhagicum was present on the caudal end of the ovary and a large preovulatory follicle was observed dorsal to the corpus hemorrhagicum (Figure 3A). The right ovary contained a regressed corpus albicans. The marmoset was laparoscoped the following day, day 56, to determine whether ovulation of the follicle had occurred. On day 56 the estrogen concentration was 517 micrograms/liter. The corpus luteum on the left ovary was lighter in color and more vascular than it was the previous day. The follicle had ruptured, and a corpus hemorrhagicum was seen in its place. The corpus hemorrhagicum observed the previous day was still present (Figure 3B). It appeared that two ovulations had occurred from the left ovary.

The estrogen concentrations from marmoset 78A011 are shown in Figure 4. There were 4 cycles recorded over the 87 days of urine collection. The average cycle length was 20 $\pm$ 6 days. There was considerable intra- and intercyclic variation and the estrogen values ranged from 43 micrograms/liter to 1,260 micrograms/liter. The first estrogen cycle had a high estrogen peak while the second cycle revealed no high peak estrogen concentration. The

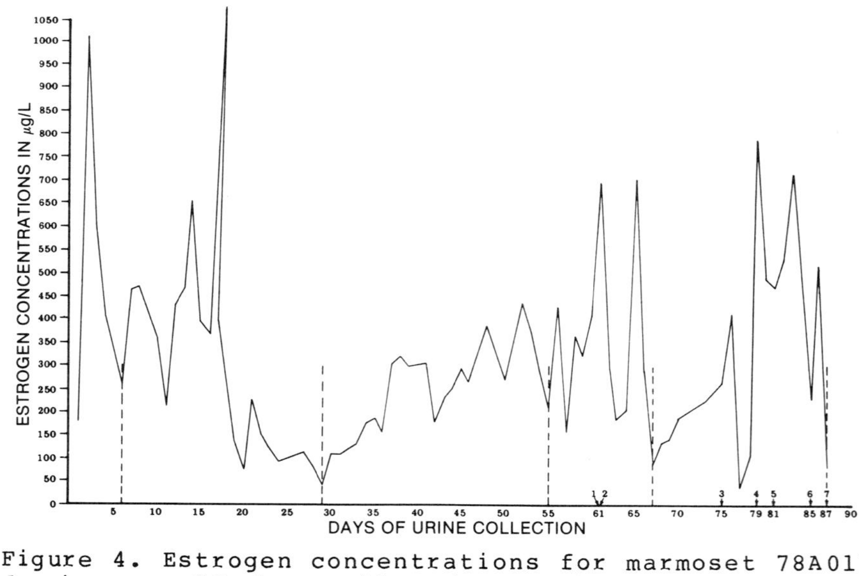

Figure 4. Estrogen concentrations for marmoset 78A011 during an 87-day collection period. 1 = day male marmosets were introduced into room, 2 = first laparoscopic examination, 3 = second laparoscopic examination, 4 = third laparoscopic examination, 5 = fourth laparoscopic examination, 6 = fifth laparoscopic examination, 7 = sixth laparoscopic examination.

third cycle consisted of two high peaks of almost equal estrogen concentration and the fourth cycle showed a high estrogen peak and 2 smaller peaks in the luteal phase.

Female 78A011 was laparoscoped 6 times (on days 61, 75, 79, 81, 85, and 87 of the urine collection). In the first laparoscopy, insertion of the trocarcannula appeared to cause excessive trauma to the female and the ovaries were not examined. The next examination occurred on day 75 when estrogen concentration was 264 micrograms/liter. This was in the follicular phase of the fourth cycle and estrogen levels were increasing. The right ovary was larger than the left and contained a very large corpus luteum assumed to be from the preceding cycle. The left ovary had a smaller corpus luteum. The marmoset was laparoscoped again on day 79, when estrogen levels peaked at 773 micrograms/liter. The right ovary was still larger and contained a large corpus luteum while the corpus luteum on the left ovary had regressed. On day 81 (2 days after the estrogen peak) estrogen levels had declined to 472 micrograms/liter, and laparoscopy was performed again. The right ovary was still larger and contained a persistent corpus luteum on its cranial portion (Figure 5A). The left ovary had enlarged and contained a corpus hemorrhagicum, indicating that an ovulation had occurred. Examination of the ovaries was performed again on day 85 (6 days following the estrogen peak) when the estrogen level was 233 micrograms/liter. Both ovaries contained large corpora lutea. The corpus luteum on the left ovary contained a prominent stigma or ovulation site (Figure 5B). The corpus luteum on the right ovary covered the entire cranial end of the ovary. The final laparoscopic examination occurred on day 87 at the end of the luteal phase of the fourth cycle. Estrogen concentrations had decreased to 74 micrograms/liter. Observation revealed the corpora lutea on both ovaries to be lighter in color but approximately the same size as observed 2 days earlier.

The 5 remaining females are described briefly. Six estrogen cycles were demonstrated for female 79B001; with an average cycle length of 15 $\pm$ 3 days. The sixth cycle was validated by observing a developing follicle in the follicular phase and a corpus luteum during the luteal phase on the left ovary. Three estrogen cycles were noted for female 80D001 with an average cycle length of 18 $\pm$ 3.6 days. Laparoscopic examinations performed during the first cycle revealed a corpus luteum on each ovary, indicating that two ovulations had occurred. Female 80D003 died during the first laparoscopic examination but had displayed one estrogen cycle. Examination of the ovaries revealed 2 large corpora lutea on the right ovary, indicating all ovarian activity had occurred on one ovary. Female 79B005 displayed irregular cyclic activity. One cycle of estrogen concentrations of 18 days was noted. Estrogen levels then decreased and remained low throughout the sampling period. Laparoscopic

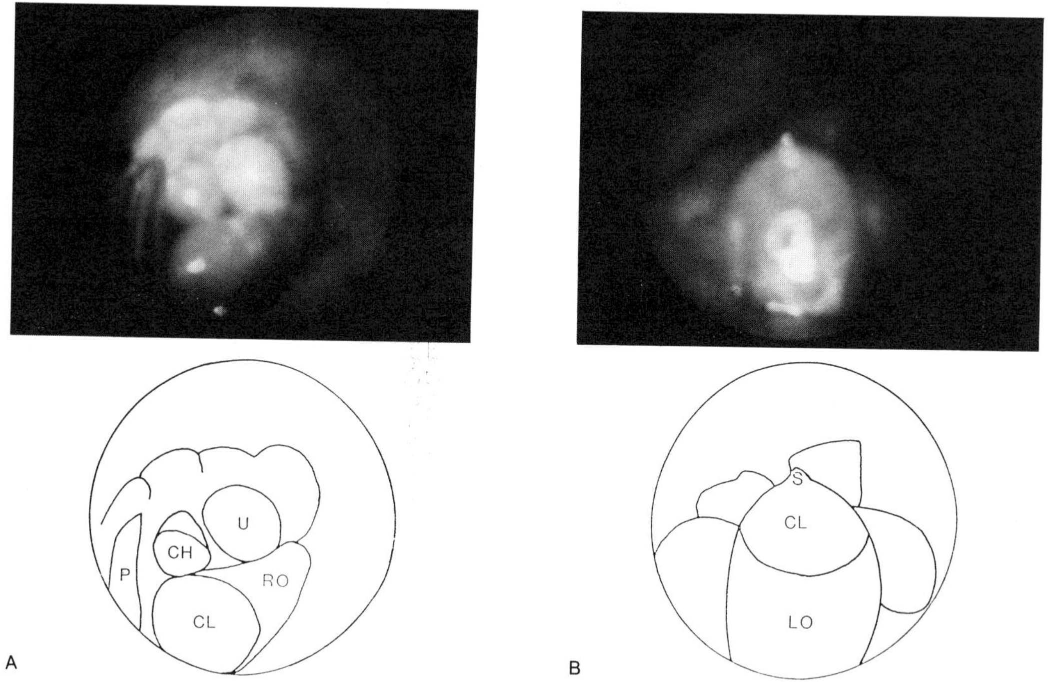

Figure 5. Laparoscopic photographs and descriptive drawings of marmoset 78A011. A: Right ovary (RO) on day 81 with a large corpus luteum (CL). The uterus (U), the probe (P), and the left ovary containing a corpus hemorragicum (CH) are in the background. B: Left ovary (LO) demonstrating a large corpus luteum (CL) and prominent stigma (S).

examination during the low estrogen period revealed no ovarian activity except a corpus albicans was noted on each ovary, presumably from the previous cycle. Female 77D003 had no clear cyclicity of estrogen concentrations throughout the sampling period, and laparoscopic examinations revealed quiescent ovaries.

It was not possible to monitor creatinine levels in this experiment as an adjustment for urine output. However, all peak levels of estrogen were four-to fivefold increases in concentration and creatinine concentrations do not vary more than twofold in marmosets and tamarins (French and Snowdon, personal communication; Ziegler, personal observation). Therefore, the cycle peaks of estrogen would still be clearly visible. It is possible, however, that baseline fluctuations would be reduced.

DISCUSSION

The estradiol enzymatic assay has been developed and used for monitoring the estrogen cycle of humans with excellent results (Nicolas et al., 1979; Descomps et al., 1980; Ziegler, personal observation). Preliminary results in the baboon (Ziegler, personal observation) indicated the estradiol enzymatic assay to be a reliable indicator of the estrogen peak in this species. This experiment indicated that the estradiol enzymatic assay was effective in monitoring estrogen cycles of the common marmoset. Even though the irregular cycles did not allow for exact prediction of the estrogen peak, ovulation appeared to be imminent 1-2 days after the estrogen peak as was noted in the 2 marmosets that were laparoscoped at that time. In marmoset 78A011 ovulation occurred 1 to 2 days after the estrogen peak. For marmoset 79A017, 1 ovulation occurred less than 24 hours after the estrogen peak and the other less than 2 days following the estrogen peak. Therefore, the estrogen peak, as determined by the estradiol enzymatic assay appears to be useful in determining the time of ovulation.

Estrogen concentrations are reported in marmosets and tamarins to be higher than values reported from other primates (Hampton and Hampton, 1974; Hearn and Lunn, 1975). The marmosets in this study were in the range of urinary estrogen concentrations reported for marmosets and tamarins; with estrone plus estradiol ranging from 2 micrograms/liter to 3,326 micrograms/liter. Lunn (1975) reported urinary estrogens measured by a fluorimetric assay for common marmosets to be 5.2-1,039.9 micrograms/24 hours. Recently, Hodges and Eastman (1984) found urinary estrogens to be slightly higher than reported in this study. Estrone ranged between 0.17-9.5 micrograms/mgCr and estradiol-17beta ranged from 0.29-48.9 micrograms/mgCr. Through differential hydrolysis and solvolysis, they found that the highest concentration of estrogens appeared in the sulfate fraction. For the cotton-topped tamarin (Saguinus oedipus), Brand (1981) reported urinary estrogens ranging 2 to 60 micrograms/mgCr and French et al. (1983) reported 3.8-10 micrograms/mgCr. Hodges and Eastman (1984) reported 2.16-9.68 micrograms/mgCr for urinary estrone and 1.02-7.35 micrograms/mgCr for urinary estradiol-17beta. Hodges et al. (1981) reported total urinary estrogens for a single white-lipped tamarin (Saguinus fuscicollis) to range between 20-90 micrograms/mgCr and Katz and Epple (1980) reported estrone levels to range between 1-10 micrograms/mgCr and estradiol-17beta to range between 0.3-0.9 micrograms/mgCr for the same species. Although these urinary estrogen levels are within a similar range, it is difficult to compare these results because of the different assays and antiserum used. Also there appears to be considerable variability between females even within the same species.

For the most part estrogen concentrations and ovarian events observed during laparoscopy correlated very well. During laparoscopic examinations of the marmosets' ovaries, a persistent opened stigma was noted on many of the corpora lutea. This occurrence of a persistent opened stigma has been reported for the rhesus monkey (Corner et al., 1963), the bonnet monkey (Barnes et al., 1978), and the cebus monkey (Nagle et al., 1980), although a functional significance has not been indicated.

It was also noted that ovarian activity could take place on one or both ovaries during any cycle. In 3 marmosets activity was observed in both ovaries. In 2 marmosets ovarian activity was limited to 1 ovary. The other 2 marmosets had inactive ovaries.

It was not possible to determine from the marmoset cycles whether the addition of males into the room affected their cycles. However, all females showed great interest in the males. Dominant female marmosets are known to affect the reproductive cycles in subordinate females (Epple, 1967; Abbott and Hearn, 1978; Katz and Epple, 1980) by causing a suppression of steriod hormones in the subordinate female (Abbott and Hearn, 1978; Katz and Epple, 1980). This has usually been associated with females sharing the same cage. In this study the cages were placed as far as possible from each other to avoid any cycle suppression and there did not appear to be any dominance effects on ovulation.

The estradiol enzymatic assay offers an alternative method for monitoring estrogen cycles without the hazards of radioactive isotopes. Considering the low cost and relatively short time requirement of this assay, it could prove to be a valuable test for monitoring the estrogen levels and hence for ovulation detection in nonhuman primates.

The marmoset urine samples were analyzed using a 20-hour incubation time in the enzymatic assay. Because of this, the method is approximately a 24-hour test. Nicolas

(1979) indicated that this test could be made into a 9-hour test by reducing the incubation time to 5 hours at 37°C, but the assay sensitivity would be reduced by a factor of 2. In this experiment we did not alter the incubation time since we were concerned with high sensitivity and determining if the test would work using marmoset urine. The test could be adapted to automated spectrophotometers, which would allow many samples to be assayed in a short time. In order to further validate the results of this assay in the common marmoset, it would be advantageous to correlate results of this assay with other hormonal profiles on the same samples.

ACKNOWLEDGMENTS

The authors wish to thank Dr. F. J. Stein and Dr. R. F. Sis for providing the marmosets for this study from the marmoset colony at Texas A&M University which is funded by the National Institutes of Health/DHEW RF-79-599(3523). We thank Dr. D. Kraemer who very graciously provided the equipment for the laparoscopies. Appreciation is extended to Dr. D. H. Lewis and Dr. J. E. Martin for counsel and advice throughout the study, to Dr. M. L. Davis for her photographic assistance with the laparoscopy, to Dr. R. W. Lewis for the use of the spectrophotometer in his laboratory, and to Nancy Arnold and Leo Busch for their laparoscopic assistance. We offer special appreciation to Mr. F. Febuay of BioMerieux for his helpful advice and supplying the enzyme kits.

REFERENCES

Abbott, D. H. and Hearn, J. P.: Physical, hormonal and behavioral aspects of sexual development in the marmoset monkey, Callithrix jacchus. J. Reprod. Fertil. 53:155-166. (1978).

Barnes, R. D., Lasley, B. L., and Hendrickx, A. G.: Midcycle ovarian histology in the bonnet monkey, Macaca radiata. Biol. Reprod. 18:537-553. (1978).

Brand, H. M.: Urinary oestrogen excretion in the female cotton-topped tamarin (Saguinus oedipus). J. Reprod. Fertil. 62:467-473. (1981).

Corner, G. W., Bartelmez, G. W. and Hartmen, C. G.: On normal and aberrant corpora lutea of the rhesus monkey. Amer. J. Anat. 59:433-457. (1963).

Descomps, B., Nicolas, J. C., Chikhaoui, Y., and Crastes de Paulet, A.: Prediction and detection of ovulation by hormonal measurements: Contribution of a new enzymatic method. J. Steroid Biochem. 12:385-393. (1980).

Epple, G.: Vergleichende untersuchungen uber sexual-und sozialverhaldten der Krallenaffen (Hapalidae). Folia Primatol. 7:37-65. (1967).

French, J. A., Abbott, D. H., Scheffler, G., Robinson, J. A., and Goy, R. W.: Cyclic excretion of urinary oestrogens in female tamarins (Saguinus oedipus). J. Reprod. Fertil. 68:177-184. (1983).

Hampton, J. K., and Hampton, S. H.: Some expected and unexpected characteristics of reproduction in Callitrichidae. Lab. Anim. Handbooks 6:235-240. (1974).

Harding, R. D., Hulme, M. J., Lunn, S. F., Henderson, C., and Aitken, R. J.: Plasma progesterone levels throughout the ovarian cycle of the common marmoset (Callithrix jacchus). J. Med. Primatol. 11:43-51. (1982).

Harrison, R. M.: Laparoscopy in monkeys and apes. In Harrison and Wildt. Animal Laparoscopy. pp. 73-93. Waverly Press, Baltimore. (1980).

Hearn, J. P., and Lunn, S. F.: The reproductive biology of the marmoset monkey, Callithrix jacchus. Lab. Anim. Handbooks. 6:191-202. (1975).

Hearn, J. P.: The endocrinology of reproduction in the common marmoset, Callithrix jacchus. In Kleiman. Biology and Conservation of the Callitrichidae. pp. 163-171. Smithsonian Institution Press, Washington. (1977).

Hearn, J. P., Abbott, D. H., Chambers, P. C., Hodges, J. K., and Lunn, S. F.: Use of the common marmoset, Callithrix jacchus, in reproductive research. Prim. in Med. 10:40-49. (1978).

Hodges, J. K., Gulick, B. A., Czekala, N. M., and Lasley, B. L.: Comparison of urinary oestrogen excretion in South American primates. J. Reprod. Fertil. 61:83-90. (1981).

Hodges, J. K., Brand, H., Henderson, C., and Kelly, R. W.: Levels of circulating and urinary oestrogens during pregnancy in the marmoset monkey (Callithrix jacchus). J. Reprod. Fertil. 67:73-82. (1983).

Hodges, S. K., and Eastman, S. A. K.: Monitoring ovarian function in marmosets and tamarins by the measurement of urinary oestrogen metabolites. Amer. J. Primatol. 6(3):187-197. (1984).

Katz, Y., and Epple, G.: Social influences on urinary estradiol cyclicity of female Saguinus fuscicollis (Callitrichidae). Antropol. Contemp. 3:219. (1980).

Lunn, S. F.: Urinary oestrogen excretion in the common marmoset, Callithrix jacchus. In Rothe, Wolters, and Hearn. Biology and Behavior of Marmosets. pp. 67-73. Eigenverlag Hartmut Rothe, Gottingen. (1978).

Nagle, C. A., Riarte, A., Quiroga, S., Azorero, R. M., Carril, M., Denari, J. H., and Rosner, J. M.: Temporal relationship between follicular development, ovulation and ovarian hormonal profile in the capuchin monkey (Cebus apella). Biol. Reprod. 23:629-635. (1980).

Nicolas, J. C., Boussious, A. M., Descomps, B., and Crastes de Paulet, A.: Enzymatic determination of estradiol and estrone in plasma and urine. Clin. Chim. Acta 92: 1-9. (1979).

Preslock, J. P., Hampton, S. H., and Hampton, J. K., Jr.: Cyclic variations of serum progestins and immunoreactive estrogens in marmosets. Endocrin. 92:1096. (1973).

5

Effects of Lactate and Pyruvate on *In Vitro* Fertilization of Squirrel Monkey Oocytes

P. J. Chan and W. R. Dukelow

Squirrel monkeys were induced to ovulate and follicular oocytes, recovered laparoscopically, were inseminated and cultured in medium with combinations of the three levels of lactate and 0, 0.25, 0.5, and 1.05 mM pyruvate. Pyruvate at a level of 1.05 mM significantly increased in vitro fertilization of squirrel monkey oocytes but no effects were noted on subsequent cleavage.

INTRODUCTION

Energy metabolism and substrate utilization by naturally fertilized embryos has been reported for a number of laboratory species (Kane, 1979; Brinster, 1965). This paper is concerned with the effects of various substrates on the maturation, fertilization and development, in an in vitro environment, of squirrel monkey oocytes. The energy sources interact to provide an optimum environment for embryonic growth (Brinster, 1965). Early mouse embryos utilize pyruvate as the major energy source (Brinster, 1971) although a synergistic effect is noted with the inclusion of lactate (Brinster, 1965; Wales and Whittingham, 1973; Quinn and Wales, 1973). The dividing mouse embryo at the 8-cell stage and beyond attains the ability to metabolize glucose (Brinster and Thomson, 1966). Glucose uptake by bovine blastocysts has been reported (Renard et al., 1980). Other energy sources such as oxaloacetate, malate and succinate and their utilization have been reported (Brinster and Thomson, 1966), but most work has focused on glucose, lactate and pyruvate, as substrates in culture systems.

The objective of this study was to determine the effects of added energy substrates on the maturation, fertilization and cleavage of in vitro cultured squirrel monkey oocytes.

METHODS AND MATERIALS

Squirrel monkeys (Saimiri sciureus) of Bolivian origin (Primate Imports Corp., Port Washington, New York) were maintained on a 12 hour:12 hour light/dark cycle at 21 ± 2°C. They were fed a commercial monkey feed with fresh water ad libitum. The squirrel monkeys were administered four daily intramuscular injections of follicle stimulating hormone (1 mg, FSH-P, Burns-Biotic Labs, Inc., Omaha, NE) and a single injection of HCG (250 I.U., APL[R] Ayerst Labs, Montreal) on the fourth day (Dukelow, 1970) 9 hours after the last FSH treatment. The anesthetized monkeys were laparoscoped (Dukelow et al., 1971) and follicular oocytes recovered (Kuehl and Dukelow, 1979) 15-16 hours after HCG treatment. The oocytes were incubated at 37°C (5% CO_2 in air) in supplemented TC 199 medium containing treatment combinations of 0, 0.25, 0.5, and 1.05 mM pyruvate and 0, 0.25, and 50 mM lactate. Semen was collected by means of electroejaculation (Kuehl and Dukelow, 1974) and added to the cultures.

At approximately 24 hour intervals the cultures were examined. The presence of a polar body indicated a mature oocyte, and the appearance of two or more polar bodies or two pronuclei and/or equal sized blastomeres indicated a fertilized oocyte. Statistical evaluation of the data was performed using multidimensional contingency analysis to assess interactions and Bonferroni chi-square to compare pooled totals.

RESULTS

The interaction of pyruvate and lactate was not significant for oocyte maturation (Table 1). The maturation rates of the 50 mM (45.6%) and the 25 mM lactate groups (27.8%), were not significantly different from the control group (37.1%).

The maturation rates in the control and ascending pyruvate concentrations were not significantly different. The overall maturation rate was 36.6%.

The absence or presence of lactate had no effect on in vitro fertilization rates (Table 2). However, the presence of pyruvate increased fertilization rates by 17.2 to 35.5 percentage units over the control group, with a significant increase ($p < 0.05$) at the 1.05 mM pyruvate level.

The number of embryos continuing development in culture (Table 3) were limited and no statistical differences were found in the treatment groups.

TABLE 1. Effect of pyruvate and lactate supplementation on *in vitro* maturation of squirrel monkey oocytes

No. of Matured/No. of Recovered Oocytes (%)

Lactate Concentration (mM)	Pyruvate Concentration (mM) 0	0.25	0.50
0	4/15 (26.7)	6/15 (40.0)	8/21 (38.1)
25	4/17 (23.5)	5/19 (26.3)	7/25 (28.0)
50	10/23 (43.5)	9/17 (52.9)	9/23 (39.1)
Total	18/55 (32.7)	20/51 (39.2)	24/69 (34.8)

Lactate Concentration (mM)	Pyruvate Concentration (mM) 1.05	Total
0	25/65 (38.5)	43/116 (37.1)
25	11/36 (30.6)	27/97 (27.8)
50	13/27 (48.1)	41/90 (45.6)[a]
Total	49/128 (38.3)	111/303 (36.6)

[a] Significantly different from the 25 mM level of lactate, but not from the control ($p<0.05$).

TABLE 2. Effect of pyruvate and lactate supplementation on *in vitro* fertilization of squirrel monkey oocytes

No. of Fertilized/No. of Matured Oocytes (%)

Lactate Concentration (mM)	Pyruvate Concentration (mM) 0	0.25	0.50
0	0/4 (0.0)	0/6 (0.0)	3/8 (37.5)
25	2/4 (50.0)	4/5 (80.0)	3/7 (42.9)
50	3/10 (30.0)	5/9 (55.5)	6/9 (66.7)
Total	5/18 (27.8)a	9/20 (45.0)	12/24 (50.0)

	1.50	Total
0	18/25 (72.0)	21/43 (48.8)
25	5/11 (45.5)	14/27 (51.9)
50	8/13 (61.5)	22/41 (53.7)
Total	30/49 (63.3)b	57/111 (51.4)

[a,b] Different superscripts indicate a significant difference ($p<0.05$).

TABLE 3. Effect of pyruvate and lactate supplementation on clevage of squirrel monkey oocytes

No. of 2 Cell Embryos/No. of Fertilized Oocytes (%)

Lactate Concentration (mM)	Pyruvate Concentration (mM) 0		0.25		0.50	
0	-		-		0/3	(0.0)
25	2/2	(100.0)	0/4	(0.0)	0/3	(0.0)
50	0/3	(0.0)	0/5	(0.0)	1/6	(16.7)
Total	2/5	(40.0)	0/9	(0.0)	1/12	(8.3)

Lactate Concentration (mM)	Pyruvate Concentration (mM) 1.05		Total	
0	3/18	(16.7)	3/21	(14.2)
25	2/5	(40.0)	4/14	(28.6)
50	2/8	(25.0)	3/22	(13.6)
Total	7/31	(22.6)	10/57	(17.5)

DISCUSSION

The results of the present paper indicate no influences of lactate on the in vitro fertilization process. In contrast, the addition of pyruvate to the medium increased the fertilization rate above the control. This may be a result of pyruvate acting to sustain sperm motility. Pyruvate is the major energy source of preimplantation mouse embryos (Biggers et al., 1967; Brinster, 1971) and thus it is possible that added pyruvate enters the oocyte mitochondria to produce energy for growth.

ACKNOWLEDGMENTS

The authors thank Mrs. L. M. Cleeves and Dr. Takehito Asakawa for their help in this paper. This work was supported by research grants from the March of Dimes Birth Defects Foundation and the National Institutes of Health.

REFERENCES

Biggers, J. C., Whittingham, D. G., and Donahue, R. D.: The pattern of energy metabolism in the mouse oocyte and zygote. Proc. Nat. Acad. Sci. 58:560-567. (1967).

Brinster, R. L.: Studies on the development of mouse embryos in vitro. J. Reprod. Fertil. 10:227-240. (1965).

Brinster, R. L., and Thomson, J. L.: Development of eight-cell mouse embryos in vitro. Exp. Cell Res. 42:308-315. (1966).

Brinster, R. L.: Oxidation of pyruvate and glucose by oocytes of the mouse and rhesus monkey. J. Reprod. Fertil. 24:187-191. (1971).

Chan, P. J., Hutz, R. J., and Dukelow, W. R.: Nonhuman primate in vitro fertilization: seasonality, cumulus cells, cyclic nucleotides, ribonuclei acid, and viability assays. Fertil. Steril. 38:609-615. (1982).

Dukelow, W. R.: Induction and timing of single and multiple ovulations in the squirrel monkey (_Saimiri sciureus_). J. Reprod. Fertil. 22:303-309. (1970).

Dukelow, W. R., Jarosz, S. J., Jewett, D. A., and Harrison, R. M.: Laparoscopic examination of the ovaries in goats and primates. Lab. Animal Sci. 21:594-597. (1971).

Kane, M. T.: Fatty acids as energy sources for culture of one-cell rabbit ova to viable morulae. Biol. Reprod. 20:323-332. (1979).

Kuehl, T. J., and Dukelow, W. R.: A restraint device for electroejaculation of squirrel monkeys (_Saimiri sciureus_). Lab. Animal Sci. 24:364-366. (1974).

Kuehl, T. J., and Dukelow, W. R.: Maturation and in vitro fertilization of follicular oocytes of the squirrel monkey (_Saimiri sciureus_). Biol. Reprod. 21:545-556. (1979).

Mizoguchi, H., and Dukelow, W. R.: Effect of timing on HCG injection on fertilization in superovulated hamsters. Biol. Reprod. 23:237-241. (1980).

Quinn, P., and Wales, R. G.: Uptake and metabolism of pyruvate and lactate during preimplantation development of the mouse embryo in vitro. J. Reprod. Fertil. 35:273-287. (1973).

Renard, J. P., Philippo, A., and Menezo, Y.: In vitro uptake of glucose by bovine blastocysts. J. Reprod. Fertil. 58:161-164. (1980).

Wales, R. G., and Whittingham, D. G.: The metabolism of specifically labelled lactate and pyruvate by two-cell mouse embryos. J. Reprod. Fertil. 33:207. (1973).

6

Cryopreservation of Squirrel Monkey Ova

F. J. DeMayo, R. J. Hutz,
and W. R. Dukelow

The squirrel monkey is the only nonhuman primate in which, to date, reliable and constant *in vitro* fertilization has been achieved. The present study concerned the feasibility of the preservation of squirrel monkey ova in liquid nitrogen. Preliminary studies were carried out on the more readily obtainable oocytes of the golden hamster. Cryopreservation was accomplished by placing ova in media containing 1.5 M or 2 M DMSO and slowly cooling the ova (-0.24°C/min) to a temperature of -80°C. The ova were then immersed in liquid nitrogen (-196°C) and stored. Ova were recovered by slow thawing (1 to 2°C/min) or rapid thawing (100°C/min). Ovum viability was judged by morphological appearance, trypan blue exclusion and fluorescein diacetate uptake. Hamster ova survived at a rate of 64.2% when frozen and slow thawed in 1.5 M DMSO in Dulbeccos phosphate buffered saline. Changing the medium, rapid thawing or increasing the DMSO concentration to 2 M had no effect on survival after cryopreservation. Squirrel monkey ova were judged morphological viable (37.5%) after cryopreservation and slow thawing. After 3 hours in culture, 11.8% of the frozen thawed monkey ova survived as judged by trypan blue exclusion. This is a significant decrease when compared to controls. Rapid thawing did not increase viability but increasing the DMSO concentration from 1.5 M to 2 M increased the viability of cryopreserved squirrel monkey ova after 3 hours in culture.

INTRODUCTION

The science of *in vitro* fertilization in humans is approaching widespread clinical applicability. This generation of human embryos emphasizes the need for a nonhuman primate model to study the events that occur immediately at the time of fertilization.

At the Endocrine Research Unit of Michigan State University, we have developed a repeatable and highly efficient *in vitro* fertilization system for the squirrel monkey (*Saimiri sciureus*; Kuehl and Dukelow, 1975a, 1979; Asakawa and Dukelow, 1982; Chan et al., 1982).

The productivity of this system, however, would be enhanced by a readily available pool of ova which would be used in *in vitro* fertilization trials. The intent of the present study was to develop a system for the cryopreservation of squirrel monkey ova. However, before attempts were made to freeze squirrel monkey ova, the procedures and variables involved in cryopreservation such as the optimal medium, thawing rate and concentration of cryoprotectant were tested on hamster ova. The procedures for the cryopreservation of hamster ova were based on the techniques described by Quinn et al. (1982).

MATERIALS AND METHODS

Hamster ova were collected by flushing the oviducts of euthanized female hamsters which had received a superovulating regimen of gonadotropin. Thirty IU of pregnant mares serum (PMS) given by intraperitoneal injection on Day 1, the day of postovulatory vaginal plug and (72 to 76 hours later) an injection of 30 IU human chorionic gonadotropin (HCG) intraperitoneally. The ova were collected 17 hours later (Mizoguchi and Dukelow, 1981).

Squirrel monkey ova were collected by the laparoscopic technique of Dukelow and Ariga (1976) and aspirating the contents of the follicle into 0.1 ml of culture medium using a 25 gauge needle and tuberculin syringe. Prior to the collection of ova, the squirrel monkeys received an ovulatory regimen of gonadotropin consisting of an intramuscular injection of 1 mg follicle stimulating hormone (FSH) per day for four days followed by an intramuscular injection of 250 IU of HCG on the fourth day (Dukelow, 1979). The HCG was administered 17 hours prior to ovum collection. During the summer, July through September, the FSH regimen was extended to five days due to the seasonal follicular response of the squirrel monkey (Kuehl and Dukelow, 1975b),

Hamster ova were collected in either Dulbeccos phosphate-buffered saline (PBS) or tissue culture media 199 modified with 20% fetal bovine serum, pyruvate and either gentamicin or penicillin-streptomycin. Squirrel monkey ova were collected in this modified tissue culture medium 199 (mTC 199).

As the ova were collected, they were placed in an ampule containing 0.1 ml of the freezing medium. The ampule was held in an ice bath until all the ova were collected. After the completion of ovum collection, 0.1 ml of either 3 M DMSO or 4 M DMSO was added to the ampule to bring the final concentration of this cryoprotectant to either 1.5 M or 2 M. The ampule was then placed in a 95% ethanol bath at -7°C -10°C and then seeded by touching the side of the ampule with forceps which had been immersed in liquid nitrogen. The seeded ampule was allowed to cool in the freezing apparatus at an average rate of 0.24°C/min. When the temperature of the apparatus reached -80°C the ampule was plunged in liquid nitrogen and stored at -196°C.

The freezing apparatus was an unsilvered Dewar flask which has been drawn to a vacuum of 6 Torr; a well insulated reservoir of liquid nitrogen into which the Dewar flask was placed; a laboratory elevator jack to regulate the depth of the Dewar flask in liquid nitrogen and a thermometer. The Dewar flask contained 350 ml of 95% ethanol. The freezing rate was determined by the slope of the least squares fit regression line of the changing temperature with time. This rate is a function of the vacuum in the unsilvered flask and its submersion in liquid nitrogen.

Recovery of these ova, was by either slow or rapid thawing of the ampules. Slow thawing was accomplished by placing the frozen ampule in a 600 ml beaker containing 200 ml of 95% ethanol, which had been cooled to -110°C and which was allowed to equilibrate to 20°C. The warming rate was estimated by least squares regression analysis and averaged 1 to 2°C/min.

Rapid thawing was accomplished by transferring the ampules from liquid nitrogen directly to an ice bath. The rapid thawing rate approximated 100°C/min.

Once the contents of the ampule had thawed, 1.8 ml of media was added stepwise to dilute the DMSO. This was done by adding 0.2 ml, 0.2 ml, 0.4 ml and 1 ml of media at 1 minute intervals.

The recovered ova were examined for morphologic normality as described by Quinn et al. (1982), for the ovum's ability to exclude the vital stain, trypan blue and by viewing their ability to incorporate fluorescine diacetate (FDA), as described by Chan et al. (1982). The ova were then cultured for three hours in mTC199 at 37°C and their viability again assessed by the same three procedures.

Cryopreserved, squirrel monkey ova and in frozen control ova were cultured with 20microCi/ml of ^{3}H leucine. After culture, they were fixed, embedded in parafin, sectioned, and examined by autoradiography to estimate rates of protein synthesis as described by Chan et al. (1982).

RESULTS

Hamster ova responded well when frozen in either PBS or mTC199 as shown in Table 1.

Although the viability of cryopreserved hamster ova was significantly greater in mTC199 immediately after recovery (X^2=14.6), the advantage disappeared after 3 hours in culture as shown by Table 2.

Since the *in vitro* fertilization medium used for the primate was mTC199, this confirmed the applicability of this medium for freezing. All subsequent freezing procedures used this medium.

The effects of rapid and slow thawing on hamster ova are shown in Table 3.

Rapid thawing gave no significant advantage over slow thawing (Table 4) and no significant difference was noticed after three hours in culture between rapid and slow thawing.

The effect of increasing the concentration of DMSO from 1.5 M to 2 M is shown in Table 5. This did not effect the viability of hamster ova after freezing and thawing and no difference in viability after 3 hours of culture was seen.

The response of squirrel monkey ova to these freezing and thawing regimens is shown in Table 6. Viability of ova recovered from storage in 1.5 M DMSO and following slow thawing was significantly lower than nonfrozen controls as measured by all three procedures (trypan blue, FDA, morphology). There was a sharp decrease in the viability after 3 hours in culture which was not seen in controls as shown in Table 7.

Control ova (14) and frozen squirrel monkey ova (13) were recovered and cultured with 20micro-Ci/ml of ^{3}H-leucine. After culturing, they were fixed, embedded in parafin, sectioned and examined by autoradiography to estimate rates of protein synthesis. There was an abundance of grains in autoradiographs of control ova (359 $\pm$ 21 grains/1000 micro-M^2) while cryopreserved ova showed a sparcity of grains 5 $\pm$ 1.51 g/1000 microM^2. This would indicate a diminished abiility of cryopreserved ova to incorporate ^{3}H-leucine and a decrease in protein synthesis.

Changing from slow to rapid thawing (Table 8) did not increase the viability of cryopreserved squirrel monkey ova, but increasing the concentration of DMSO from 1.5 to 2 M, as shown in Table 9, showed a slight but not significant increase in viability of recovered ova. However, the viability of ova after 3 hours in culture was significantly increased (X^2=7.02) when the DMSO concentration was increased to 2 M.

DISCUSSION

The cryopreservation of hamster ova can be accomplished with high efficiency as demonstrated above, and as previously described by Quinn et al. 1982. Under these procedures, changing from PBS to the modified TC199 medium had no effect on the viability of frozen-thawed hamster ova. Since the modified TC199 had no adverse effects on the viability of hamster ova after cryopreservation it should serve as a suitable medium for the cryopreservation of squirrel monkey ova. This would represent a convenience, since the modified TC199 is the

TABLE 1. Media effects on the viability of hamster ova frozen in 1.5M DMSO and slowly thawed

Medium	Mean warming rate (^{o}C/min)	Ova deposited	Ova recovered (%)	Morpho-logically viable	Trypan blue negative	FDA+
PBS	1.88	415	338 (81.4)	228 (67.4)	217 (64.2)	215 (63.6)
mTC199	1.53	424	346 (81.0)	276 (79.8)	268 (77.5)	252 (72.8)

TABLE 2. Media effects on the viability of hamster ova frozen and thawed after 3 hours in culture

Medium	Mean warming rate ^{o}C/min	Ova cultured	Trypan Blue negative before culture (%)	Trypan Blue negative after culture (%)
PBS	1.84	390	253 (64.9)	247 (63.3)
mTC199	1.53	346	368 (77.4)	228 (65.4)

TABLE 3. The effects of rapid and slow thawing on hamster ova frozen in 1.5M DMSO in mTC199

Mean thawing rate ^{o}C/min	Ova frozen	Ova recovered (%)	Morphol-logically viable (%)	Trypan blue negative (%)	FDA positive (%)
1.53	427	346 (81.7)	276 (79.8)	268 (77.5)	252 (72.8)
100.0	157	133 (84.7)	107 (80.5)	112 (84.2)	111 (83.5)

TABLE 4. The effects of rapid and slow thawing on hamster ova frozen in 1.5M DMSO in mTC199 after 3 hours in culture

Mean thawing rate $^{o}C/min$	Ova cultured	Trypan blue negative prior to culture (%)	Trypan blue negative after 3 hours in culture (%)
1.53	346	268 (77.5)	228 (65.9)
100.0	133	112 (84.2)	93 (69.9)

TABLE 5. The effects of concentration of DMSO on the viability of frozen/thawed hamster ova

Concentration of DMSO (M)	Mean thawing rate	Ova frozen	Ova recovered (%)	Morphologically viable (%)	Trypan blue negative (%)	FDA+ (%)
1.5	1.53	427	346 (81.7)	276 (79.8)	268 (77.5)	252 (72.8)
2.0	100.0	477	450 (94.3)	358 (79.5)	345 (78.7)	327 (72.6)

TABLE 6. The viability of squirrel monkey ova after cryopreservation in 1.5M DMSO in mTC 199

Treatment	Mean thawing rate ($^{o}C/min$)	Ova frozen	Ova recovered	Morphologically viable (%)	Trypan blue negative (%)	FDA+ (%)
Control	-	-	16	15 (93.8)	15 (93.8)	15 (93.8)
Cryopreserved	1.83	51	48	18 (37.5)	17 (35.4)	6 (12.5)

TABLE 7. The viability of frozen/thawed squirrel monkey ova after 3 hours in culture

Treatment	DMSO concentration (M)	Ova frozen	Trypan blue negative before culture (%)	Trypan blue negative after culture (%)	FDA+ before culture (%)	FDA+ after culture (%)
Control	-	16	15 (93.8)	15 (100.0)	15 (100.0)	13 (86.7)
Frozen	1.5	48	17 (35.4)	2 (11.8)	6 (12.5)	0 (0.0)
Frozen	2.0	20	10 (50.0)	6 (60.0)	5 (25.0)	3 (60.0)

TABLE 8. The effect of thawing on squirrel monkey ova cryopreserved in 1.5M DMSO

Mean thawing rate °C/min	Ova deposited	Ova recovered (%)	Morphologically viable (%) (%)	Trypan blue negative (%)	FDA+ (%)
1.83	51	48 (94.1)	18 (37.5)	17 (35.4)	6 (12.5)
100.0	31	28 (90.3)	6 (21.4)	2 (7.1)	1 (3.6)

TABLE 9. The effects of concentration of DMSO on the viability of cryopreserved squirrel monkey ova

Concentration DMSO (M)	Ova deposited	Ova recovered	Morphologically viable (%)	Trypan blue negative (%)	FDA+ (%)
Control	-	16	15 (93.8)	15 (93.8)	15 (93.8)
1.5	51	48 (94.1)	18 (37.5)	17 (35.4)	6 (12.5)
2.0	21	20 (95.2)	11 (55.0)	10 (50)	5 (25)

medium already used for the *in vitro* fertilization of squirrel monkey ova. Changes of other variables, such as slow to rapid thawing, or increasing the concentration of DMSO from 1.5 M to 2 M, had no effect on the success of cryopreservation of hamster ova.

The viability of squirrel monkey ova after cryopreservation was much less than that of hamster ova. Thawing at 100°C/min or 1.83°C/min did not alter the viability. This decrease in viability is seen not only by vital staining with trypan blue and FDA (which measure membrane integrity) but also in the ovum's ability to synthesize protein, as determined by autoradiographic analysis.

Increasing the concentration of DMSO from 1.5 M to 2 M slightly increased cryopreserved squirrel monkey ovum viability, although not to control levels. This could indicate a difference in permeability of the squirrel monkey ovum's membrane compared to the hamster. Since cooling rate lethality is a function of membrane permeability, a decrease in the cooling rate may improve cryopreservation of squirrel monkey ova.

REFERENCES

Asakawa, T., and Dukelow, W.R.: Chromosomal analysis after *in vitro* fertilization of squirrel monkey (*Saimiri sciureus*) oocytes. Biol. Reprod. 26:579-585. (1982).

Chan, P., Hutz, R., and Dukelow, W.R.: Non-human primate *in vitro* fertilization, seasonality, cumulus cells, cyclic nucleotides, ribonucleic acid and viability assays. Fertil. Steril. 38(5):616-620. (1982).

Dukelow, W.R.: Human chorionic gonadotropin: Induction of ovulation in the squirrel monkey. Science 206:234-235. (1979).

Dukelow, W.R., and Ariga, S.: Laparoscopic techniques for biomedical research. J. Med. Primatol. 5:89-99. (1976).

Kuehl, T.J., and Dukelow, W.R.: Fertilization *in vitro* of *Saimiri sciureus* follicular oocytes. J. Med. Primatol. 24:209-216. (1975a).

Kuehl, T.J., and Dukelow, W.R.: Ovulation induction during the anovulatory season in *Saimiri sciureus*. J. Med. Primatol. 4:23-31. (1975b).

Kuehl, T.J., and Dukelow, W.R.: Maturation and *in vitro* fertilization of follicular oocytes of the squirrel monkey (*Saimiri sciureus*). Biol. Reprod. 21:545-556. (1979).

Quinn, P., Barros, C., and Whittingham, D.G.: Preservation of hamster oocytes to assay the fertilizing capacity of human spermatozoa. J. Reprod. Fert. 66:161-168. (1982).

7

Primate Colony Management of Harem Breeding Groups of Rhesus Monkeys (*Macaca mulatta*)

G. P. Goo

The harem breeding system has been the sole breeding mode employed at the Texas Primate Center since its establishment in 1974. By January 1982, the population exceeded 6,400 animals, distributed across six discrete colonies. Rhesus monkeys comprised 91% of the population. Of these, 51% were adults and the remainder were offspring under 3 years of age. All social groups are housed outdoors in corn crib type cages. Colony management practices include: an average adult breeder ratio of one male to seven females, weaning of infants of 6 months of age to peer groups of 10-14 animals, formation of breeder replacement groups at 3 years of age for females and 5 years of age for males, a fostering program for rejected or orphaned infants, permanent removal from their social group and relocation after hospitalization of animals requiring treatment for traumatic injuries. Colony-born females are retained for future breeding stock and colony-born males are retained on a ratio of one male for every 10 females. Reproductive rates (after Taub, 1982) from the 1981 breeding season yielded a pregnancy rate of 84.3% and production birth rate of 68%. Colony-born females were found to be as reproductively proficient as wild-caught females. The mortality rate for 1981 was 8.5% and the morbidity rate was 590 clinical episodes per 1,000 animals. Adult females had a higher proportion of traumatic injuries than the population as a whole.

INTRODUCTION

At the Texas Primate Center (TPC), the harem breeding concept has been the sole system of breeding rhesus monkeys (Macaca mulatta) since its establishment in 1974. The harem system, which houses one male with multiple females for random breeding, has advantages over other housing systems including: low initial caging costs, production of a multipurpose animal of known parentage which is suitable as an experimental subject or as a breeder, ease of clinical surveillance, and ready access to individual animals. However, with respect to the other ways of housing breeding colonies that are being discussed in this symposium, there are also disadvantages. Unlike animals housed indoors, there is a distinctly seasonal breeding pattern. Labor costs are higher than those associated with island or corral operations, and clinical problems are more frequent due to higher levels of aggression and stress that result from a more restrictive environment. This paper presents an overview of the housing and management practices at the TPC and summarizes recent data on reproduction, mortality, and morbidity in rhesus monkeys.

METHODS AND MATERIALS

Facility layout and animal housing

The facility is located on 84.2 hectares in south Texas. Contract breeding programs are maintained as autonomous colonies, each composed of animal units, animal and personnel support buildings, and a work force which is spatially separated from other colonies. Administrative functions, animal records, and a central supply area are housed in a single office building.

All social groups of monkeys are housed outdoors in modified corn crib type units ("F-C" Model Series, Behlen Mfg. Co., Columbus, NB). The basic unit is 3.86m in diameter and is constructed of prefabricated welded wire panels that are 2.39m in height. The panels are topped by a conical roof which is comprised of sheet metal sections, bringing the overall height of each unit to 4.83m. The unit rests on concrete blocks (41 by 20 by 20 cm) which enclose a gravel base. Each unit is equipped with a two level wooden scaffold which provides 15.54m of lineal perch space. The scaffold is located 1.83m above the gravel base, with the upper level, at 2.44m, protected from the elements by the roof. During the winter season, the cage is sheltered from wind and rain by a galvanized metal screen. The screen is composed of overlapping sheet metal panels (2.39 by 0.66m) secured to the north side of each cage with banding material.

Animals

The animal population at the beginning of 1982 was about 6,400 nonhuman primates (Table 1). The majority were rhesus monkeys. The remainder were crab-eating macaques (*M. fascicularis*), squirrel monkeys (*Saimiri*

TABLE 1. Animal population as of January 1, 1982

Total Animals 6,398	Average per Cage 11 animals		Total Cages 566
	SPECIES DISTRIBUTION		
91%	*Macaca mulatta*	5,831	
	Macaca fascicularis	568	
9%	*Saimiri sciureus*	7	
	Hylobates lar	1	
	MACACA MULATTA 5,831 animals		
51.5% Adults	3,004	(Males)	406
		(Females)	2,598
48.5% Offspring	2,827		

sciureus), and a white handed gibbon (*Hylobates lar*). Virtually all the original breeding stock for each breeding program was composed of wild-caught animals. By 1982, the rhesus population was evenly divided between adult breeders and subadult offspring.

Management practices

Basic management practices are straightforward and have remained consistent over the years.

1. Adult breeders are housed in harem groups with an average ratio of one male to seven females. Weanlings are housed with 10-14 age peers per unit.
2. Animals are observed three times daily for animal health problems and reproductive events. Animals may receive treatment in their social groups or in separate treatment facilities. Those hospitalized for trauma are permanently removed from their original social groups and reformed into new groups upon discharge from the hospital.
3. Rejected or orphaned infants are fostered onto other female breeders.
4. With a few exceptions, all colony-born females are retained and enter the breeding colony at 3 years of age. Males are retained at a ratio of one male for every 10 females and are placed into breeding situations at 5 years of age.
5. Routine procedures include: quarterly tuberculin testing of all animals and abdominal palpatation of female breeders to detect pregnancies; semiannually relocating groups to facilitate cleaning of gravel bases; identification and sexing of newborns; tattooing infants at 4 months of age; weaning at 6 months; and consolidation of small groups of breeders.

6. Husbandry procedures include: feeding twice daily with a commercial monkey diet (Monkey Chow #5038, Ralston-Purina Co., St. Louis, MO), water is available <u>ad libitum</u>; and animals are captured using long handled nets.

Personnel

At the beginning of 1982, 43 employees were required to support the operation. Since 25 of these were full-time animal technicians, the animal to technician ratio was 256:1.

Data collection

The 1981 reproductive data were collected between October 1, 1981 through September 30, 1982 from birth logs. Critical reproductive rates were calculated from these data, after Taub (1982), to facilitate comparisons with the other breeding programs. The 1981 mortality data were tabulated from death logs between January 1 and December 31. Causes of deaths were summarized from gross necropsy reports. Histopathological examinations of tissue samples were limited to those cases in which death due to an infectious agent was suspected. The morbidity data represent a tabulation of all health data through June 1982 that was accessible by our microcomputer (IBM 5110 Model B32, IBM Corporation, Armonk, NY). Animal records for each of the breeding programs have been computer-maintained for variable periods since 1981.

An in-house survey was done in 1981 of the health and reproductive histories of a representative subgroup of female breeders (n = 767). Harem breeding groups of seven or more females were identified. The records of those groups comprised solely of females acquired as adults were selected for further examination. Their health and reproductive data were tabulated in order to characterize the average female breeder experience at this facility. In addition, the reproductive performance of colony-born female breeders (n = 228) was determined by examination of animal records and their performance was compared with that of wild-caught females.

RESULTS

Reproduction

Reproductive data from the 1981 breeding season are shown in Table 2. The effective breeding population, those females that were at least 4 years old as of October 1, totalled 2,046. The pregnancy rate of 84.3% was based on confirmed pregnancy terminations. This

TABLE 2. Reproductive rates (after Taub, 1982) from the 1981 breeding season

	Rate		
Effective Breeding Population (4+ years on Oct. 1) 2,046			
Total Pregnancies		1,725	
Pregnancy Rate	84.3%		
256 Fetal Deaths . . .	14.8%		
1,469 Live Births . . .	85.2%		100%
78 Neonatal Deaths* .	5.3%		
1,391 Viable Live Births	94.7%		100%
Production Births		1,391	
Production Birth Rate**	68.0%		

* Infant deaths occurring ≤ 30 days of age.
** Production Birth Rate = Viable Live Births divided by Effective Breeding Population.

figure did not include previously detected pregnancies that were resorbed or miscarried without evidence. The rate did include the bonus babies of 3-year-old females. The production birth rate for the 1981 breeding season was 68%.

Table 3 shows the reproductive performance of female breeders acquired as adults compared with colony-born female breeders. Both rates, unlike those above, were calculated as a proportion of the number of female years. The average conception rate of the acquired females initially appeared to be higher than that of the colony-born females. A dissection of this lower rate into separate breeding years showed that the poor average performance of colony-born females was due to a 50% rate of the first year when they were 3 and 4 years of age. The conception rates of colony-born females for their second and third years of breeding (years 2 and 3) were

TABLE 3. Reproductive comparison of wild-caught females and colony-born females

	Females acquired as adults	TPC-born females Average*	Yr1	Yr2	Yr3
Conception Rate**	70%	60%	57%	67%	65%
Live Birth Rate***	61%	54%	50%	60%	65%
Female Years	2,883	329	228	81	20

* Average of reproductive performances in Yr1, Yr2, and Yr3.
** Conception Rate = Number of pregnancies divided by number of female years.
*** Live Birth Rate = Number of live births divided by number of female years.

statistically indistinguishable from the rates of females acquired as adults.

The reproductive rates of the females acquired as adults were below the 1981 breeding season rates. The 70% conception rate for the wild-caught females was calculated in the same manner as the pregnancy rate, which was 84.3%. The 61% live birth rate had no comparable Taub statistic but, as recalculated from Table 2 (number of live births divided by effective breeding population), was 71.7% in 1981. The differences between the 1981 and pre-1981 rates may be due, in part, to a change in weaning age that was instituted in November 1981. The weaning age was advanced by 2 months, from 8 months to 6 months of age. This change was made as a result of a recent study done at this facility (Goo et al., 1981) which showed that offspring weaning age was inversely associated with maternal reproductive performance in the subsequent year.

Mortality

Mortality rates for 1981 are shown in Table 4. Unweaned offspring showed the highest mortality rate of all age groups; 8% of all deaths were due to infant malnutrition. For all age groups, deaths due to enteropathies were the most common, followed by deaths resulting from trauma. The trauma to enteropathy ratio was 1:1.7. Pneumonias, which tended to afflict debilitated animals, were the third major cause of death. Shock, anemia, peritonitis, and gastric dilatation were among several causes of death which occurred at low frequencies and were, therefore, grouped under the heading "all others." The lack of routine histopathological examination of all deaths accounted for the relatively high proportion of deaths due to undetermined causes.

TABLE 4. 1981 mortality rates and causes of death

Overall Mortality Rate	8.5%
Adults (≥ 4 years)	8.5%
Weaned offspring (> 6 months & ≤ 3 years)	6.8%
Unweaned offspring (< 6 months)	10.5%
Causes of Death	
Enteropathies	40%
Trauma	24%
Pneumonia	11%
Infant malnutrition	8%
Undetermined	6%
All others	11%

TABLE 5. Morbidity patterns

	All animals (8,465 animal years)	Adult females (2,883 animal years)
Clinical Episode Rate per animal year*	0.59	----
Hospitalization Rate per animal per year	0.45	0.64
Enteropathies	0.25 (56%)**	0.14 (22%)
Trauma	0.14 (31%)	0.45 (70%)
Others	0.06 (13%)	0.05 (8%)
Average Hospital Stay	16 days	

* Clinical Episode: any health-related incident for which an animal required a physical examination or was rendered treatment.
** Proportion of hospitalizations.

Morbidity

Morbidity rates are shown in Table 5. Clinical episodes were tabulated for 8,400 animal years and were found to occur at an average rate of 590 clinical episodes per 1,000 animals per year. Approximately 75% of these involved hospitalization. Enteropathies represented over half the causes for hospitalization. The trauma to enteropathy ratio, at 1:1.8, was nearly identical to that seen for mortalities. Adult females had a higher hospitalization rate than did the population as a whole, and the profile of causes of morbidity for the females was substantially different as well. Trauma represented the cause of 70% of their hospitalizations, a rate of 0.45 clinical episodes per female per year. This rate was equivalent to the overall hospitalization rate for all animals.

DISCUSSION

The harem mode of breeding rhesus monkeys can be an effective method which, in terms of cost and environment, steers an intermediate course between the individual caging mode and corral or free ranging modes. As with individual caging of breeders, animals of known parentage and with known birth dates are produced. In addition, unweaned offspring are socialized with continuous exposure to adults and other infants. As is the case with both corral and free ranging modes, precise conception dates are unknown. In harem groups, however, individuals are highly visible at close range and are more accessible than animals in other housing environments. These factors allow earlier detection of health problems and an accurate accounting of reproductive performance.

Harem breeding is not without liabilities, as evidenced by the high rates of treatment and hospitalization. Traumatic injury is a particular problem among the adult females. The impact of this on reproduction is unknown.

Management practices can serve to ameliorate or exacerbate problems related to animal health and reproduction. At most facilities, it is difficult to elucidate population level responses due to small numbers of animals and the large degree of individual variation. At the TPC, the large number of animals provides the opportunity to obtain species level information on the responses of rhesus monkeys to harem group management practices.

ACKNOWLEDGMENTS

The Texas Primate Center operations have been supported by the Animal Resources Branch, Division of Research Resources, National Institutes of Health; by the Bureau of Biologics, Food and Drug Administration; and by Lederle Laboratories Division, American Cyanamid Co., Pearl River, NY. The author would like to acknowledge the dedication and commitment shown by Dr. D. D. Boyd, Director, Texas Primate Center, and Dr. D. A. Valerio, President, Hazelton Research Animals, Inc., in the establishment of the largest primate breeding facility in the world (animal population, as of October 1, 1983, was 7,794). The author also gratefully acknowledges the programming assistance of Dr. J. K. Fugate, Corpus Christi State University, Corpus Christi, TX, in the retrieval and analysis of the computerized morbidity data.

REFERENCES

Goo, G. P., Fugate, J. K., and Boyd, D. D.: Effect of infant weaning age on subsequent reproduction in rhesus monkeys (*Macaca mulatta*). Paper presented at the fourth meeting of the American Society of Primatologists. San Antonio, Texas. June 2-6, (1981).

Taub, D. M.: Large scale breeding and husbandry of nonhuman primates. Symposium presented at the IXth Congress of the International Primatological Society. Atlanta, Georgia. August 8-13, (1982).

8

Captive Breeding in a Source Country

J. G. Else

Data on the captive breeding of vervet monkeys (Cercopithecus aethiops) in Kenya are presented and discussed in the context of the establishment of a primate captive breeding program in a source country. Once conditioned and established in stable breeding groups, the vervet monkey breeds well and annual pregnancy and livebirth rates of 90% and 85%, respectively, can be expected. Production results and other parameters were equivalent to those reported from non-source countries. This program was undertaken on a small scale, but production costs were smaller than those reported elsewhere and compare favorably with the cost of primates captured from the wild. Our experience with source country captive primate breeding is that it is a feasible and economically viable approach to supplying user countries with healthy and relatively inexpensive animals for research purposes.

INTRODUCTION

Because nonhuman primates have a close evolutionary kinship to man, they have become crucial to many human health research programs. Their anatomical and biochemical similarities make them indispensable for many biomedical studies, and they are ideal as well for the production of biologics and for the testing of compounds for toxicity. However, declines in natural populations over the past several years in many countries have led to increasing difficulties in obtaining primates from their countries of origin. This decline in primate numbers in the wild has resulted in considerable concern and has led to increased efforts to both conserve primates and to breed them in captivity (Committee on Conservation of Nonhuman Primates, 1975). Existing captive breeding programs are primarily to provide animals for biomedical needs and are the only ultimate answer for this requirement (World Health Organization, 1971). The concept of captive primate production in source countries has been considered but, to date, programs have been undertaken almost entirely in non-source countries such as the United States and to a lesser extent in Europe. Although these have been quite successful, the high overheads involved have limited their scope and it has not been feasible to breed a variety of species (Held, 1981, 1982).

Source country captive breeding has several inherent advantages such as ready availability of primary and replacement stock, indigenous climatic conditions and low overhead. The Institute of Primate Research (IPR), National Museums of Kenya, has established a small vervet monkey (*Cercopithecus aethiops*) captive breeding program to partially supply its own needs as well as other research organizations in Kenya. The vervet was selected because of its small size, rapid maturity and high reproductive potential. The program has been very successful and could easily be expanded to a larger scale to supply other countries.

METHODS AND MATERIALS

Standard IPR Procedures

Animals are either individually caged or kept in group enclosures, depending on the study. They are given commercial monkey chow (Unga Foods Ltd., Nairobi) twice daily fresh fruit, and vegetable supplementation thrice weekly. All animals are examined visually each day and data relative to health or reproduction are entered into individual files by technicians and veterinarians. The animals receive medical treatments as required and are routinely processed twice yearly. During processing each animal is anesthetized with ketamine hydrochloride, TB tested, given a physical exam, tattooed and biological samples required for research studies collected.

To calculate production costs, precise records are kept of all animal related expenditures such as labor, food and supplies, laboratory and medical expenses and protective clothing. Forty percent is added for routine institution operating costs; 30% of this total is added as administrative overhead. Costs were not itemized by species; therefore, the total expenditures calculated by the above formula were prorated for each primate species by weight, and a per diem rate was calculated (Table 1).

Procedures specific to the Vervet Breeding Program

Animals that were to comprise the basic captive breeding colony were trapped from various locales within

TABLE 1. Calculation of Animal Per Diem at the Institute of Primate Research

I. Expenditures
- A. Animal Related Expenses (100%)
  1. Personnel
  2. Food
  3. Supplies
  4. Cage Maintenance
  5. Animal Room Maintenance
  6. Protective Clothing
  7. Laboratory and Medical
- B. Institute Running Costs (40%)
  1. Daily Maintenance
  2. General Expenditures
- C. Administrative Overhead
  30% of A + B

II. Per Diem Calculation
- A. Classes of Animals
  1. Class 1: > 12kg (3x Factor)
  2. Class 2: 5-12kg (2x Factor)
  3. Class 3: < 5 kg (1x Factor)
- B. Individual Animal Calculation
  1. Expenditure per Annum/(Average No. of Animals times Class Factors) all divided by 365 = Per Diem Unit
  2. Per Diem Unit times Animal's Class Factor = Animal's Per Diem.

Kenya by IPR personnel. While in the field they were given a cursory physical exam and TB tested (intrapalpebrally). Upon negative TB reaction through 72 hours they were transported to IPR quarantine facilities. During the 3 month quarantine period the animals were caged individually, given water and monkey chow ad libitum, with fruit and vegetable supplements each afternoon. During the latter half of quarantine the animals were TB tested every 2 weeks and treated with antihelminthics twice.

Upon completion of quarantine the animals were formed into 10 single-male harem social groups with 5-10 females per group. The outdoor enclosures in which each harem was housed measured 3m by 6m by 2.5m, had concrete flooring and was covered partially on the top and on one side.

Every attempt was made to keep the original social groups intact but if females were seriously injured or excessively harassed they were removed, treated and, if necessary, placed into new harem groups. Male progeny were removed from the natal social group when 3-6 months of age and placed in same sex "weanling" groups; female progeny were allowed to remain in their natal groups so that harem group size could increase through internal recruitment. The single, sexually mature harem male was replaced when his female progeny reached 2 years of age.

RESULTS

Conditioning

Vervets newly acquired from the field were very excitable and at first they were extremely reluctant to adapt to commercial monkey chow. Consequently, they rapidly lost condition over the first 4-6 weeks. Slowly thereafter they began to calm down, eat more regularly and regain their former condition, so that by the end of the 3-month quarantine period they ate the diet readily, were in good condition and gaining weight. However, total conditioning required a minimum of 6 months, and more frequently 12 months.

Routine manipulation, such as ketamine anesthesia, TB testing, tubing and blood collecting during the first two months of quarantine resulted in mortality rates up to 80%. When such procedures were delayed for 6 weeks, mortality rates for the 3-month quarantine period dropped to 5-10%.

Group formation and reproduction

The establishment of stable breeding groups initially formed with adult animals was extremely difficult because of fighting among the females. On the average, groups remained stable after the first year, although by this time there were only about half (2-4) of the original females in each of the 10 groups. Exchanging breeding males when required has not caused problems.

Colony reproductive data over the past two years are summarized in Table 2. Just under 90% of the adult females became pregnant each year with 85% live births. Eight of the 45 offspring died as infants, resulting in an annual animal production rate of 70% for the two year period.

Costs

Total expenditures for vervet upkeep for the past 2 years has been $0.20 per animal per day, or $73 per year. Of this, 40% has been for food, slightly less for labor and other animal related expenses and the balance overhead. Thus, with an animal production rate of 70% and a 1:5 adult male:female ratio, the production cost for a vervet monkey was $125. If an animal production rate of 80% is used, as was obtained during the second year of this study, the cost becomes $110.

TABLE 2. Two Year Summary of Vervet Breeding

Year	No. of Breeding Females	Pregnant No.	Pregnant %	Stillbirths[a]
1980/81	29	26	89.7	2
1981/82	24[c]	21	87.5	0
TOTAL	53	47	88.7	2

Year	Livebirth Rate (%)	Infant Deaths	Animal Production Rate (%)[b]
1980/81	82.2	6	62.1
1981/82	87.5	2	79.2
TOTAL	84.7	8	69.8

a No abortions were noted but they could easily be missed due to outdoor conditions.
b Number born per annum and surviving 6 months expressed as percentage of total females.
c Of the 5 missing females, 2 died, 1 was medically culled and 2 assigned to another study.

DISCUSSION

The establishment of a source country vervet captive breeding program in Kenya proved quite feasible. IPR trapping of stock animals enabled precise information on origin, screening of animals while still in the field and subsequent selection of only quality animals.

The vervet monkey has a reputation as a poor experimental animal because of excitability and high mortality. We have found that recently trapped vervets are extremely prone to stress and that any early manipulations have disastrous results. It is therefore little wonder that wild caught vervets which are immediately shipped to facilities in other countries and placed under strict quarantine or subjected to clinical manipulative procedures fare poorly. In our experience, this species requires at least a year to truly stabilize and adapt to captive conditions, which is in line with a previous report that stabilization of a vervet breeding colony may require 6-12 months or more after a 3-month quarantine period (Johnson et al., 1973).

Establishing breeding groups with adult vervets was difficult because of fighting among the females as they established a stable hierarchy. The animals must be monitored closely and careful decisions taken regarding when to treat or remove an animal and whether to disband a group. With the procedures used, and by weaning away male and retaining female progeny in family groups, we hope to eventually establish larger breeding groups but this is a long and laborious process. In retrospect, intact social troops could have been trapped and put directly into large enclosures which would have undoubtably reduced considerably the stress of adapting to captive conditions. This method would have also helped alleviate the fighting occurring during group formation and could have resulted in large stable breeding groups from the onset. This would have dramatically reduced initial expenditures.

Once stable harem groups are established, vervets breed readily. The 90% annual pregnancy rate and 85% livebirth rate compare favorably with rates reported for other Old World primates such as the rhesus (Wilson et al., 1978), cynomolgus (Kaplan et al., 1981), baboon (Hendrickx and Kriewaldt, 1967; Moore, 1975) and patas (Kaplan, 1981) as well as for the vervet (Kushner et al., 1982). The overall annual production rate of 70% was strongly influenced by the 6 infant deaths (25%) during the first year. Refinement of husbandry techniques combined with increased colony stability resulted in a reduction of the infant death rate to 10% during the second year, which resulted in a production rate of 80% which is probably more indicative of what can be expected from this type of captive vervet breeding program (and similar to what has been reported previously for this species; Kushner et al., 1982).

Assuming an animal production rate of 80%, and a 1:5 adult male:female ratio, production costs are about $110 per animal or $182 for an animal 1 year of age. This is 10-20% of estimated rhesus captive breeding costs in the United States (Soave, 1981) and is competitive with the cost of wild trapped animals, especially when shipping and quarantine losses of wild trapped animals are taken into account (Hobbs and Bleby, 1976).

This study was done on an extremely small scale as it was designed for the purpose of developing the techniques, and to supply colony bred animals for intramural uses. If it was undertaken on a larger scale and a more commercial basis, costs could be reduced and high quality, colony born animals could be supplied to other countries at competitive prices.

ACKNOWLEDGMENTS

This investigation received financial support from the Special Programme of Research, Development and Research Training in Human Reproduction, World Health Organization. The author is grateful to Dr. Mbaruk Suleman and John Njuguna for animal care and data collection, to Dr. Robert Eley and Holly Wall for editorial advice and to Margaret Omoto for secretarial assistance. We further acknowledge the support of the Yerkes Primate Research Center.

REFERENCES

Committee on Conservation of Nonhuman Primates. Nonhuman primates: usage and availability for biomedical programs. National Academy of Sciences. Washington, D. C. (1975).

Held, J. R.: Breeding and use of nonhuman primates in the USA. International Journal for the Study of Animal Problems. 2:27-37. (1981).

Held, J. R.: Conservation and management of nonhuman primates. Microbiologica. 5:161-169. (1982).

Hendrickx, A. G., and Kriewaldt, F. H.: Observations on a controlled breeding colony of baboons. pp. 69-83. International Symposium on the Baboon and its Use as an Experimental Animal. San Antonio, Texas. (1967).

Hobbs, K. R., and Bleby, J. Jr.: Laboratory nonhuman primates for biomedical research in the United Kingdom: A report to the Medical Research Council on the existing provision and future needs. Medical Research Council Laboratory Animals Centre, Carshalton, Survey, U. K. (1976).

Johnson, P. T., Valerio, D. A., and Thompson, G. E.: Breeding the African green monkey, Cercopithecus aethiops, in a laboratory environment. Laboratory Animal Science. 23:355-359. (1973).

Kaplan, J. R., Anthony, M., and Wood, L.: Domestic breeding of patas monkeys (Erythrocebus patas). Laboratory Animal Science. 31:409-412. (1981).

Kushner, H., Kraft-Schreyer, N., Angelakos, E. T., and Wundarski, E. M.: Analysis of reproductive data in a breeding colony of African green monkeys. Journal of Medical Primatology. 11:77-84. (1982).

Moore, G. T.: The breeding and utilization of baboons for biomedical research. Laboratory Animal Science. 25:798-801. (1975).

Soave, O. A.: The cost of domestic breeding of rhesus monkeys. Interagency Primate Steering Committee, NIH. Bethesda, Maryland. (1981).

Timmermans, P. J. A., Schouten, W. G. P., and Krinjnen, J. C. M.: Reproduction of cynomolgus monkeys (Macaca fascicularis) in harems. Laboratory Animals. 15:119-123. (1981).

Wilson, M. E., Gordon, T. P., and Bernstein, I. S.: Timing of births and reproductive success in rhesus monkey social groups. Journal of Medical Primatology. 7:202-212. (1978).

World Health Organization. Health, aspects of the supply and use of nonhuman primates for biomedical purposes. Technical Report Series No. 470, p. 11. WHO, Geneva. (1971).

PART II

Physical Anthropology

9

Systematic Position and Evolution of Primates Within Eutheria: Amino Acid and Nucleotide Sequence Findings

J. E. Beeber, J. Czelusniak, and M. Goodman

The cladistic position of Primates within higher level eutherian systematics and processes of molecular evolution in primate phylogeny were investigated utilizing amino acid and nucleotide sequence data. Genealogical trees constructed from these data by the maximum parsimony method indicate that Lagomorpha is the closest order to Primates; Tupaioidea and Rodentia may also be relatively close to Primates, whereas Paenungulata (Proboscidea, Sirenia, and Hyracoidea), Tubulidentata, and Edentata appear most distant. Perissodactyla, Cetacea, and Artiodactyla group together, constituting another distant clade from Primates. An additional clade consists of a branch of Insectivora (Lipotyphla) and Chiroptera joined to Carnivora, with Pholidota possibly joining such a clade. The nucleotide replacements between ancestral and descendant sequences in the reconstructed genealogies provided evidence that natural selection caused recurring speed ups and slow downs in rates of codon sequence evolution with accelerated rates in basal eutherians and early primates followed by markedly slowing rates in later higher primates.

INTRODUCTION

Any consensus drawn from the views of systematists concerning the higher level phylogeny of placental mammals woud depict a bush-like pattern of origin and radiation for the 17 or more orders in the infraclass Eutheria. This pattern reflects an ancient initial burst of cladogenetic change during which a series of presumed dichotomous branchings occurred in rapid succession. The challenge of resolving the "consensus" bush into a more discriminating tree remains. So challenged we have been constructing, from protein and DNA sequence data, genealogical trees which provide new evidence on eutherian phylogeny. In addition to indicating the cladistic position of Primates within Eutheria, these trees revealed that natural selection caused recurring speed ups and slow downs in rates of codon sequence evolution with accelerated rates in basal eutherians and early primates followed by markedly decelerating rates from the catarrhine-platryrrhine divergence node to present.

The main new findings presented here are based on cladistic analysis of amino acid sequence data representing 664 polypeptide chains of 273 species. In order to carry out this analysis, the genealogical trees were constructed by the maximum parsimony method (Goodman, 1981; Goodman et al., 1982a). A genealogical tree for 44 hemoglobin genes and pseudogenes of 10 species (9 vertebrates and 1 plant), constructed by this maximum parsimony method modified to analyze nucleotide sequences (Czelusniak et al., 1982), also yielded results (Goodman et al., 1982b) which we discuss. Other findings discussed concern genealogical trees (Czelusniak and Goodman, 1982; Goodman et al., 1982b) constructed for nucleotide sequence data on corresponding sections of the mitochondrial DNA of hominoid genera (Brown et al., 1982), mouse (Bibb et al., 1981) and ox (Anderson et al., 1982).

METHODS AND MATERIALS

The maximum parsimony method (Moore et al., 1973; Goodman, 1981) accounts for evolutionary descent of amino acid sequences by the fewest possible genic changes. In order to accomplish such, a genealogical arrangement must be found which <u>maximizes genetic likenesses</u> associated with common ancestry, while <u>minimizing</u> incidences of parallel and back mutations. This is done by using the genetic code to represent amino acid sequences as <u>mRNA</u> sequences and by then seeking a tree with a minimum number of nucleotide replacements, or lowest <u>NR</u> length. Since common ancestry, as opposed to convergent evolution, is most likely responsible for extensive matches of nucleotide sequences between species, the tree reconstruction approach based on maximum parsimony is a sensible means of arriving at a <u>preferred</u> genealogical hypothesis.

The amino acid sequence data for the parsimonious genealogical reconstructions came from 94 cytochromes *c*, 46 lens alpha crystallins, 47 combined fibrinopeptides A and B sequences, 21 carbonic anhydrases, 46 members of the protein family containing calmodulin, 321 globins, of which those from the jawed vertebrates subdivide into myoglobin, alpha-hemoglobins, and beta-hemoglobin, and 36 ribonucleases. The vast majority of sequences (587 of 664) and a majority of species (209 of 273) are from vertebrates. The vertebrate class best represented by the sequences is the Mammalia (156 species), in particular its subclass Eutheria (148 species) within which the order Primates is well sampled (38 species). Catalogues of most of these amino acid sequences and species from which they originated are given for ribonucleases in Bientema and Lenstra (1982) and for the remaining protein chains in Goodman (1981) and Goodman et al. (1982a). Table 1 lists, for the eleven kinds of protein chains most widely sequenced in mammals, the orders of Eutheria represented.

TABLE 1. Protein sequences utilized in the study of ordinal relationships: a=alpha hemoglobin; B=beta hemoglobin; m=myoglobin; l=lens alpha-crystallin A; f^A=fibrinopeptide A; f^B=fibrinopeptide B; c=cytochrome *c*; CaI=carbonic anhydrase I; CaII=carbonic anhydrase II; CaIII=carbonic anhydrase III; r=ribonuclease.

	a	B	m	l	f^A	f^B	c	CaI	CaII	CaIII	r
Primates	X	X	X	X	X	X	X	X	X	X	
Lagomorpha	X	X	X	X	X	X	X	X	X		
Tupaioidea	X	X	X	X							
Rodentia	X	X		X	X	X	X		X		X
Insectivora	X	X	X	X							
Chiroptera			X	X			X				
Carnivora	X	X	X	X	X	X	X				
Pholidota				X							
Perissodactyla	X	X	X	X	X	X	X	X	X		X
Cetacea			X	X			X				X
Artiodactyla	X	X	X	X	X	X	X	X	X	X	X
Edentata	X	X		X							X
Proboscidea		X	X	X	X	X	X				
Sirenia				X							
Hyracoidea				X							
Tubulidentata			X	X							

Two kinds of most parsimonious genealogical trees were constructed: gene phylogenies for separate collections of sequences (each collection consisting only of evolutionarily related sequences, e.g., hemoglobin genes and pseudogenes); and species phylogenies from sequences of separate collections combined in extended tandem alignments as though encoded for by giant genes. While the sequence file for a gene phylogeny can have orthologous sequences aligned against paralogous sequences, the sequence file for a species phylogeny should have only orthologous sequences aligned against other orthologous sequences. Orthologous sequences represent gene-lineages which split from common ancestors simultaneously with separation of the species-lineages, whereas paralogous sequences represent gene-lineages which arose from gene duplication prior to species separation (Fitch, 1970).

It has been found (Goodman et al., 1979; Maeda and Fitch, 1981) that different proteins can yield lowest NR length trees with non-concordant branching arrangements for the same animal species; thus each such reconstruction disagrees with some of the features of the species phylogeny supported by evidence from other proteins. The disagreements could be due to real differences in branching arrangements between gene and species phylogenies. Alternatively, such disagreements could indicate incorrect groupings of sequences that happen to have excesses of convergent residues causing these sequences to be represented as sharing a more recent common ancestor than they actually do. In order to overcome this source of error, the parsimony criterion was expanded to encompass not only NRs, but also those categories of genic changes which can account for a different branching order of the lineages in a putative gene phylogeny (e.g., a tree of lowest NR length) from the branching order of the species-lineages. As discussed in detail elsewhere (Goodman et al., 1979; Goodman, 1981; Goodman et al., 1982a), these categories are namely gene duplications (GDs) and deletions and regulatory mutations affecting gene expression, i.e., gene expression events (GEs). By utilizing this expanded parsimony criterion...in which the aim is to minimize the sum of NRs+GDs+GEs, we can then construct more accurate genealogies. GDs and GEs may be viewed as "extra-curricular genetic events" because the maximum parsimony programs employed in the repetitive computer searches for trees of lower length only count NRs.

When the computer search revealed a range of lowest length trees, consensus branching arrangements (Adams, 1972; Penney et al., 1982) were constructed.

RESULTS

Figure 1 (A&B) shows an Adams consensus tree derived from the 20 trees of the lowest NR length (each requiring

1950 NRs) found after extensive branch-swapping for 70 vertebrate taxa using tandemly aligned alpha- and beta-hemoglobin sequences. This Adams tree shows 7 disagreements (indicated by heavy dots) with well supported traditional evidence on monophyletic vertebrate groupings. Among the 20 lowest NR length trees one is found which...by grouping the frog *Xenopus* with other tetrapods, duck with other Ansiformes (i.e., with geese and swan), and marmoset and spider monkey with capuchin monkey (the other ceboid)...shows only 3 disagreements: (1) The failure of marsupials to join Eutheria and thereby yield a monophyletic Theria; (2) failure of coatimundi to share a monophyletic origin with raccoon, both members of subfamily Procyoninae; and (3) failure of rat to join monophyletically with mouse in the subfamily Murinae.

Figure 2 is the Adams consensus tree derived with the 14 trees of lowest NR length (each requiring 603 NRs) found for 26 vertebrate species using tandemly aligned myoglobin and lens alpha-crystallin A sequences. The wide separation of pig from the other member of Artiodactyla (ox) represented in this data set is the only flagrant phylogenetic "error".

Results of an extended data set, one in which additional proteins are included in the tandem alignment, are shown in Figure 3. While a total of 64 taxa are involved only 10 could be represented by all 7 polypeptide chains. Nevertheless, the results obtained using this data set presently serve as our best synthesis of branching patterns in higher level eutherian phylogeny. Figure 3 depicts an Adams tree derived from the trees of lowest NR length, of which 13 were found each requiring 2811 NRs. This consensus tree shows 7 disagreements with prior well supported monophyletic groupings. However, among the 2811 NR trees, there is one which groups duck with geese and swan, llama with camel, and hedgehog with mole and musk shrew; thus, this tree shows 4 rather than 7 disagreements. But before pointing to these remaining disagreements, we can note certain positive areas of the Adams 2811 NR tree which seem to hold up under the expanded parsimony criterion ...the ancient separation of Proboscidea and Edentata holds, in spite of the surrounding marsupials and monotremes; rabbits stay next to primates; Cetacea remains with ungulates; and the Insectivora-Chiroptera branch consistently joins Carnivora. Since our "bat" (Chiroptera) is a hybrid taxon represented by a megachiropteran for myoglobin, a microchiropteran for lens alpha-cystallin A, and yet a different microchiropteran for cytochrome *c*, we were precluded from investigating the question of Chiroptera's monophyletic versus diphyletic origins (Smith, 1976). However, once orthologously related proteins are sequenced in several or more species of these two suborders, such species will be treated as separate taxa in future reconstructions of mammalian phylogeny.

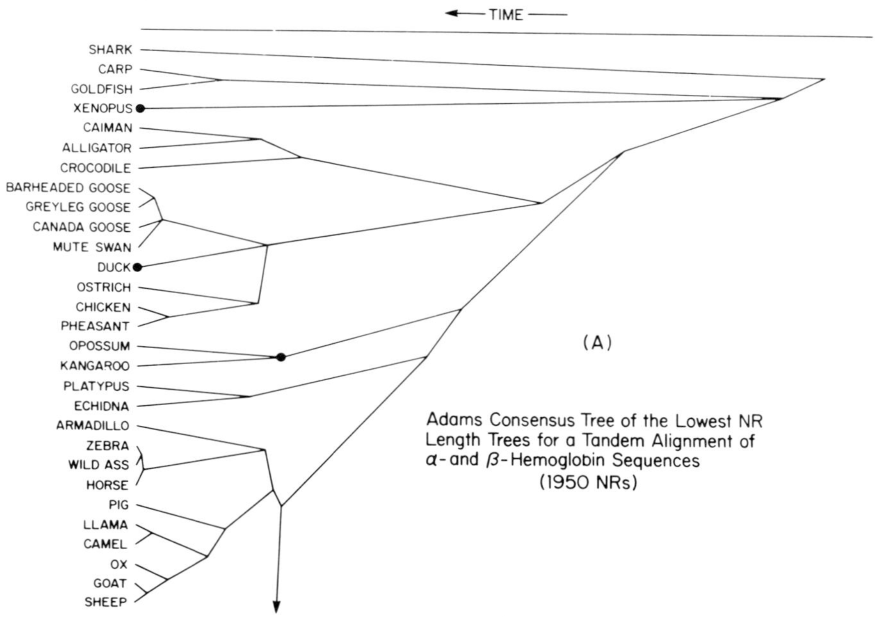

Figure 1. In addition to the alpha- and beta-hemoglobin sequences referenced in either Goodman (1981) or Goodman et al. (1982a) or Czelusniak et al. (1982), or in references cited therein, alpha- and beta-hemoglobin sequences from the following mammals and birds were employed: Musk shrew/*Suncus murinus* (Maita et al., 1981); European mole/*Talpa europaea* (Kleinschmidt et al., 1981); Zebra/*Equus zebra* and wild ass/*Equus hemionus kulan* (Mazur and Braunitzer, 1982); Armadillo/*Dasypus novemcinctus* (Kleinschmidt et al., 1982); Edentata of Figure 1 is represented not only by armadillo hemoglobin sequences but also by two-toed sloth lens alpha-crystallin A. Canada goose/*Branta canadensis* and mute swan/*Cygnus olor* (Oberthur et al., 1982); Pheasant/*Phasianus colchichus colchicus* (Braunitzer and Godovac, 1982).

Aside from the broad cladistic pattern of vertebrate and mammalian branches noted above, other noteworthy features in the Adams 2811 NR tree which do not change, even when the expanded parsimony criterion is applied, concern the primate branch. Within Primates, Strepsirhini (lorises and lemurs) separates from Haplorhini (tarsier and Anthropoidea). Within Anthropoidea, Platyrrhini (marmosets and new world monkeys) separates from Catarrhini. The latter divides into Cercopithecoidea (Old

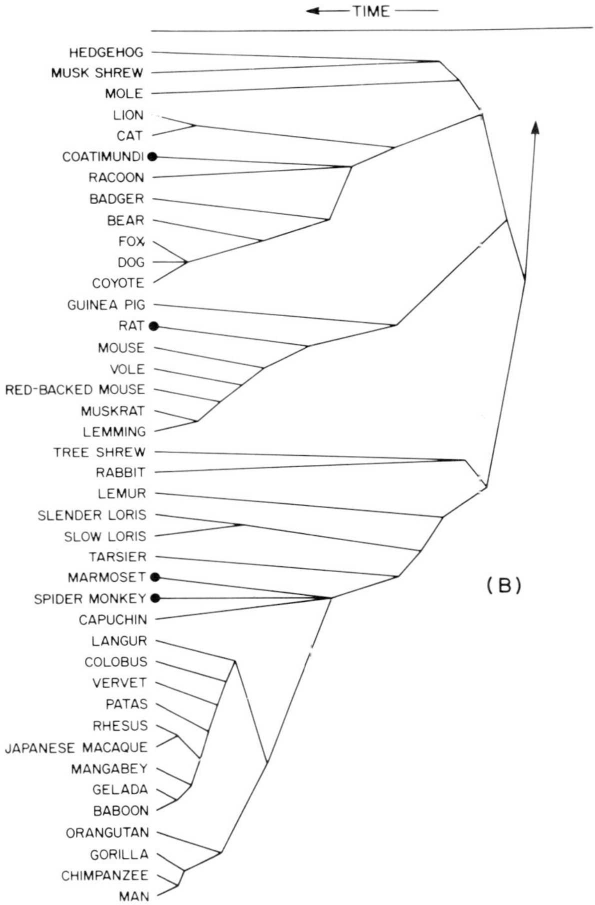

(B)

World monkeys) and Hominoidea and within Hominoidea the two most closely related genera are _Homo_ and _Pan_. This human-chimpanzee branch is joined by gorilla to form the genealogical subfamily Homininae (Goodman and Moore, 1971). Thus the two African ape genera are cladistically closer to _Homo_ than to the Asiatic ape branches: Ponginae (_Pongo_ or orangutan) and Hylobatinae (_Hylobates_ or gibbon). Although the present amino acid sequence evidence (e.g., Figure 3b) shows _Pongo_ to be as distant from Homininae as is _Hylobates_, the cladistic evidence (Goodman et al., 1982b) from mitochondrial DNA sequences indicates strongly that Ponginae is the sister group of Homininae within Hominidae and that this family then has Hylobatidae as its sister group.

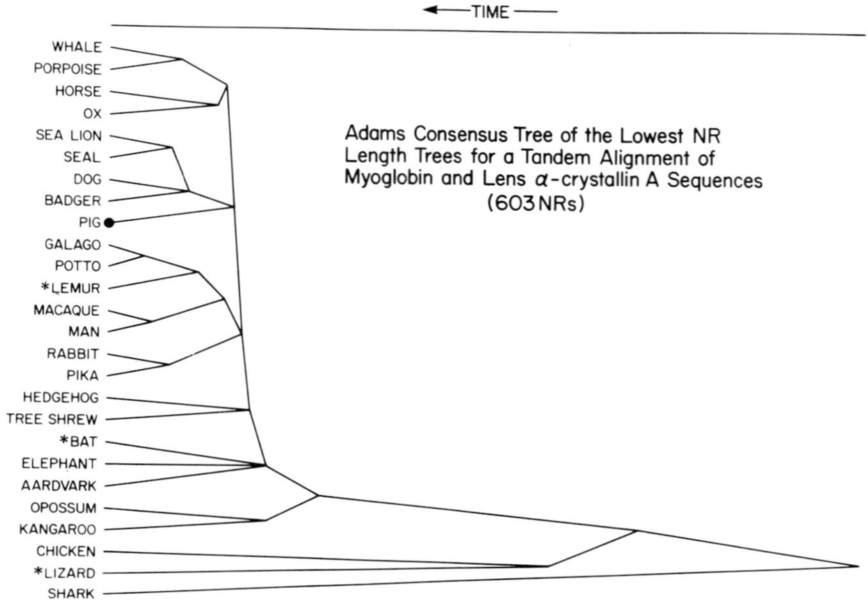

Figure 2. This lowest NR length Adams tree for tandemly aligned myoglobin and lens alpha-crystallin A sequences fails to support the inclusion of aardvark (Tubulidentata) in a monophyletic Paenungulata indicated by alpha-crystallin sequences alone (de Jong and Goodman, 1982). However, with myoglobin sequences alone, aardvark does have an ancient ancestral origin within Eutheria (H. Dene, M. Goodman, D. Walz, and A.E. Romero-Herrera, unpublished data).

At least 18 extra curricular genetic events (4GDs+14GEs) are required for the 2811 NR tree which shows the least number of disagreements (4) if we choose not to challenge the substantial traditional evidence for the monophyletic origin of theriological mammals and, later in descent, the monophyletic origin of artiodactyls. In other words, this tree then has a total length of 2829 genic events (2811 NRs+4GDs+14GEs). Certain trees which do not oppose such well substantiated monophyletic groupings prove more parsimonious than the 2811 NR trees. Indeed trees were found with scores of 2819 NRs but requiring no postulation of extra curricular genetic events. By the expanded parsimony criterion, these 2819 NR trees are in fact the lowest length trees which we found. Trees almost as parsimonious, requiring 2820 and 2821 NRs but no GDs and GEs, were also found. From the range of these most- and nearly-most parsimonious trees a consensus tree after Penney et al. (1982) was choosen, i.e., a tree in which its branching features were the ones repeated most often in the aforementioned range of trees. The mammalian portion of this "majority" consensus tree is shown in Figure 4, and salient features of the main most-parsimonious (the 2819 NR length) trees of the range from which the consensus tree was chosen are depicted in Figure 5. The Adams tree, for these data, is shown in Figure 6; it depicts a similar arrangement.

From the full body of relevant amino acid sequence data analyzed (the 7 polypeptide chains individually and in tandem alignment, plus three kinds of carbonic anhydrases and ribonuclease) tentative conclusions can be drawn about cladistic relationships among eutherian orders and major vertebrate groups: (1) Vertebrata separates into Agnatha and Gnathostomata with the latter subdividing into Elasmobranchii and a sister-branch which in turn subdivides into Teleostei and Tetrapoda, (2) within Tetrapoda, Amniota is a monophyletic branch, and within Amniota, Archosauria (Aves and Crocodilia) and Mammalia are each monophyletic, (3) within Eutheria the following cladistic pattern is suggested. Lagomorpha appears to be closest to Primates, whereas Paenungulata (Proboscidea, Sirenia, Hyracoidea), Tubulidentata, and Edentata appear most distant. Perissodactyla, Cetacea, and Artiodactyla group together, constituting another distant clade from Primates. An additional clade consists of a branch of Insectivora (Lipotyphla) and Chiroptera joined to Carnivora, with Pholidota joining such a clade as suggested by the a-crystallin A sequence tree which depicts pangolin aslpha the sister group of Carnivora.

DISCUSSION

The close grouping of Primates to Lagomorpha, which is neither supported nor opposed by any overriding paleontological views, is one of the more consistent findings from maximum parsimony analysis of sequence data (e.g., Figures 3-6). In contrast, the suggested superordinal clade containing Carnivora (along with Insectivora and Chiroptera and possibly Pholidota as well) is ambiguously placed, sometimes next to ungulates and cetaceans (e.g., Figure 3a) and other times with a primate-lagomorph-tupaiid-rodent assemblage (Figures 4-6). Such ambiguity is not surprising when we consider that the fossil record of the late Cretaceous and early Paleocene indicates that most orders of eutherian mammals separated from one another over a relatively short period of evolutionary time (see review by Novacek, 1982). Clearly the course of branching of Eutheria into its major monophyletic clades will be clarified as more extensive sequence data (as well as denser data in terms of taxa represented within each order) are gathered and

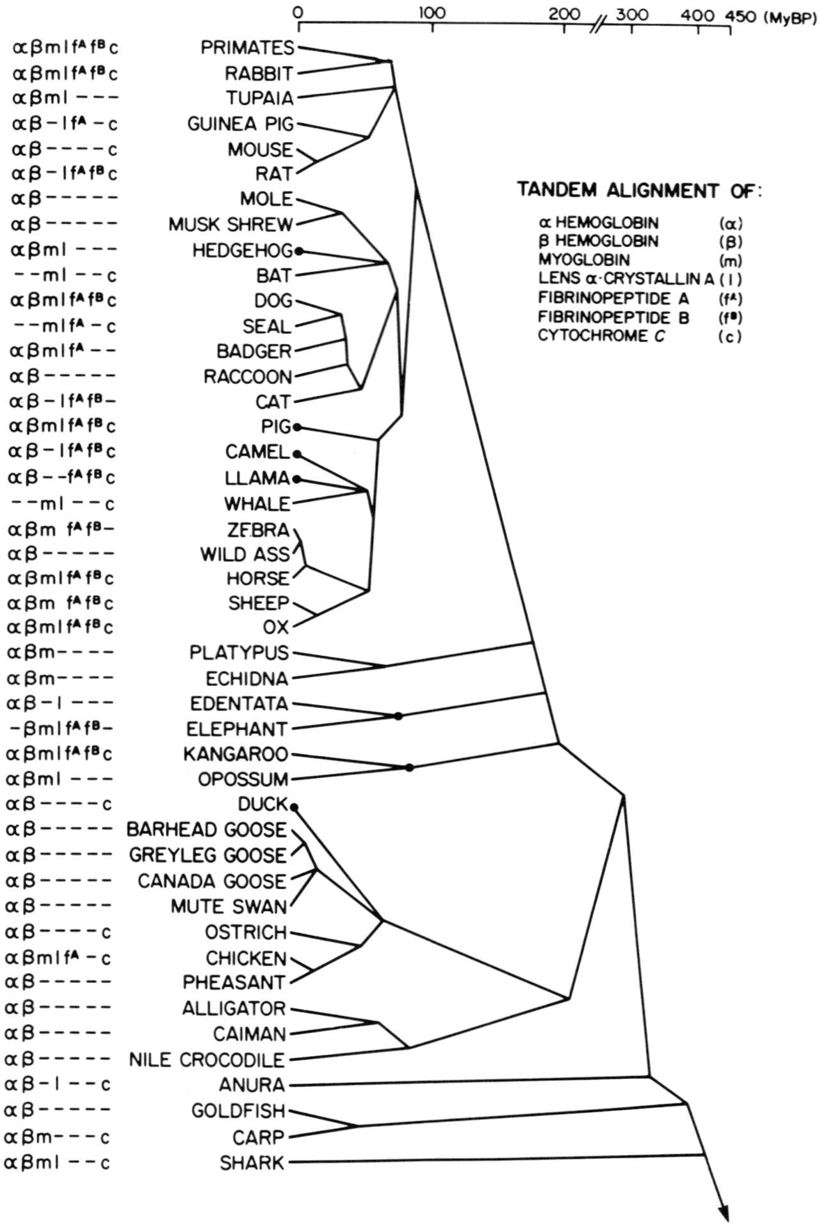

Figure 3. Adams consensus tree derived from the 13 trees of lowest NR length (each requiring 2811 NRs) found for 64 taxa using a tandem alignment of up to 7 polypeptide chains. A) Branching arrangement of all taxa except for subbranching of Primates. B) Subbranching of Primates.

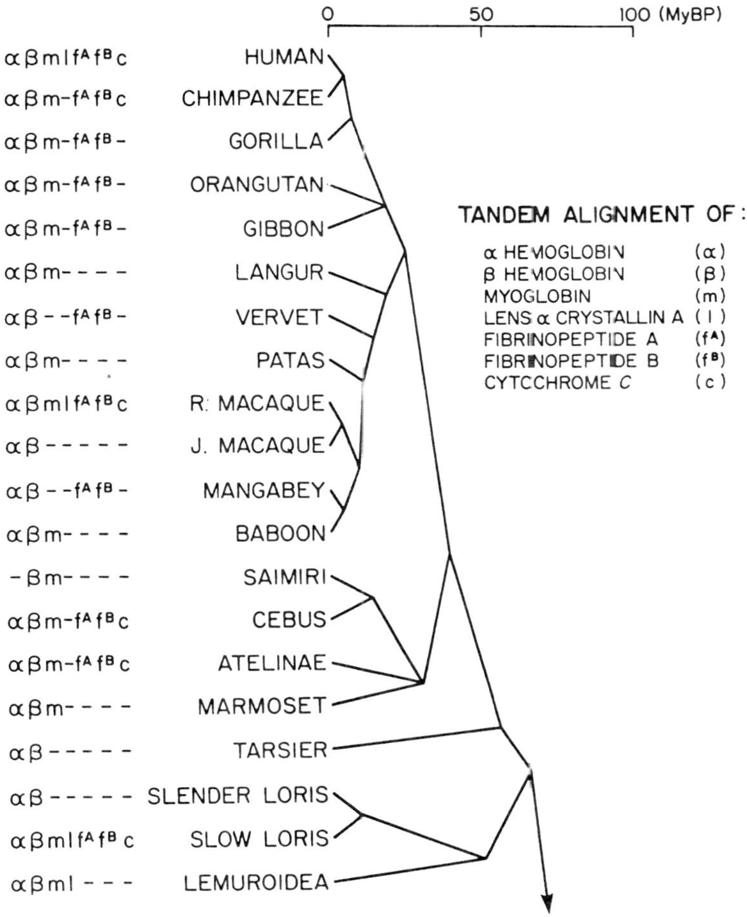

used for the reconstruction of phylogenetic history. In the meantime, despite uncertainties as to the correct branching pattern, some striking trends have been revealed (Goodman, 1981) on calculating rates of protein evolution from the numbers of nucleotide replacements (NRs) on the branches of the individual genealogical trees constructed for globins, cytochrome c, lens alpha-crystallin, fibrinopeptides A and B, and carbonic anhydrases I, II, and III. The same general trends in evolutionary rates were observed when the individual genealogical trees followed a common branching arrangement or when the individual trees had been constructed by our maximum parsimony approach without being constrained to follow a common branching arrangement for the species.

Apparently there was a general trend (see Table II

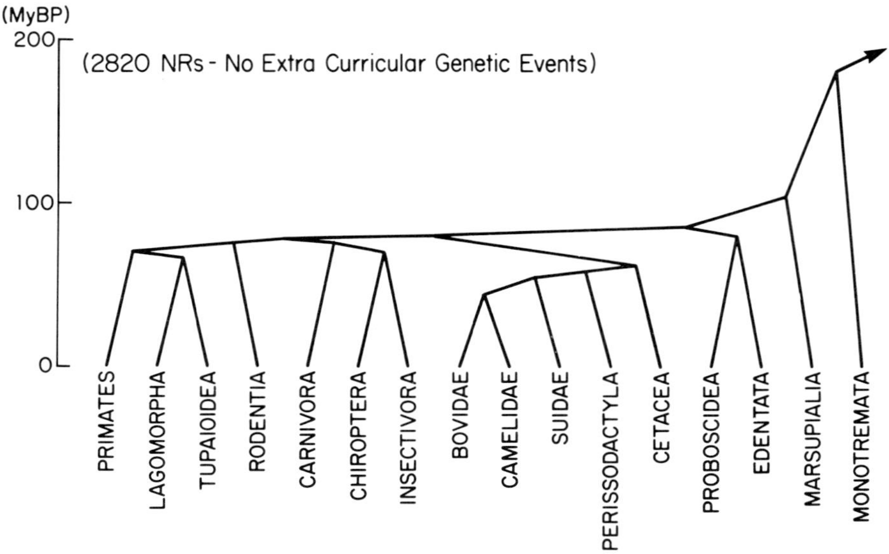

Figure 4. Mammalian portion of "majority" tree derived from lowest and near-lowest length trees utilizing the expanded maximum parsimony criterion. Portions not shown (the non-mammalian region and most subbranchings within mammalian orders) are identical to corresponding portions in the 2811 NR lowest length tree of Figure 3; the general pattern of mammalian branchings is similar in the two trees.

of Goodman, 1981) towards highly elevated rates of protein evolution between 500 and 300 MyBP (millions of years before present) during the emergence of Gnathostomata and Tetrapoda. Myoglobin, alpha-hemoglobin, beta-hemoglobin, cytochrome _c_, and lens alpha-crystallin each evolved 5 to 7 times faster than during the following 210 or so million years in the lineage from the bird-mammal ancestor to basal eutherian mammals. Then, during the emergence of primates, lens alpha-crystallin A and cytochrome _c_ evolved 10 to 20 times faster than in the previous period. Furthermore, myoglobin, alpha- and beta- hemoglobins and carbonic anhydrase I evolved two to five times faster. After this upsurge in rates, globins, lens alpha-crystallin A, fibrinopeptides A and B, and carbonic anhydrases eventually all evolved at markedly decelerated rates in the lineage to man. Indeed, the deceleration became especially steep within Homininae, with the evolutionary rate of the sequences in the protohuman lineage being at least 9-10 times slower than the average eutherian rate. (This assumes that the last _Pan-Homo_ common ancestor preceded rather than followed the appearance of _Australopithecus_ in the fossil record; Goodman, 1982). From the Homininae ancestral node to

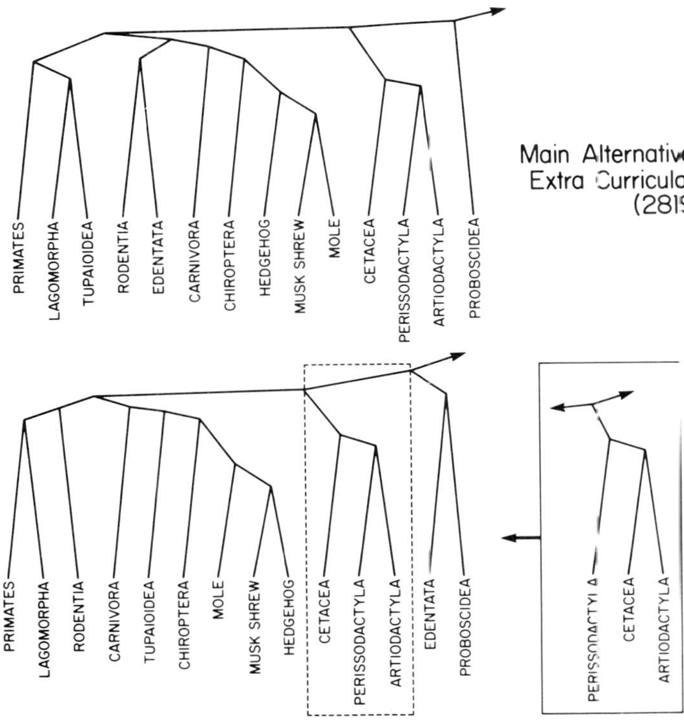

Figure 5. Maintenance alternatives of the most-parsimonious trees. These trees have a score of 2819 NRs and require no extra curricular genetic events.

present, not only did the protein sequences evolve slower on the line to Homo than on the lines to Pan and Gorilla (Goodman, 1981; Goodman, 1982), but mitochondrial DNA sequences also evolved slower (Brown et al., 1982; Goodman et al., 1982b), as illustrated in Figure 7, and apparently chromosomes too (Yunis and Prakash, 1982).

It may be noted that the two periods of highly elevated protein evolution described above, first in basal vertebrates and early tetrapods and second in basal eutherians and early primates, were periods of large scale adaptive radiation. Thus these bursts of rapid protein evolution correspond to the bursts of rapid morphological evolution observed in the fossil record. Similarly, just as in the model of punctuated equilibrium (Eldridge and Gould, 1972; Stanley, 1979) or the earlier Darwinian model of Rensch (1959) in which the explosive phase of adaptive radiation is followed by slowing rates of morphological change and even in some later lines of descent...morphological stasis, molecular evolution decelerated in later clades. The genealogical reconstructions from sequence data suggest (Goodman,

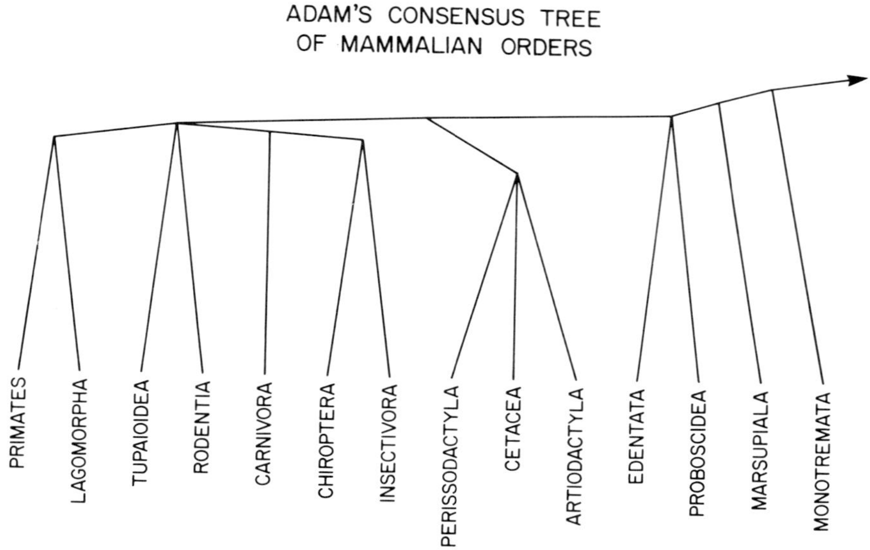

Figure 6. From mammalian portion of Adams tree derived from lowest and near-lowest length trees utilizing expanded maximum parsimony criterion. It may be noted that just the most parsimonious trees alone (those in Figure 5 at 2819 NRs requiring no "extra-curricular" genetic events) yield this same Adams consensus tree.

1981) that, at the molecular level, the earlier period of accelerated rates often involved gene duplications and selection on expressed genes among the duplicates to encode molecular sites with new or modified functions which in turn led to the later period of decelerated rates when selection then protected many functionally coadapted molecular sites from further changes.

The reconstructions of phylogenetic history carried out on sequence data from globins, cytochrome c, and calmodulin and its relatives (proteins with well characterized three-dimensional structures and functional sites) yielded in each case a detailed correlation between rates of evolution in different parts of the protein molecule and change of function in these parts (Goodman, 1981). The main finding was that during periods of rapid evolution of a protein the most rapidly evolving sites were those acquiring new important functions, and later in the decelerated phase of the protein's evolution these strong functional sites were among the most conservative.

In addition to rate of amino-acid substitutions and distribution of substitutions in functionally different parts of the protein molecule, a third parameter for

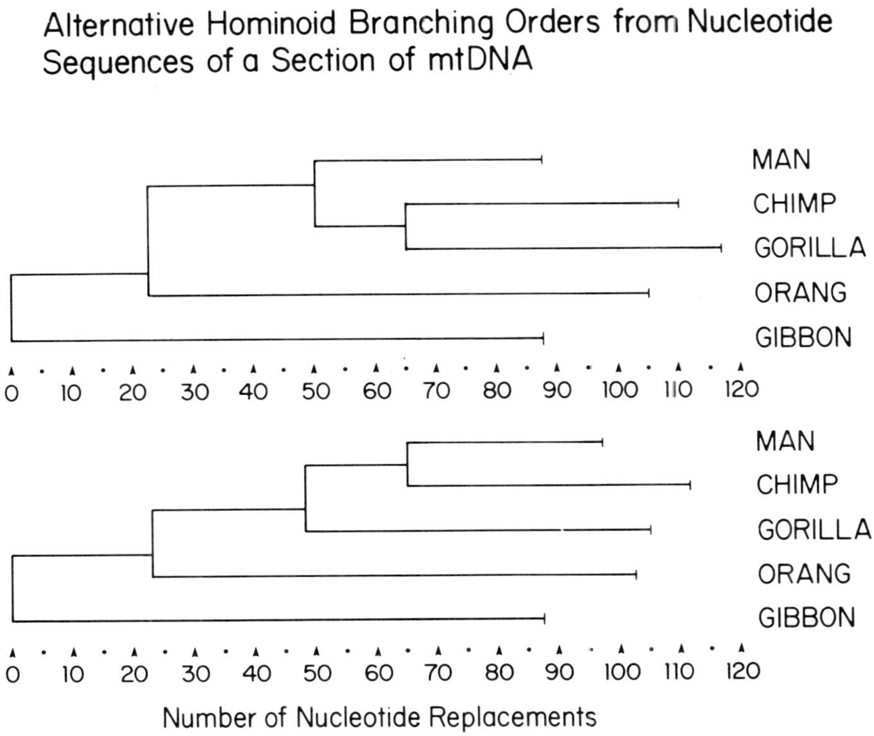

Figure 7. Alternative hominoid branching order of the two lowest NR length trees (each with score 750 NRs) found for the mtDNA sequences. The length of each branch equals the number of NRs calculated to have occurred along that line of evolution.

evaluating the role of natural selection in molecular evolution has been provided by the nucleotide sequence data on exon (coding) regions of genes. This parameter, R_{cs} value, is the ratio of amino acid-changing to silent substitutions, a value that was calculated for each internodal link of the hemoglobin gene nucleotide sequence tree (Czelusniak et al., 1982). The changes in values of the three parameters during different evolutionary periods provided evidence that a substantial proportion of amino acid substitutions were adaptive. For example, in the line of descent to catarrhine gamma-hemoglobin chains the rate of evolution of amino acid changing base replacements from the primate-lagomorph to catarrhine ancestral node (about 70-25 MyBP) was 6 to 7 times faster than from the catarrhine ancestral node to present. The amino acid substitutions which occurred during the earlier faster period of evolution in this primate gamma-line allowed fetal hemoglobin to strip greater amounts of oxygen from maternal hemoglobin. This may have helped make possible the relatively long period of fetal life coupled with extensive brain development distinguishing catarrhine primates from other eutherian mammals. In correlation, the R_{cs} values revealed that most base replacements on the primate gamma-line were amino acid changing ones rather than silent (Czelusniak et al., 1982). That these were adaptive base replacements is further indicated by the very slow rate of evolution of the catarrhine gamma-chains, averaging only 6 NR% (6 NRs per 100 codons per 10^8 years) and, in fact, being 0 NR% from the _Pan-Homo_ ancestor to present. Since this decelerated rate was probably caused by natural selection in its stabilizing form, the previous faster rate of base replacements can logically be attributed to natural selection, in its transforming mode, choosing the amino acid replacement residues which thereafter were conserved.

ACKNOWLEDGMENT

This work was supported by NSF grant DEB7810717.

REFERENCES

Adams, E.N.: Consensus techniques and the comparison of taxonomic trees. Syst. Zool. 21:390-397 (1972).

Anderson, S., de Bruihn, M.H.L., Coulson, A.R., Eperon, I.C., Sanger, F., and Young, I.G.: The complete sequence of bovine mitochondrial DNA: conserved features of the mammalian mitochondrial genome. J. Mol. Biol. 6(4):683-717. (1982).

Beintema, J.J., and Lenstra, J.A.: Evolution of mammalian pancreatic ribonucleases. In: M. Goodman (Ed.). Macromolecular Sequences in Systematic and Evolutionary Biology. Plenum Press, New York. (1982).

Bibb, M.J., Van Etten, R.A., Wright, C.T., Walberg, M.W., and Clayton, D.A.: Sequence and gene organization of mouse mitochondrial DNA. Cell 26:167-180 (1981).

Braunitzer, G., and Godovac, J.: Hemoglobins, XLV: The amino acid sequence of pheasant (_Phasianus colchicus colchicus_) hemoglobins. Hoppe-Seyler's Zschr. Physiol. Chem. 363:229-238 (1982).

Brown, W.M., Prager, E.M., Wang, A., and Wilson, A.C.: Mitochondrial DNA sequences of Primates: tempo and mode of evolution. J. Mol. Evol. 18:225-239 (1982).

Czelusniak, J., Goodman, M., Hewett-Emmett, D., Weiss, M.L., Venta, P.J., and Tashian, R.E.: Phylogenetic origins and adaptive evolution of Asian and mammalian haemoglobin genes which are expressed differentially during ontogeny. Nature 298:297-300 (1982).

Czelusniak, J., and Goodman, M.: Methods of utilizing all information generated in optimal tree searches. Amer. Zool. 22:976 (1982).

de Jong, W.W., and Goodman, M.: Mammalian phylogeny studied by sequence analysis of the eye lens protein a-crystallin. Z. Saugetierkunde 47:257-276 (1982).

Eldridge, N., and Gould, S.J.: Punctuated equilibrium: an alternative to phyletic gradualism. In: J.M. Schapf (Ed.). Models in Paleobiology. Freeman & Co., San Francisco. (1972).

Fitch, W.M.: Distinguishing homologous and analogous proteins. Syst. Zool. 19:99-113 (1970).

Goodman, M.: Decoding the pattern of proteins evolution. Progr. Biophys. Mol. Biol. 38:105-164 (1981).

Goodman, M., and Moore, G.W.: Immunodiffusion systematics of the Primates: I. The Catarrhini. Syst. Zool. 20:19-62 (1971).

Goodman, M.: Biomolecular evidence on human origins from the standpoint of Darwinian theory. Human Biol. 54:247-264 (1982).

Goodman, M., Czelusniak, J., Moore, G.W., Romero-Herrera, A.E., and Matsuda, G.: Fitting the gene lineage into its species lineage: a parsimony strategy illustrated by cladograms constructed from globin sequences. Syst. Zool. 28:132-163 (1979).

Goodman, M., Romero-Herrera, A.E., Dene, H., Czelusniak, J., and Tashian, R.E.: Amino acid sequence evidence on the phylogeny of primates and other Eutherians. In: M. Goodman (Ed.). Macromolecular Sequences in Systematic and Evolutionary Biology. Plenum Press, New York. (1982a).

Goodman, M., Olson, C.B., Beeber, J., and Czelusniak, J.: New perspectives in the molecular biological analysis of mammalian phylogeny. Acta Zool. Fennica 169:19-35 (1982b).

Kleinschmidt, T., Jelkmann, W., and Braunitzer, G.: Hamoglobine, XLIII: Die Primarstruktur des Hamogloblins des Maulwurfs (Talpa europaea). Hoppe-Seyler's Zschr. Physiol. Chem. 362:1263-1272 (1981).

Kleinschmidt, T., de Jong, W.W., and Braunitzer, G.: Haemoglobine, KXVI: Die Primarstruktur der a-Ketten des Hamoglobins vom Gurteltier Dasypus novemcinctus, Edentata). Hoppe-Seyler's Zsch. Physiol. Chem. 363:239-245 (1982).

Maeda, N., and Fitch, W.M.: Amino acid sequence of a myoglobin from lace monitor lizard, Varanus varius, and its evolutionary applications. J. Biol. Chem. 256:4301-4309 (1981).

Mazur, G., and Braunitzer, G.: Hamoglobine XLIV: Perissodactyla: Die Sequenz der Hamoglobine von Wildesel (Equus hemionus kulan) and Zebra (Equus zebra). Hoppe-Seyler's Zschr. Physiol. Chem. 363:59-71 (1982).

Maita, T., Matsuda, G., Takenaka, O., and Takahashi, K.: The primary structure of adult hemoglobin of musk shrew (Suncus murinus). Hoppe-Seyler's Zschr. Physiol. Chem. 362:1465-1474 (1981).

Moore, G.W., Barnabas, J., and Goodman, M.: A method for constructing maximum parsimony ancestral amino acid sequences on a given network. J. Theor. Biol. 38:459-485 (1973).

Novacek, M.H.: Information for molecular studies from anatomical and fossil evidence on higher eutherian phylogeny. In: M. Goodman (Ed.). Macromolecular Sequences in Systematic and Evolutionary Biology. Plenum Press, New York. (1982).

Oberthur, W., Godovac-Zimmermann, J., and Braunitzer, G.: The amino acid sequence of Canada goose (Branta canadensis) and mute swan (Cygnus olor) hemoglobins. Hoppe-Seyler's Z. Physiol. Chem. 363:777-787 (1982).

Penney, D., Foulds, L.R., and Hendy, M.D.: Testing the theory of evolution by comparing phylogenetic trees constructed from five different protein sequences. Nature 297:197-201 (1982).

Rensch, B.: Evolution Above the Species Level. Columbia Univ. Press, New York. (1959).

Smith, J.D.: Chiropteran evolution. Biology of bats of the new world family Phyllostomatidae, I. Special Museum Publications, Texas Tech Univ., Lubbock. (1976).

Stanley, S.M.: Macroevolution. Freeman & Co., San Francisco. (1979).

Yunis, J.J., and Prakash, O.: The origin of man: a chromosomal pictorial legacy. Science 215:1525-1530 (1982).

10

Dermaglyphics and Subspecific Systematics of Mantled Howler Monkeys (*Alouatta palliata*)

J. W. Froehlich and P. H. Froehlich

Previously, dermal ridge patterns on the hands and feet of mantled howler monkeys, Alouatta palliata, have been used effectively to study the population genetic structure of neighboring troops on Barro Colorado Island, Panama, and in Guanacaste, Costa Rica, with interpretations ranging from isolation by distance or vicariance to historically corroborated migrations. Larger differences between Panama and Costa Rica than within the two samples suggested that the same traits may also have resolving power at the subspecies level. Pursuing that hypothesis, we recorded data from museum specimens collected between the Colombian border and Lake Nicaragua; the resulting eight regional samples represent four previously identified subspecies. Discriminant analysis confirmed the distinctness of isolated populations on the Azuero Peninsula and on Coiba Island, Panama, although the latter still shows affinities to western Panama despite geographical separation of the island from the mainland since the last Pleistocene sea level transgression. This genetic proximity of A. coibensis to mainland howlers after 15,000 years suggests that fingerprint genes are relatively stable over time. The most distinct subspecies is A. trabeata on the Azuero Peninsula. It is suggested that the present isolation of this taxon by savanna has been long-standing, perhaps for tens of thousands of years. Our data seem to indicate continuous intergradation, however, between A. palliata and A. aequatorialis, the principal two subspecies of Central America. The subspecific boundary appears to lie near the Isthmus of Panama, rather than along the Costa Rican border as was previously thought. In sum, dermatoglyphic data have proved to be a useful tool for subspecific systematics in the context of the complex biogeographical history of Central America.

INTRODUCTION

The dermatoglyphic study of nonhuman primates has progressed little since the pioneering work of Midlo and Cummins (1942); this is in dramatic contrast with both the quantity and sophistication of population (Froehlich and Giles, 1981b) or clinical and genetic studies (Holt, 1968) of human fingerprints (cf., Mavalwala, 1978). Many nonhuman primate studies have been primarily descriptive (e. g., Brehme, 1965a, 1965b; Iwamoto, 1964); systematic comparisons have been largely limited to summary indices (e. g., Brehme and Wiersema, 1973; Brehme, 1975) or simple graphics (e. g., Brehme, 1968a; Pereira da Silva, et al., 1972).

In most of these systematic studies, the focus has been at the generic or specific level, with conclusions relating primarily to questions of classification (e. g., Brehme, 1968b). In a few instances involving macaques, there have been statistical, quantitative comparisons of subspecific taxa (Cauble and Mavalwala, 1973; de Stefano, et al., 1981). These have shown that intraspecific differences may overpower those between geographically distant species, a possibility suggested by Biegert (1963). Morris and Kerr (1974) have concluded that insular microevolution of _Macaca fuscata_ populations has produced this result. Presumably, at higher taxonomic levels conservatism or convergence in functional adaptations may obscure phylogenetic relationships. Although some of these functional adaptations have been discussed by Biegert (1963), it remains difficult to distinguish the effects of function and phylogeny. For perspective on understanding this distinction, the range and significance of intraspecific variation requires further study.

In previous work (Froehlich and Thorington, 1982) on dermatoglyphic variation between contiguous troops of mantled howler monkeys, _Alouatta palliata_, on Barro Colorado Island, Panama and at Finca La Pacifica, Guanacaste, Costa Rica, we used multiple discriminant analysis to show a close congruity between fingerprint distances and geographical vicariance or historical relationships. Moreover, a relatively large multivariate distance between the two populations, representing different subspecies, suggested that the same data and analytic techniques should have significant resolving power at the subspecific level.

Except for a summational, Euclidean distance parameter used by Morris and Kerr (1974) in a study of genetic variability in macaques, we know of no other nonhuman primate fingerprint study where multivariate analysis has been utilized. As shown in previous work (Froehlich and Giles, 1981a), discriminant analysis is an effective method for reducing and interpreting dermatoglyphic data; moreover, it eliminates the confounding redundancy of correlated variables characteristic of simple Euclidean distance measures.

The success of previous subspecific investigations of macaque fingerprints and our own study of genetic structure in howler monkey populations encouraged us to pursue the problem of subspecific relationships within the context of the very complex, recent biogeographical history of Central America (Marshall et al., 1982). We report here the results from a study of available museum samples of howler monkeys with provenances from the Darien of eastern Panama to the shores of Lake Nicaragua, encompassing four previously identified subspecies (Lawrence, 1933).

METHODS AND MATERIALS

Beginning in 1973, we have been involved in a study of howler monkey population dynamics and genetic structure on the northeast corner of Barro Colorado Island (BCI), Panama (Froehlich et al., 1981). After capturing the animals from eight contiguous troops with CO_2 propelled anesthetic darts, we collected a variety of data, including impressions of the dermal ridge patterns on the left hand, foot, and the ventral surface of the tail. We used a silicone rubber casting compound, which provided an undistorted, three dimensional record of each individual's dermatoglyphs. Similar techniques were used in Guanacaste Province, Costa Rica for sampling five contiguous troops of howlers (Scott et al., 1976)

With a total of nearly 200 animals from these two geographical samples, we developed techniques of data analysis for optimally recording the variability of dermal ridge patterns. These scoring methods were based on, but not limited to, the techniques of Midlo and Cummins (1942). In particular, we found the previously undescribed patterns on the proximal and medial phalanges to be extremely complex, while the distal digital pads and the tail prints were essentially identical in all animals sampled. We designed an ordinal coding system for each pattern area (see Figure 1), based on our perception of increasing pattern complexity in *Alouatta palliata*. Our system is species specific; we made no attempt to make it applicable to other taxa, because of the necessary simplifications this would have entailed and the consequent loss of systematic information concerning *A. palliata* subtaxa. In addition to the codes for each pattern area, we devised summary indices for the complexity of plantar and palmar surfaces, based on the number of ridge recurvatures.

From the mammalogy collections of the American Museum of Natural History, the Museum of Comparative Zoology, and the United States National Museum, we recorded dermatoglyphic data from additional Central American howler monkey populations (see Figure 2). In order to avoid damage to the skins from the rubber molding material, we scored the data directly, using a magnifying light apparatus. This procedure frequently

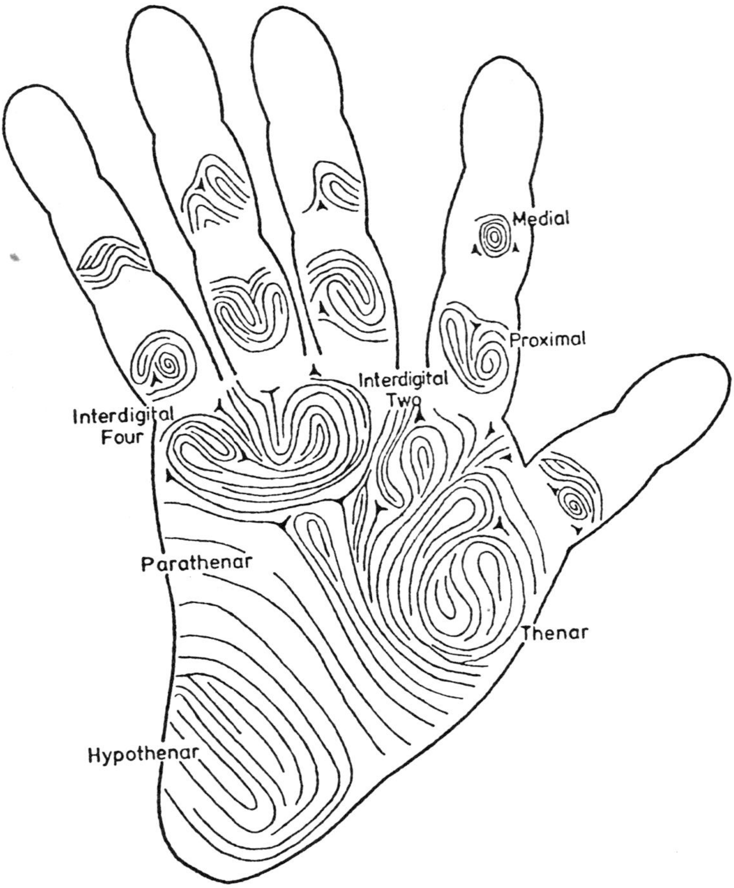

Figure 1. Schematic drawing of the principal dermatoglyphic pattern areas in Central American howler monkeys.

proved difficult or even impossible in practice, because some collectors had made no attempt to preserve the digits or palmar surfaces. In many cases, only partial data could be obtained; the digits, where the most systematic variability had been observed in the previous study, were frequently amputated, incised, or everted. These difficulties greatly reduced sizes of some samples, and some large geographical collections could not be included because of their poor preservation.

Nevertheless, we succeeded in obtaining eight regional samples, which were of sufficient size for analysis and had little or no missing data. Two, from Nicaragua and Costa Rica, complemented our large sample of the A. palliata subspecies from Guanacaste. According

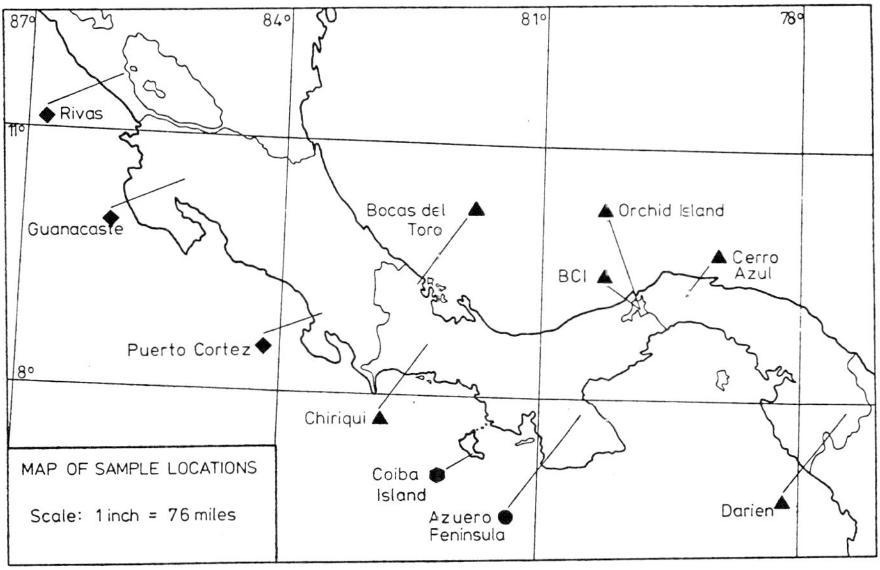

Figure 2. Map of Central America, showing the locations of 11 regional sampling areas. The symbols in this and subsequent figures denote traditional subspecies: diamonds = A. palliata; triangles = A. aequatorialis; circles = A. trabeata; hexagons = A. coibensis.

to Lawrence (1933), animals of this taxon are distinguished from neighboring animals in Panama by long cinnamon flank hairs and short and broad, flattened skulls. The contiguous A. aequatorialis subspecies of Panama is represented by four samples, in addition to that from BCI; this subspecies is characterized by a tendency to lighter coloration than that of A. palliata, as well as relatively short, golden ochraceous mantle hairs. The A. trabeata subspecies, isolated historically on the Azuero Peninsula of Panama by a wide strip of savanna, is notable for its long, silky golden flanks. The subspecies A. coibensus was proposed by Lawrence (1933) for the isolated Coiba Island population because of their small size and dull coloration. Coiba is separated today by a 25-mile wide channel at least 42 fathoms deep. Interpolating from data presented by Bartlett and Barghoorn (1976), the island was last connected to western Panama approximately 24,000 to 15,000 years ago.

To obtain an idea of the magnitude of intrapopulation variability and to test the ability of discriminant analysis to cluster samples according to phenotypic similarity, the BCI data were subdivided into three samples on the basis of geography and previous analysis (Froehlich and Thorington, 1982). The second sample (BCI_2) comes from peninsular populations historically somewhat isolated from the area around the laboratory clearing (BCI_1), while the small Orchid Island population is completely isolated from, but contiguous with, one of the peninsulas. With this subdivision, a total of 12 samples representing four subspecies was available for subsequent statistical analysis.

The initial step of the analysis was a study of the strength of single variables for distinguishing among groups. This was done by appropriate parametric and nonparametric statistical tests for the significance of intergroup differences. These were performed for both subspecific and geographic groupings. In both cases, a majority of the dermatoglyphic variables differed significantly among groups, but the patterns of intergroup variability were inconsistent from trait to trait. Multiple discriminant analysis was thus used to sort our the complex systematic relationships reflected in the single variables. Because of the extensive missing data from the digits in some of our museum samples and because discriminant analysis requires subjects without missing variables, only data from the plantar and palmar surfaces were employed in the initial multivariate analyses. This provided a check on the effects of the small sizes of some samples when we proceeded to an analysis of the complete data set.

RESULTS

Statistical analyses of the 38 fingerprint variables revealed significant variation among subspecies and among the individual geographical populations. Table 1 shows results of *F* tests for quantitative palmar and plantar complexity indices in 11 howler regional samples, with the BCI data only subdivided into Orchid and main island samples. Chi-square statistics for qualitative traits are summarized in Table 2. Over 60% of the traits vary significantly among subspecies; the same is true of the 11 regional samples.

TABLE 1. *F* statistics for the summary complexity indices of dermatoglyphic patterns in 11 howler monkey samples from Central America (*df* 10, 226)

Trait	*F* Value	Significance Level
Total palmar complexity	2.7	0.005
Palmar complexity without thenar	4.8	0.0001
Total plantar complexity	2.3	0.01
Plantar complexity without thenar	2.8	0.005

TABLE 2. Significance levels for chi-square comparisons of 34 dermatoglyphic traits in Central American howler monkey subspecies and regional population samples

Number of Groups	Number of Traits at or below Significant Levels: Non-significant	0.05	0.01
4 subspecies	13	6	15
11 population samples	13	3	18

Two of the 11 sample chi-square analyses are presented in Tables 3 and 4. With respect to patterning on the thenar pad of the palm in Table 3, the Azuero Peninsula and Coiba Island populations share a high degree of complexity, along with two of the more northern A. palliata populations. At the opposite extreme, the Cerro Azul animals have significantly reduced complexity. In sharp contrast, a different pattern of relationships is suggested by the data from the medial phalangeal pad on the fifth toe in Table 4. Here the Azuero animals join Cerro Azul with high complexities, not shared by several other samples, albeit one must also note the much smaller sample sizes in certain cases.

The difficulties of interpreting population relationships in all 21 significant variables, when each shows idiosyncratic clusterings such as in these two examples, are resolved by a multiple discriminant analysis (Figure 3). In order to maintain the largest possible sample sizes, this analysis excludes the digital data. With the BCI data subdivided into three samples,

TABLE 3. Percentages of simple and complex palmar thenar patterns in 11 Central American howler monkey samples ($p < 0.001$, 10 df)

Sample	Direction of Trend	Pattern Type Simple	Pattern Type Complex	Sample Size
Darien		73	27	15
Cerro Azul	Down	91	9	11
Barro Colorado Island		82	18	92
Orchid Island		80	20	10
Chiriqui		69	31	16
Bocas del Toro		71	29	14
Azuero Peninsula	Far up	21	79	14
Coiba Island	Up	50	50	16
Puerto Cortez		67	33	9
Guanacaste	Up	54	46	48
Rivas	Up	50	50	10

TABLE 4. Percentages of pattern types for the medial fifth phalanx of the foot in 11 Central American howler monkey population samples (*p* < 0.001, *df* 10)

Sample	Direction of Trend	Pattern Type Arch	Pattern Type Loop	Sample Size
Darien		64	36	14
Cerro Azul	Far up	27	73	11
Barro Colorado Island	Down	86	14	92
Orchid Island		60	40	10
Chiriqui		56	44	9
Bocas del Toro	Down	77	23	13
Azuero Peninsula	Up	42	58	12
Coiba Island		60	40	5
Puerto Cortez	Far down	100	0	8
Guanacaste		58	42	48
Rivas		50	50	10

this graph shows the relationships among 12 populations. In this analysis, the first function accounted for 35% of the variance, and the second, 20%. Both were highly significant by Wilks' criterion; indeed, the first four extracted functions had null probabilities below 0.0002. Another measure of the significance of this discriminant display is the fact that classification of individual animals into the 12 samples was 37.1% accurate, considerably above the 8.3% correct assignments expected from random classification.

In this analysis of a reduced data set, a pattern of population relationships not discernible in univariate analyses emerges. The *A. trabeata* sample from the Azuero Peninsula is the most distinct, while the geographically isolated Coiba animals are intermediate, falling closer to a Costa Rican *A. palliata* sample. The three *A. palliata* samples are broadly grouped, but their internal relationships are not congruent with geography. Moreover, they are joined by an intergrading *A. aequatorialis* sample from western Panama. For central Panama, the BCI animals are adjacent both to Chiriqui and to eastern Panamanian samples, but not closely so. Finally, the relationships within the three BCI samples accord with expectations, the Orchid Island sample showing its close affinity with the peninsular BCI_2 sample.

With more resolving power, but with sample sizes as small as 5 for Coiba, a second discriminant analysis was performed on the complete data set, including the digital pads with their significant intergroup variability. Table 5 presents an abbreviated summary of this analysis; for clarity of interpretation, only variables with canonical coefficients above 0.25 are shown. As noted, the first and second functions are similar to those of the preceding analysis, with a total of 53.1% of the overall

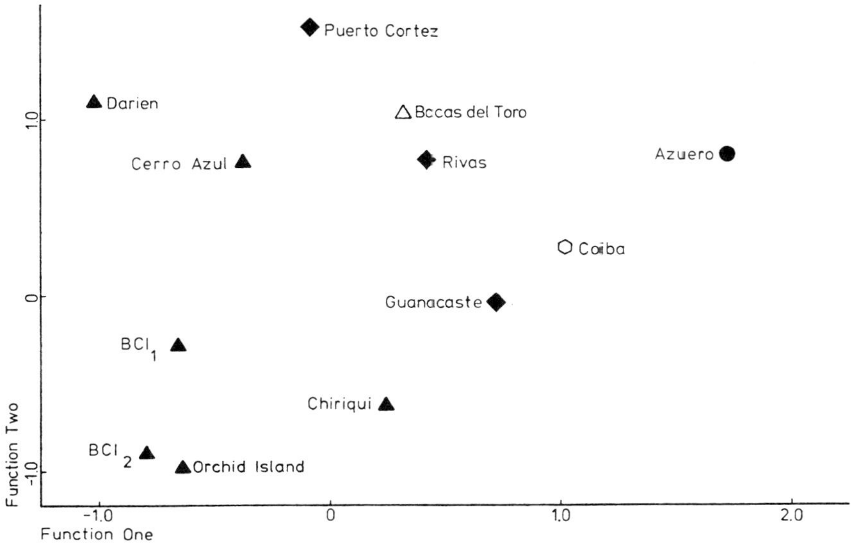

Figure 3. Centroids of 12 howler monkey samples from Central America on the first and second discriminant functions of a discriminant analysis based on 20 palmar and plantar dermatoglyphic traits. Subspecies symbols as in Figure 2; a lack of shading indicates a deviation from expectation.

dermal ridge variance accounted for by these two vectors. A total of five functions were extracted with null probabilities below 0.0004 by Wilks' Lambda statistic. The increased discriminatory power of this analysis can also be discerned in an assignment accuracy of 53%, well above the 37% of the previous discrimination.

These increases of significance and discriminatory power underscore the systematic importance of the dermal patterns on the proximal and medial phalanges, at least in howler monkeys. Comparing the plot of the first two functions in Figure 4 with the previous discriminant display in Figure 3, it is also clear that the smaller sample sizes in the former are not a critical factor. Although the spatial patterns are partially rotated in this analysis, the small Coiba sample still has Rivas and Guanacaste as its closest neighbors.

The pattern of population relationships is also more clearly interpretable in Figure 4 than in the previous analysis. Again the Azuero _A. trabeata_ sample is separated as an isolate, validating Lawrence's (1933) subspecific designation. Similarly, Coiba Island animals appear isolated in discriminant space as they have been isolated for 15,000 years by water. Their closest neighbors appear to be the populations of western Panama and Costa Rica. The latter form a fairly clear cluster,

TABLE 5. Variables making a major contribution to discrimination among 12 Central American howler monkey regional samples. Of the 25 variables included in the analysis, only those with standardized coefficients greater than 0.25 on functions one and/or two are shown

Dermatoglyphic Variable	Discriminant Functions One	Two
Medial index finger	-0.41	-0.28
Medial fifth finger		-0.41
Proximal third finger	0.27	
Palmar third interdigital		0.58
Palmar fourth interdigital		0.36
Parathenar		0.43
Thenar	0.27	
Palmar complexity without thenar		-0.51
Medial fifth toe		-0.41
Proximal third toe	0.33	
Proximal fifth toe		0.33
Plantar second interdigital	0.26	
Plantar fourth interdigital	0.48	
Plantar complexity	0.32	0.58
Plantar complexity without thenar	-0.72	-0.26
Eigenvalue	1.68	0.60
Percent of variance	39.21	13.93
Significance level	<0.00005	<0.00005

with spatial patterns congruent with their actual geographical relationships. These three *A. palliata* samples from Costa Rica and Nicaragua are now in a distance pattern nearly identical with geography (cf., Figure 2), while the western Panamanian samples of *A. aequatorialis* approach the southern end of this distribution. On the basis of these data, it is tempting to redefine the subspecies boundary eastward toward the isthmus, where the BCI samples aggregate quite closely with Cerro Azul and Darien. Again the internal relationships of the BCI samples are according to expectations, with the distances between them being the smallest in the entire figure. Thus the same data which previously elucidated genetic relationships among contiguous troops (Froehlich and Thorington, 1982) are relevant to subspecific patterns at a different magnitude of variance.

DISCUSSION

Clearly, dermatoglyphic data, which have previously been shown to be so highly heritable in humans (for

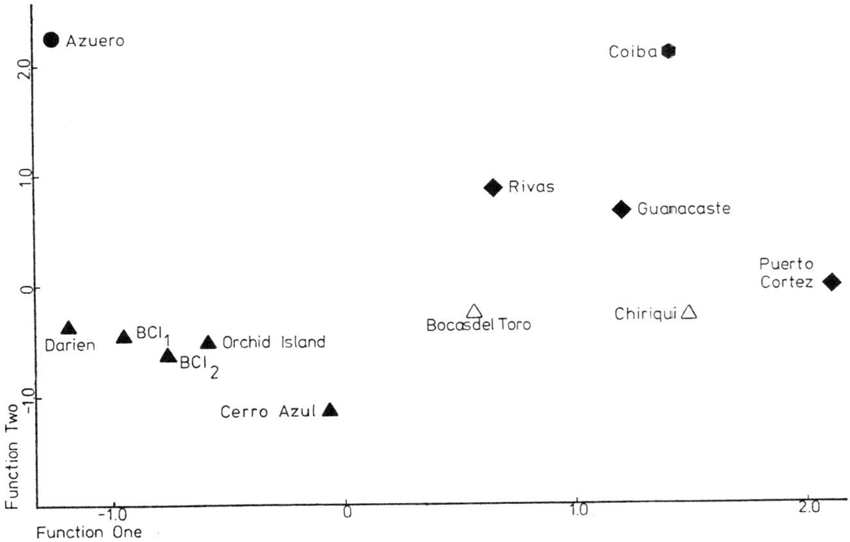

Figure 4. Centroids of 12 howler monkey samples from Central America on the first and second axes of a discriminant analysis based on a comprehensive set of 25 dermatoglyphic traits. The most important traits are listed in Table 5. Subspecies symbols as in Figure 2; a lack of shading indicates a deviation from expectation.

review, see Froehlich, 1976), are of value in elucidating systematic relationships between subspecies, perhaps even preferable to conventional pelage data or the environmentally influenced metric traits. On the supposition of greater stabilizing selection for characters which play a role in locomotion, it is likely that the heritabilities for dermatoglyphic traits in nonhuman primates are as high or higher than those for human fingerprints. It remains to be seen what resolving power these genetic data have for systematic relationships among species or higher taxa. Certainly, quantitative assessment of one of the most unique, derived traits shared by all living primates has been too long neglected.

For the moment, the patterning of dermatoglyphic relationships among Central American subspecies of howler monkeys suggests a number of new conclusions concerning the biogeography of these animals. In general, according to the discriminant vectors in Table 5, howler monkeys inhabiting Coiba Island, western Panama, Costa Rica, and Nicaragua show complex patterns on the thenar of the hand and in the thenar and interdigital areas of the foot, in conjunction with proximal third digit complexity and medial index finger simplicity. The Coiba animals, however, are isolated from the others on the second axis by overall plantar and palmar complexity, coupled with simple patterns on the medial fifth digits. Depth soundings and the reconstructions of Pleistocene sea level transgressions and recessions (Bartlett and Barghoorn, 1973) support the assumption that Coiba Island was last connected to western Panama and southern Costa Rica prior to 15,000 years ago and that this connection had endured for more than 5,000 years. During this time, our data suggest that howler monkeys from western Panama north to Nicaragua formed a continuous, intergrading population, with Coiba animals being closely related to or part of this population.

After 15,000 years and roughly 3,500 generations since the vicariance event of sea level transgression separated Coiba Island from the mainland, the isolated Coiba population has evolved the smaller size and drab coloration typical of island populations. The dermatoglyphic data, however, still reveal affinities to what we presume to be the ancestral population, although some secondary distinctions have evolved. This evidence corroborates an earlier conclusion that genes controlling dermatoglyphic characters are relatively stable in microevolutionary time (Froehlich and Giles, 1981b). Our data, however, do not support the suggestion of Goldman (1920) that the Coiba animals are a separate species.

The highly distinctive _A. trabeata_ of the Azuero Peninsula, with their uniform golden flanks, are uniquely isolated in discriminant space. Although they share with Coiba animals the overall plantar and palmar complexity in conjunction with simplicity on the medial fifth digits (see Table 5), their characteristics on the first axis are totally opposite to those of the Coiba sample and align them with eastern Panama. Azuero population affinities possibly lie to the east, but their apparent long isolation leaves this in question. Judging by the time Coiba animals seem to have been isolated from the mainland and the relative distinctness of the two groups, the wide strip of grassland, which presently separates the Azuero animals from the rest of Panama, should be of great antiquity. It has been suggested that the majority of the migrants involved in the Great American Interchange, beginning about 3 million years ago, were members of a savanna or grassland fauna (Marshall et al., 1982). Moreover, Janzen and Martin (1982) have recently argued that the megafauna of Central America helped to maintain a forest characterized by patchiness punctuated with grasslands until about 10,000 years ago. Thus, the Azuero _A. trabeata_ subspecies may be the product of tens of thousands of years of relative isolation.

The samples from the isthmus and eastern Panama form a tight cluster in Figure 4, fairly widely separated from the two western Panama populations. In terms of the vectors in Table 5, the trait complex which clusters BCI and eastern Panama is the antithesis of the pattern complexities which distinguished the Coiba animals, described earlier. The wide separation and tight clustering of these animals supports a separate subspecies designation, but in contrast to Lawrence's (1933) conclusions, the two samples from western Panama are at best intermediate between these and *A. palliata*, if not actually a part of the *A. palliata* subspecies. In her defense, it should be noted that she acknowledged considerable intergradation in coat color between *A. palliata* and *A. aequatorialis*, with total overlap in skull characters. Indeed, her Chiriqui sample, which is the same as ours, was the most intermediate. These factors and the dermatoglyphic results suggest a subspecific dividing line closer to the Isthmus of Panama. The *A. aequatorialis* subspecies may be more allied to South America.

Following the geological reconstructions of Lloyd (1963), which show eastern Panama to be a province of South America, howler monkeys of *A. aequatorialis* ancestry may have lived in eastern Panama since at least the Pliocene. The Central American *A. palliata* subspecies may represent a howler population which migrated and became isolated during the Great American Interchange around 3 million years ago. Whitmore and Stewart (1965) have argued that intermittent connections between western Panama and the Talamanca Range of Costa Rica have existed since the Miocene, when the isthmic connection was only a filter bridge for island hoppers. Our data would seem to corroborate these ancient geographical relationships by demonstrating genetic differences between eastern and western Panama. In sum, the fingerprint data of primates have provided a useful tool for the study of Central American howling monkey biogeography.

ACKNOWLEDGMENTS

We are grateful for the cooperation and assistance we received from the staffs of the American Museum, the Museum of Comparative Zoology, and the U. S. National Museum of Natural History. Travel was partially supported by the University of New Mexico, where computational time was also provided. Finally, we wish to thank Manzano Day School for the use of its word processor. Comments on the text were generously provided by R. W. Thorington, Jr., and by others attending the IXth Congress of the International Primatological Society.

REFERENCES

Bartlett, A. S., and Barghoorn, E. S.: Phytogeographic history of the Isthmus of Panama during the past 12,000 years (A history of vegetation, climate, and sea-level change). In Graham. Vegetation and Vegational History of Northern Latin America. pp. 203-299. Elsevier, New York. (1973).

Biegert, J.: The evaluation of characteristics of the

skull, hands, and feet for primate taxomony. In Washburn. Classification and Human Evolution. pp. 116-145. Aldine, Chicago. (1963).

Brehme, H.: Uber das Hautleistensystem der Palma und Planta von *Nasalis larvatus*. Homo. Suppl. -Bd.: 244-248. (1965a).

Brehme, H.: Uber die Variabilitat des Hautleistensystems der Palma und Planta von *Chiropotes satanas* (Satansaffe). Z. Morph. Anthrop. 56:206-216. (1965b).

Brehme, H.: Uber das Hautleistensystem der Palma und Planta einiger seltener Schlankaffen (*Presbytis* und *Pygathrix*). Mitt. Anthrop. Ges. Wien. 98:1-14. (1968a).

Brehme, H.: Zur Variabilitat des Hautleistensystems der Meerkatzen (*Cercopithecus*) und des Husarenaffen (*Erythrocebus patas*). Folia Primatol. 9:41-67. (1968b).

Brehme, H.: Epidermal patterns of the hands and feet of the pygmy chimpanzee (*Pan paniscus*). Am. J. Phys. Anthrop. 42:255-262. (1975).

Brehme, H., and Wiersema, H. D.: Finger- und Zehenbeerenmuster von Pongiden. Folia Primatol. 20:391-409. (1973).

Cauble, R. G., Mavalwala, J.: The palmar dermatoglyphics of *Macaca fascicularis*: A comparison with *Macaca fuscata*. J. Hum. Evol. 2:137-152. (1973).

Froehlich, J. W.: The quantitative genetics of fingerprints. In Giles and Friedlaender. The Measures of Man. pp. 260-320. Peabody Museum Press, Cambridge. (1976).

Froehlich, J. W. and Giles, E.: A multivariate approach to fingerprint variation in Papua New Guinea: Implication for prehistory. Am. J. Phys. Anthrop. 54:73-91. (1981a).

Froehlich, J. W. and Giles, E.: A multivariate approach to fingerprint variation in Papua New Guinea: Perspectives on the evolutionary stability of dermatoglyphic markers. Am. J. Phys. Anthrop. 54:93-106. (1981b).

Froehlich, J. W. and Thorington, R. W., Jr.: The genetic structure and socioecology of howler monkeys (*Alouatta palliata*) on Barro Colorado Island. In Leigh and Rand. Ecology of a Neotropical Forest: Seasonal rhythms and long-term changes. Smithsonian Institution Press, Washington. (1982).

Froehlich, J. W., Thorington, R. W., Jr., and Otis, J. S.: The demography of howler monkeys (*Alouatta palliata*) on Barro Colorado Island, Panama. Int. J. Primatol. 2:207-236. (1981).

Goldman, E. A.: Mammals of Panama. Smithsonian Misc. Coll. 69:1-309. (1920).

Holt, S. B.: The genetics of dermal ridges. Charles C. Thomas, Springfield. (1968).

Iwamoto, M.: Morphological studies of *Macaca fuscata*. I

Dermatoglyphics of the hand. Primates 5:53-73. (1964).

Janzen, D. H., and Martin, P. S.: Neotropical anachronisms: the fruits the gomphotheres ate. Science 215:19-27. (1982).

Lawrence, B.: Howler monkeys of the _palliata_ group. Bull. Mus. Comp. Zool. Harvard. 75:315-354. (1933).

Lloyd, J.: Tectonic history of the south Central American orogen. Am. Assoc. Petrol. Geologists. Mem. 2:88-100. (1963).

Marshall, L. G., Webb, S. D., Sepkoski, J. J., and Raup, D. M.: Mammalian evolution and the Great American Interchange. Science 215:1351-1357. (1982).

Mavalwala, J. Ed. : Dermatoglyphics: an International Perspective. Mouton, The Hague. (1978).

Midlo, C., and Cummins, H.: Palmar and plantar dermatoglyphics in primates. Amer. Anat. Mem. 20:1-198. (1942).

Morris, L. N., and Kerr, B. A.: Genetic variability in macaques: Correlations between biochemical and dermatoglyphic systems. J. Hum. Evol. 3:223-235. (1974).

Periera da Silva, M. A., Petit-Marie, N., and Rakotosamimanana, B. R.: Contribution a l'etude des dermatoglyphes palmaires et plantaires des genres _Cebus_, _Alouatta_ et _Ateles_. Mammalia 36:279-291. (1972).

de Stefano, G. F., Falcone, G. F., and Fuciarelli, M.: Finger and palmar dermatoglyphics in a troop of _Macaca fuscata_. Antropol. Contemp. 4:33-45. (1981).

Scott, N. J., Jr., Scott, A. F., and Malmgren, L. A.: Capturing and marking howler monkeys for field behavioral studies. Primates 17:527-533. (1976).

Whitmore, F. C., Jr., and Stewart, R. H.: Miocene mammals and Central American seaways. Science 148:180-185. (1965).

11

On Size of the Semicircular Canals Organ in the *Tarsius bancanus*

S. Matano, T. Kubo, T. Matsunaga, C. Niemitz, and M. Gunther

Tarsius's three dimensional movement in leaping with extraordinary rapidity is recognized as being remarkably unique. To relate this behavior to anatomical structures concerned with linear and rotational acceleration we made measurements of the osseous semicircular canals organ of *Tarsius bancanus* and compared these with those of *Nycticebus coucang*. In skulls of Tarsius and Nycticebus, the osseous semicircular canals were exposed by a surgical drill. The values of angle between the lateral canal plane and Reid's horizontal plane are 29.12° in Tarsius and 49.60° in Nycticebus. This angle may have a close relationship to normal postures in the lives of these animals. In both species the values of the angle between the plane of the anterior canal and the sagittal plane fall among those of other primates. These values may be related to the location of the eyes in each species. The consensus is that agile creatures with orientations requiring naturally occuring rapid motion demand a larger curvature of the semicircular canals. The values of the radius of curvature of canals in ratio to the body weight are significantly much larger in Tarsius than those in Nycticebus. Moreover, volumes of the vestibular nuclei in relation to the total brain volumes are about 1.7-1.8 times in Tarsiidae and Galaginae which show leaping behaviors, compared to those in Lorisinae which do not (Stephan's Table, 1981).

INTRODUCTION

In different species locomotor behavior is considered a reflection of the structure of the somatic system. That is to say, the sense organs, motor systems, and nervous system will show structural variations as morphological evidence of animal adaptation. Viewed comparatively the divergence of various prosimian locomotions offers interesting examples. Among them, Tarsius's three dimensional locomotion is recognized as being of remarkable uniqueness. Tarsius leaps with extraordinary rapidity; turns through 180° may be made in midair (Napier and Napier, 1967) and the head may rotate through 180° (Niemitz, 1979). However, the sense organ related to body equilibrium in Tarsius has received relatively little attention.

In order to attain ocular stabilization during the head movements of daily life, the vestibulo-ocular reflex must work in a coordinated fashion. Theoretically, the sensitivity of this reflex arc depends upon the size of the semicircular canals organ of the vestibular labyrinth (Jones and Spells, 1963). In the present study, we have examined the size and the angular relationship of the semicircular canals of *Tarsius bancanus* and compared these values with those of *Nycticebus coucang*. Discussion of these is made in relation to characteristics of their locomotive pattern.

METHODS AND MATERIALS

One adult skull of *Tarsius bancanus* and one of *Nycticebus coucang* were studied.

Using a surgical drill, the osseous semicircular canals on both sides of the temporal bone were carefully exposed under magnification. The skull was mounted semi-permanently on a wooden plate with epoxyresin to avoid any dislocation during measurement. The plate was fixed in a stereotaxic frame. A measuring needle attached to the micromanipulator was moved along x, y, and z axis under magnification (Figure 1). Reid's plane was used as a reference horizontal plane which passes through the center of the external auditory meatus and the inferior margin of the bony orbit.

Several points in each semicircular canal were registered in three dimensional scales. The inclination of the plane of the lateral semicircular canal relative to Reid's horizontal plane was then calculated. Angles between the plane of the anterior canal and that of the posterior one, and between the plane of the anterior or posterior canal and the stereotaxic sagittal plane, were also measured. The diameter was measured about the radius (r) of each canal and the radius of curvature (R) of each canal. Average values were taken for each of the three canals on both sides.

RESULTS

The inclination of lateral semicircular canal is shown by the angle between Reid's horizontal plane and the plane of the lateral semicircular canal. These values of Tarsius and Nycticebus are remarkably different as

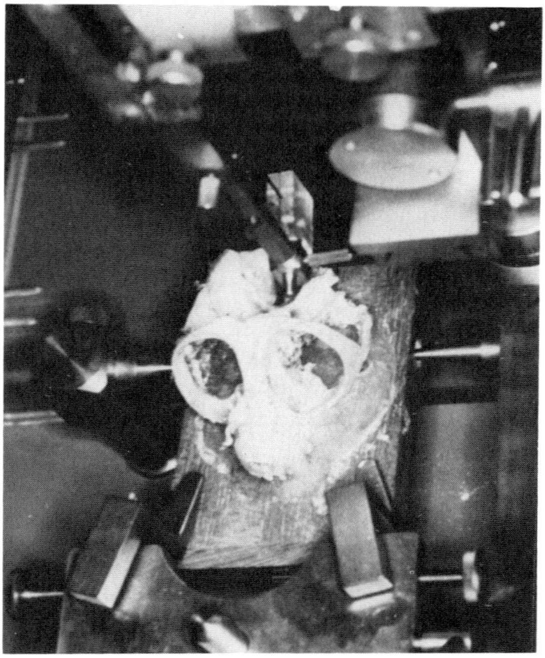

Figure 1. This photograph shows the skull of <u>Nycticebus coucang</u> attached to the stereotaxic frame.

shown in Table 1. In another example, the plane of the lateral semicircular canal of the guinea pig inclined 50.31° relative to the horizontal plane which passes through the upper incisor bar and the center of the external auditory meatus (Curthoys et al., 1975). Modest inclination of the lateral semicircular canal relative to Reid's horizontal plane was reported in the cat and man. These values are 22.92° in the cat (Blanks et al., 1975b) and 25.12° in man (Blanks et al., 1975a). From these results, the values of Tarsius and man are relatively approximate and, also, values of Nycticebus are similar to those of the guinea pig. These may have a relationship to animal posture in the usual living state. However, the biological significance of the inclination of the lateral semicircular canal relative to the skull is unclear. For this problem further measurements of various species of animals are required.

The angle between the plane of the anterior semicircular canal and that of the posterior one, as shown in Table 1, does not show strict perpendicularity. These values are also different in each animal species. For example, these are 76.71° in the guinea pig (Curthoys et al., 1975), 90.40° in the cat (Fernandez and Valentinuzzi, 1968) and 86.16° in man (Blanks et al., 1975a).

TABLE 1. Angles in degree of inclination, A-P, A-S, and P-S

	Inclination	A-P angle	A-S angle	P-S angle
Tarsius bancanus	29.12	91.78	36.87	51.35
Nycticebus coucang	49.60	88.63	34.11	57.26

Inclination: Angle between the lateral canal plane and Reid's plane. A-P: Angle between the anterior canal plane and the posterior one. A-S and P-S: Angles between the anterior or posterior canal plane and the stereotaxic sagittal plane.

The angle between the plane of the anterior semicircular canal and the stereotaxic sagittal plane (A-S angle in Table 1) is 36.87° in Tarsius and 34.11° in Nycticebus, and the angle between the plane of the posterior semicircular canal and the stereotaxic sagittal one (P-S angle in Table 1) is 51.35° in Tarsius and 57.21° in Nycticebus. These values fall among those of other primate species including man. In other mammals, such as sheep, horse, rabbit and cat, the values of P-S angle are lower than those in primates and may be related to the location of the eye in each species (Hasegawa, 1969).

In Table 2, values of the radius of the osseous canals and the radius of curvature of the osseous canals in Tarsius bancanus and Nycticebus coucang are shown with their body weight.

TABLE 2. The average values of the canal radius of curvature (R) and the radius (r) of the osseous canal shown with the body mass in kg

	R (mm)		r (mm)	body mass
Tarsius bancanus	anterior canal 1.89			
	posterior can. 1.71	1.78	0.19	0.13
	lateral canal 1.75			
Nycticebus coucang	anterior canal 1.85			
	posterior can. 1.60	1.67	0.20	1.20
	lateral canal 1.57			

Regarding the radius of the semicircular canals, the values of the osseous canals are much larger compared with those of the membranous canals. The values of the osseous canals are not sufficient for discussing the biological significance of the radius of canals. On the other hand, the radius of curvature of the osseous canals and that of the membranous canals closely approximate each other in size. Hence, a correction is not needed for the values of these sizes as described by Curthoys et al. (1976b). For this reason we focused on a comparison of the radius of curvature of the canals.

As shown in Table 2, the radius of curvature of the anterior canal is larger than that of the lateral and posterior canals in both species. This finding is in accord with the results of other investigators who also recognize this common tendency. A detailed description on this matter can be found in the reports of Gray (1951) and Curthoys et al. (1976a).

Jones and Spells (1963) and Jones (1974) measured the dimension of the semicircular canals of many animals and found that the radius of curvature of the membranous canal was proportionate to body mass. Thus, Jones and Spells showed the relationship between the radius of curvature of the canals and body mass (Figure 2). Regression line A is log 100 R = 0.1029 log m + 2.6742 in fishes; line B is log 100 R = 0.0563 log m + 2.4143 in birds; and line C is log 100 R = 0.1157 log m + 2.2466 in mammals alone (m = body mass). Then, the values of Tarsius and Nycticebus obtained by our measurements were inserted respectively in this figure. A solid rectangular mark shows the value of Tarsius (arrow) and the open one shows that of Nycticebus (arrow). A species specific tendency is clearly recognizable in both species. That is to say, the value of Tarsius is relatively larger than that of Nycticebus in relation to body weight. Therefore, Tarsius falls rather near the average range among birds, whereas Nycticebus is located among mammalian values.

DISCUSSION

It has been reported in neurophysiological studies that the sensitivity of primary canal afferents showed some differences among the neurons of the three canals in the cat; the anterior canal neurons showed the highest sensitivity, followed by the posterior and lateral canal neurons (Blanks et al., 1975b). Moreover, Curthoys et al. (1976a) compared the size of the radius of curvature of canal (R) with the mean sensitivity of canal afferents and estimated the positive correlation between them.

Therefore, from the morphological and physiological stand point, the consensus is that fast moving creatures or orientations of naturally occurring rapid motion demand the larger dimension of curvature of the semicircular canals (Gray, 1951; Hadziselimovic and Savkovic, 1964; Curthoys et al., 1976a). Our results are

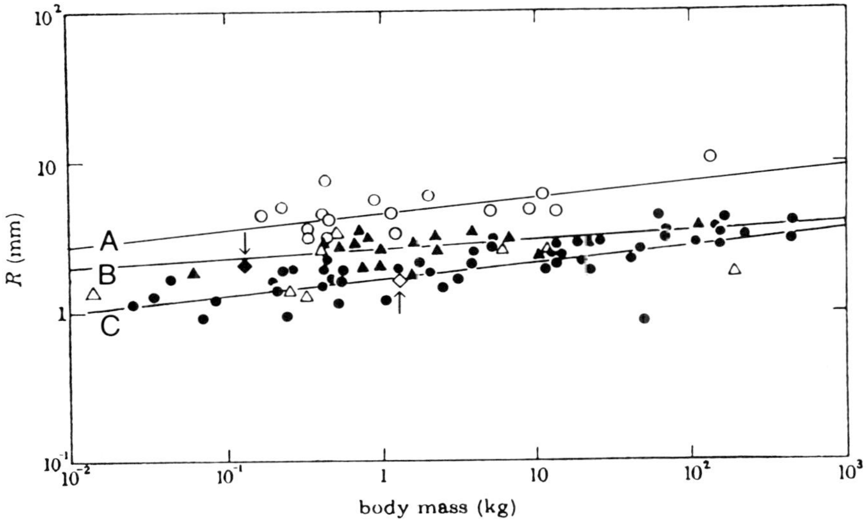

Figure 2. This figure shows change in the radius of curvature (R) with animal body mass measured from 87 species by Jones (1974). A: The regression line in fishes alone. B: The regression line in birds alone. C: The regression line in mammals alone. Black round: Mammals; Open round: Fishes; Black triangle: Birds; Open triangle: Reptiles. The values of Tarsius (black rectangle) and Nycticebus (open rectangle) obtained by our results are inserted.

also in accord with this interpretation, and are interesting in relation to the uniqueness of three dimensional behaviors in Tarsius. Jones (1974) pointed out that the suitable adjustments were brought about in both gain (canal sensitivity) and frequency response (time constant of cupular return) to maintain accurate angular velocity transduction despite changes in the likely range of head movements brought about by size, habitat and mode of movement. He postulated that the larger diameter in both radius of canal (r) and radius of curvature of canal (R) in fishes ensured the high sensitivity required for the slow head movement. This opinion about fishes, also described by Curthoys et al. (1976a), seems to contradict the consensus mentioned above.

In order to solve this problem, further evidence is needed. For example, Jones (1974) postulated that elasticity of the cupula and viscosity or density of the endolymph were identical among fishes, birds and mammals; however, there is still not an explanation for his assumption. In addition, head movements, especially as related to eye centering which ensures retinal image stabilization, are performed by an integrative mechanism through impulses from the visual system, and neck muscles, in addition to the vestibular impulses evoked by the semicircular canals stimulation. In any case, "matching of canal size" is an important problem concerned with adaptation and diversity of the various locomotive behaviors in the prosimians.

Furthermore, morphological variation may exist in the vestibular nuclei receiving sensory input from the semicircular canals. According to the data of Stephan et al. (1981), vestibular nuclei volumes were 7.69 mm^3 in Tarsius specimen, 8.45 mm^3 in *Loris tardigradus*, 17.30 mm^3 in *Perodicticus potto*, 19.90 mm^3 in *Galago crassicaudatus* and 6.65 mm^3 in *Galago demidovii*. The ratios of these values to the total brain volumes of each species are about 1.7-1.8 times in Tarsius specimen, *Galago demidovii* and *Galago crassicaudatus* compared to those in *Loris tardigradus* and *Perodicticus potto*. The latter group, as is well known, shows no leaping pattern in locomotion, whereas the former shows the outstanding leaping in normal locomotion. Furthermore, *Nycticebus coucang* is included in the latter group. There is no direct relationship between the volumes of grey matter and the activity of the specific nervous function; but the extent of the grey matter may be related to neurons in number, length and arborization of dendrites and some other factors in the structural characteristics which may give the fundamental effects upon its nervous functions. Clearly, further work is needed on this problem.

REFERENCES

Blanks, R. H. I., Curthoys, I. S., and Markham, C. H.: Planar relationship of the semicircular canals in man. Acta Otolaryngol. 80:185-196. (1975a).

Blanks, R. H. I., Estes, M. S., and Markham, C. H.: The physiologic characteristics of vestibular first order canal neurons in the cat. II. Response to constant angular acceleration. J. Neurophysiol. 38:1250-1268. (1975b).

Curthoys, I. S., Curthoys, E. J., Blanks, R. H. I., and Markham, C. H.: The orientation of the semicircular canals in the guinea pig. Acta Otolaryngol. 80:197-205. (1975).

Curthoys, I. S., Blanks, R. H. I., and Markham, C. H.: Semicircular canal radii of curvature (R) in cat, guinea pig and man. J. Morph. 151:1-16. (1976a).

Curthoys, I. S., Markham, C. H., and Curthoys, E. J.: Semicircular duct and ampulla dimensions in cat, guinea pig and man. J. Morph. 151:17-34. (1976b).

Fernandez, C., and Valentinuzzi, M.: A study on the biophysical characteristics of the cat labyrinth. Acta Otolaryngol. 65:293-310. (1968).

Gray, O.: An introduction to the study of the comparative

anatomy of the labyrinth. J. Laryngol. 65:681-703. (1951).

Hadziselimovic, H., and Savkovic, L. J.: Appearance of semicircular canals in birds in relation to mode of life. Acta Anat. 57:306-315. (1964).

Hasegawa, T.: The Labyrinth. Nagai Shoten, Osaka. In Japanese. (1969).

Jones, M. G.: The functional significance of semicircular canal size. Handbook of Sensory Physiology VI/1, 171-184, edited by H. H. Kornhuber, Springer-Verlag, Berlin. (1974).

Jones, M. G., and Spells, K. E.: A theoretical and comparative study of the functional dependence of the semicircular canal upon its physical dimensions. Proc. Roy. Soc. B. 157:403-419. (1963).

Napier, J. R., and Napier, P. H.: A Handbook of Living Primates. Academic Press, New York. (1967).

Niemitz, C.: Outline of the behavior of Tarsius bancanus. The Study of Prosimian Behavior. 631-660, edited by G. A. Doyle and R. D. Martin. Academic Press, New York. (1979).

Stephan, H., Frahm, H., and Baron, G.: New and revised data on volumes of brain structures in insectivores and primates. Folia Primatol. 35:1-29. (1981).

12

Relative Position of the Foramen Magnum in Primates

S. Matsumoto

The relative position of the foramen magnum in primates was measured on a coordinate system defined by the orbital axis. The horizontal plane was defined by the centers of right and left orbital rims and the opening of the left optic foramen; this plane is homologous and equivalent throughout primates. The position of "endobasion" was determined in the median-sagittal plane, after normalization of the sagittal distance between nasion and inion. From prosimians and catarrhines to humans, the locus of the positions of the foramen forms a quarter of an ellipse; platyrrhines are arranged in another, more superiorly situated arc. There are also specialized species whose measurements fall outside these arcs. The position of the foramen depends primarily on the degree of basicranial flexion. In recent humans, the foramen has shifted far enough anteriorly to achieve a near optimal situation; this condition had probably already been achieved at the Neanderthal stage. Sexual differences in the clivus inclination and relative location of the foramen were detected in some nonhuman primates. In almost all cases, the females have more anteriorly placed foramina. These sexual differences parallel general tendencies in the ontogenetic development of the endocranium.

INTRODUCTION

The location of the foramen magnum has received much attention in the study of human evolution. This is because the very anterior position of the foramen is one of the most remarkable characteristics of the human skull.

The skull is supported on the cervical column by nuchal musculature. Therefore, the position of the condylus occipitalis or the foramen magnum has been thought to relate to a condition of head equilibrium. In fact, the heavier the face, the stronger is the nuchal musculature which pulls on the occipital bone. Conversely, muscular development or dynamic features are reflected in skull structures such as the nuchal crest or the occipital condyles.

The special location of the foramen in humans has been thought to be associated with the evolution of the erect posture or habitual bipedalism (Lavelle et al., 1977). Whether or not this is so, the position of the occipital foramen doubtlessly reflects a species typical change in the evolution of humans.

There have been several efforts to quantify the location of the occipital condyles or foramen (Bolk, 1915; Senyurek, 1938; Schultz, 1942, 1955; Clark, 1950; Ashton and Zuckerman, 1951, 1952, 1956). There are two prerequisites to precise measurement. The first is to select the points which will define the positions of the foramen or condyles relative to the cranial structure. Although the size of the foramen and the shape or convexity of the condyles vary among the primate groups, a single point seems adequate to present the relative positions of these structures. Either basion or the midpoint of the deepest points of the condyles on both sides can be used for this purpose. The inion (I) is the best point to mark the posterior termination of the skull, because it is the posterior point from which the dorsal nuchal musculature arises. Opisthocranion is a point defined to measure maximum cranial length in humans only and should not be used for comparisons with other primates. At the anterior end three points are expected. Prosthion corresponds to inion in being the most forward point on the face in the median-sagittal plane. However, the disadvantage of this point is that its location is affected by the various directions and degrees of prognathism. Therefore, some workers have adopted nasion or glabella as the anterior terminal. Although there may not be a great difference between results obtained using nasion or glabella, the latter is used primarily for measuring maximum cranial length of the human skull; its position is affected by supraorbital development. As Schultz (1955) mentioned, nasion (N) is the most representative point of this area, because it is the anatomical border of the cerebral and visceral parts of the skull and is rather stable.

The second prerequisite to measuring the location of the foramen is to determine a base line. The position of this line should not be influenced by variation in other cranial features. The quantifications employed in prior publications are not satisfactory in this respect.

The base line has usually been defined by reference to two of the three points described above, the most anterior one and the intermediate one. The posterior point is projected onto this line (Senyurek, 1938; Schultz, 1955). Such lines are, however, not comparable, because the relative positions of these points in the median-sagittal plane vary in different primate taxa. When prosthion is adopted as the anterior termination, for instance, the course of the line is influenced by the directions of prognathism. Moreover, if the foramen magnum has shifted position in the course of human evolution, its movement must have been not only forward but also downward. Both direction and distance of the foramen from the anterior point would have been affected by this movement in two dimensions. This also holds for comparison of other primates.

Projection of points onto the ear-eye plane (Frankfurt plane) has been used to avoid such source of error (Clark, 1950; Ashton and Zuckerman, 1951, 1952). The major advantage of the Frankfurt plane is that when the human skull is oriented to place this plane in a horizontal position, natural human head posture is approximated. However, this feature is lost in other primate skulls, a loss caused, ironically, by the same factor associated with the movement of the foramen magnum (Matsumoto, 1983). Bolk (1915) used the greatest endocranial length for horizontal orientation. As pointed out by Schultz (1955), this is a poor approximation of natural head posture. Moreover, this horizon is not free of influence by endocranial shape changes. For purposes of comparative morphology, Hofer (e.g., 1965) and others have used the clivus line, while Delattre and Fenart (e.g., 1963) adopted the vestibular coordinate system. These also have shortcomings, especially for the quantification of neurocranial characters (Matsumoto, 1983).

Another major inadequacy of previous works is that the relative locations of the foramen or condyles were measured in only one direction. As mentioned above, two-dimensional shifts in the median-sagittal plane must have occurred during evolution.

In the present study, the positions of the points have been measured three-dimensionally with the aid of stereophotogrammetry, and the location of basion was computed on the coordinate system defined by the orbital axis.

METHODS AND MATERIALS

The cranial samples employed in the present study are shown in Table 1. The human skulls are from the Anatomical Institute, Goettingen, FRG, those of *Cercopithecus aethiops* from the Anatomical Institute, Frankfurt a.M. Other crania are from the Naturmuseum Senckenberg, Frankfurt a.M. and the Anthropological Institute, Goettingen. In the present study, adult skulls of both sexes with fully erupted third molars form the basis of comparison. The only exception is the inclusion of human skulls without fully erupted third molars; all such crania have fully erupted second molars, however.

Equipment restrictions prevented using small skulls; thus some prosimians and all callithricids were excluded. Three hundred fifty-one primate crania from 46 species and 27 genera are included.

The measuring method employed is almost identical with that reported by Jacobshagen (1980, 1981). In the present study, each skull in the framework was marked in advance with small black rings on the points to be measured and photographed from two directions. With a stereocomparator, the three-dimensional positions of these points were measured on the coordinate system formed by the framework.

The mean of measuring errors for coordinate values is 0 mm and the standard deviations are under 0.30 mm, confirmation that this method is sufficiently accurate for craniometric purposes. Distances or angles among points, lines, and planes are easily computed, eliminating the need for calipers or goniometers.

All coordinates were then converted to a system defined by the skull itself, as follows: (1) The origin is the midpoint of OZR and OZL; (2) The X (transverse)-axis passes through OZR and OZL; (3) The XY (horizontal)-plane includes CO. OZR and OZL are the centers of the circles defined by maxillofrontale, orbitale, and frontomalare orbitale on the right and left orbital margins, respectively. CO is the lowest point of the left optic canal opening into the orbit. This coordinate system is shown schematically in Figure 1. The Y-axis is sagittal and the Z-axis is vertical.

Because the point basion was not visible in the photograph, its coordinate values were calculated separately (see Figure 2). A rigid rod is fixed on the clivus so that the most superior point (BA) marked on the rod is at endobasion. Two other, more inferiorly situated points on the rod, BA1 and BA2, are visible in the stereophotographs, so that their coordinates can be determined. BA2 is twice as far from BA than is BA1. The coordinate values of BA are then:

$$BA_{(x,\ y,\ z)} = 2 \text{ times } BA1_{(x,\ y,\ z)} - BA2_{(x,\ y,\ z)}$$

Cranial breadths were also measured by calipers to compare skull form.

It had been previously determined that the median-sagittal plane is rather stable, and that the bilateral symmetry of the skull is well maintained (Matsumoto, personal observation). This means that lateral declinations of nasion, inion, and BA (BA1 and BA2) are negligible. Thus, in this paper, these points are considered to be in the median-sagittal plane.

RESULTS

In Table 1, the degree of inclination of the clivus line from the horizontal, the relative sagittal, and

TABLE 1. Mean values and standard deviations of clivus inclination from the horizontal(clivus), relative sagittal and vertical positions of BA ($R\text{-}BA_y$ and $R\text{-}BA_z$) and cephalic index (CI)

Species (Sex)	n	Clivus Mean	Clivus SD	$R\text{-}BA_y$ Mean	$R\text{-}BA_y$ SD	n	$R\text{-}BA_z$ Mean	$R\text{-}BA_z$ SD	CI Mean
Prosimii									
Daubentonia madag-ascariensis	2	10.93	5.04	95.44	1.29	2	8.91	4.36	60.3
Galago crassicaudatus (F)	1	10.84	---	89.79	---	1	4.07	---	56.7
Lemur macaco	7	0.60	3.06	92.05	2.53	7	7.08	2.29	57.5
L. catta	1	4.98	---	89.40	---	1	5.35	---	66.6
L. mongoz (M)	5	-0.26	1.21	92.70	1.99	5	7.81	2.85	58.8
L. mongoz (F)	5	6.41	2.84	92.86	1.54	5	3.13	1.62	57.5
Perodicticus potto (F)	1	-11.10	---	110.57	---	1	21.66	---	60.1
Cebidae									
Alouatta seniculus (M)	7	-10.98	5.29	103.81	6.61	7	22.01	4.81	64.3
Alouatta seniculus (F)	11	-2.33	3.52	95.61	3.39	11	14.93	3.03	68.6
Aotus trivirgatus (M)	1	30.28	---	70.09	---	1	-4.77	---	65.1
Aotus trivirgatus (F)	1	15.62	---	74.19	---	1	-4.44	---	68.1
Ateles geoffroyi (M)	1	20.65	---	77.68	---	1	-4.39	---	82.1
Ateles geoffroyi (F)	1	22.35	---	73.83	---	1	-3.45	---	85.1

A. paniscus (M)	1	28.06	---	73.47	---	1	-2.48	---	70.9
Cebus apella (M)	9	21.86	3.48	77.33	2.61	9	-6.66	1.17	67.9
Cebus apella (F)	6	27.58	4.39	70.99	1.32	6	-8.84	2.76	69.7
Cacajao sp.	1	20.43	---	76.21	---	1	-0.41	---	74.0
Lagothrix lagothricha (M)	2	1.37	1.73	88.68	3.94	2	6.58	4.57	68.5
Lagothrix lagothricha (F)	5	7.85	4.75	80.95	1.96	5	3.57	3.73	73.3
Pithecia monachus (F)	1	20.67	---	78.24	---	1	-5.70	---	72.5
P. pithecia (M)	1	20.68	---	82.18	---	1	-4.03	---	69.3
Colobinae									
Colobus badius (M)	5	23.83	4.37	80.80	3.94	5	-6.75	2.57	67.8
Colobus badius (F)	3	28.08	2.84	79.46	4.24	3	-9.36	1.39	68.3
Colobus guereza (M)	8	11.65	3.37	88.64	2.41	8	0.59	1.72	69.8
Colobus guereza (F)	10	17.43	5.35	84.64	2.90	10	-1.93	2.45	70.5
Colobus polycomos p. (M)	5	15.26	3.78	88.17	4.13	5	-2.60	1.39	68.5
Colobus polycomos p. (F)	3	19.90	4.75	83.12	2.38	3	-4.81	1.59	69.7
Colobus p. satanus (M)	5	19.68	4.55	82.61	2.57	5	-4.94	3.73	64.7
Colobus p. satanus (F)	4	18.80	2.86	81.53	3.24	4	-2.75	1.75	64.2
Colobus verus (M)	8	28.20	2.99	74.39	3.31	8	-8.48	2.28	69.0
Colobus verus (F)	4	27.72	2.15	75.34	2.11	4	-8.25	0.53	71.4
Nasalis larvatus (M)	5	27.16	4.13	77.81	1.19	5	-10.96	1.56	65.4

TABLE 1 continued

Species (Sex)	n	Clivus Mean	Clivus SD	$R\text{-}BA_y$ Mean	$R\text{-}BA_y$ SD	n	$R\text{-}BA_z$ Mean	$R\text{-}BA_z$ SD	CI Mean
Nasalis larvatus (F)	1	33.08	---	71.23	---	1	-12.15	---	72.2
Presbytis cristatus (M)	1	26.17	---	74.40	---	1	-8.95	---	67.4
Cercopithecinae									
Cercocebus albigena (M)	1	36.40	---	70.85	---	1	-14.20	---	74.1
Cercocebus galeritus (F)	1	32.22	---	73.28	---	1	-11.33	---	74.2
Cercocebus torquatus (M)	2	23.41	4.36	79.77	3.74	2	-7.66	2.93	67.8
Cercocebus torquatus (F)	1	25.04	---	82.58	---	1	-4.25	---	69.6
Cercopithecus aethips (M)	13	23.80	6.07	77.39	3.51	13	-7.81	3.72	70.2
Cercopithecus aethips (F)	15	26.18	5.38	73.80	2.13	15	-8.40	2.88	71.6
Cercopithecus mona (M)	4	22.78	3.33	79.91	2.23	4	-6.25	1.76	71.3
Cercopithecus mona (F)	11	27.81	4.47	74.46	3.32	11	-6.77	2.10	74.1
Cercopithecus nictitans (M)	11	21.04	3.79	79.10	2.96	11	-4.80	1.71	71.3
Cercopithecus nictitans (F)	13	26.57	3.94	75.38	2.53	13	-7.61	1.70	72.3
Cercipithecus talapoin	2	28.92	4.69	70.65	0.00	2	-6.83	3.41	75.9
Cynopithecus niger (M)	3	31.07	5.57	70.93	---	3	-8.70	---	69.6

Cynopithecus niger (F)	1	39.30	---	68.56	---	1	-16.49	---	70.9
Erythrocebus patas (M)	5	34.36	4.15	73.52	2.34	5	-14.26	1.96	65.3
Erythrocebus patas (F)	2	39.34	---	66.56	---	2	-15.42	---	68.5
Macaca fascicularis (M)	11	25.59	5.14	76.62	2.97	11	-9.27	3.23	69.0
Macaca fascicularis (F)	2	24.09	4.46	77.58	7.31	2	-7.52	6.29	72.5
Macaca maurus (M)	3	25.16	5.58	19.92	1.15	3	-10.99	2.87	70.3
Macaca maurus (F)	3	31.66	6.70	72.75	6.40	3	-12.90	3.30	72.3
Macaca sylvana (M)	1	20.03	---	76.95	---	1	-11.62	---	62.3
Mandrillus sphinx (M)	3	16.26	6.52	101.62	1.07	3	-3.54	0.47	61.2
Mandrillus sphinx (F)	2	28.75	7.36	81.99	7.31	2	-12.59	1.41	72.5
Papio hamadryas (M)	10	32.49	5.08	71.48	4.50	10	-16.10	2.77	67.7
Papio hamadryas (F)	4	39.45	8.15	69.42	2.91	4	-16.02	4.59	73.1
P. papio (M)	3	33.76	7.65	73.96	5.09	3	-16.62	3.62	68.0
P. papio (F)	1	38.85	---	69.13	---	1	-14.83	---	69.4
P. ursinus (M)	1	24.20	---	83.08	---	1	-11.08	---	69.2
Theropithecus gelada (M)	3	33.89	5.33	78.72	0.83	3	-20.68	2.19	69.7
Theropithecus gelada (F)	1	38.32	---	70.70	---	1	-18.33	---	71.4
Hylobatidae									
Hylobates agilis (F)	1	19.05	---	81.22	---	1	-1.27	---	74.6

TABLE 1 continued

Species (Sex)	n	Clivus Mean	Clivus SD	R-BA$_y$ Mean	R-BA$_y$ SD	n	R-BA$_z$ Mean	R-BA$_z$ SD	CI Mean
H. moloch (M)	3	20.44	6.74	79.56	5.52	3	-5.38	3.31	78.0
Symphalangus syndactylus (M)	6	9.69	4.51	84.68	2.13	6	0.51	2.52	66.1
Symphalangus syndactylus (F)	2	14.80	6.46	82.94	---	2	0.28	---	69.5
Pongidae									
Gorilla gorilla (M)	5	16.69	2.84	83.52	9.69	5	-5.78	2.35	50.9
Gorilla gorilla (F)	3	20.99	4.46	75.10	1.87	3	-8.10	1.84	65.5
Pan troglodytes (M)	3	33.67	2.89	74.98	1.96	3	-13.64	1.42	69.5
Pan troglodytes (F)	3	21.16	4.73	77.56	2.02	3	-6.20	0.02	74.7
Pongo pygmaeus (M)	4	15.00	4.84	85.72	6.47	4	-5.05	3.85	68.6
Pongo pygmaeus (F)	6	30.06	4.10	76.85	2.88	6	-10.02	1.11	79.2
Hominidae									
Homo sapiens (M)	25	68.46	7.30	52.42	2.51	25	-22.44	3.64	83.0
Homo sapiens (F)	13	68.90	4.47	51.95	2.29	13	-22.86	2.02	85.1
Homo sapiens subadults (M)	7	70.10	5.01	50.74	1.95	7	-21.42	1.63	83.8
Homo sapiens subadults (F)	6	65.65	3.68	52.12	1.36	6	-21.23	3.68	82.9

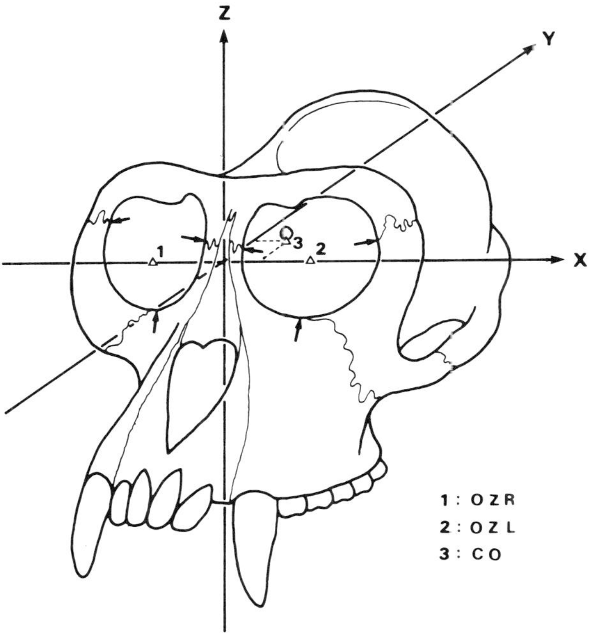

Figure 1. Definition of the coordinate system: the origin is the midpoint of OZR and OZL; the X-axis passes through OZR and OZL, and the XY-plane includes CO. Arrows indicate the points defining the orbital circles on the right and left sides.

vertical positions of BA and the cephalic index are shown.

The first of these variables is the angle formed by the intersection of the sagittal axis and the clivus line. The next two are calculated as follows:

$$\text{R-BAy} = \frac{\text{BA}(y) - \text{N}(y)}{\text{I}(y) - \text{N}(y)} \times 100$$

and

$$\text{R-BAz} = \frac{\text{BAz}}{\text{I}(y) - \text{N}(y)} \times 100$$

where BA(y) is the Y-coordinate of BA, etc. As the equations indicate, the sagittal distance between nasion and inion is normalized.

The cephalic index is then:

$$\text{C.I.} = \frac{\text{cranial breadth}}{\text{cranial length}} \times 100$$

where the denominator is the distance between nasion and inion. These strictly homologous points are used instead

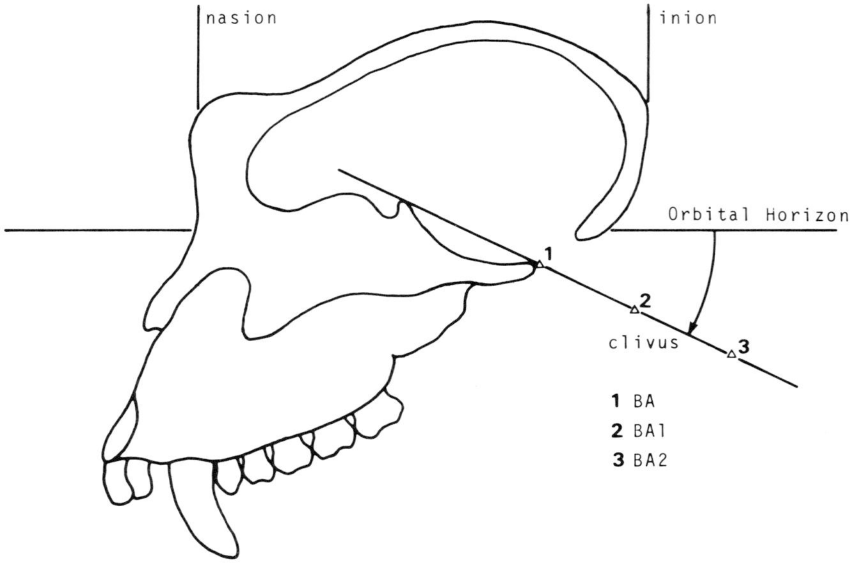

Figure 2. Median-sagittal section of a skull and relationships among points and lines. BA, BA1, and BA2 are at equal intervals on the rod fixed on the clivus. The sagittal distance between nasion and inion is normalized.

of glabella or opisthocranion even for humans. This index in humans is, therefore, somewhat smaller than the conventional one. Almost all samples used in the present study are thought to belong to the mesocran.

The degree of clivus inclination, which is closely associated with the flexion of the cranial base, ranges among prosimians from -11° in potto to 11° in aye-aye. In lemurs the clivus is almost horizontal.

Among cebid monkeys, as is well known, the howler is strongly aberrant. Because of overdevelopment of the hyobranchial apparatus, especially in males, its clivus inclines upward. The most extreme downward inclination, about 30°, occurs in _Aotus_ and _Ateles_.

The distribution in colobines is 10-30°, in cercopithecines 15-40°, in hylobatids 5-20°, and in pongids 15-35°. By contrast, it is 65-70° in humans.

In Figure 3, species values for R-BAy are shown in conjunction with the degree of clivus inclination. These variables are tightly significantly correlated (r = 0.94); only _Mandrillus_ and _Daubentonia_ display unusual relationships between the two measures. It is evident that basion moves forward, away from inion as clivus becomes more steeply declined downward.

The relative sagittal and vertical locations of BA are plotted in Figure 4. As mentioned above, the horizontal direction and height are defined by the

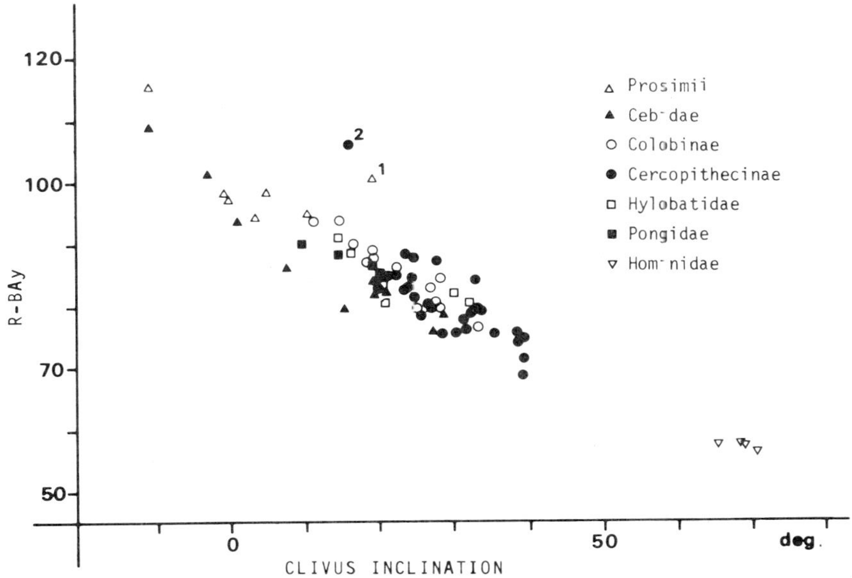

Figure 3. Distribution of values for the relative sagittal position of BA (R-BAy) arranged according to the degree of clivus inclination from the horizontal (r = 0.94). 1 = Daubentonia, 2 = male Mandrillus.

orbital axis or the three points, OZR, OZL, and CO, and the sagittal distance between nasion and inion is normalized. This figure makes clear that the foramen magnum among taxa follows certain trends.

The foramen magnum in Homo sapiens is located not only most anteriorly but also most inferiorly of all primates, and there is a remarkable gap between humans and other primates. This sample of human skulls shows no significant sexual or age difference between groups with completely erupted third molars and those in which only the second molars are completely erupted.

Aside from Homo, most cercopithecine and colobine measurements are arranged in an arc. The foramen is situated relatively inferiorly in the long-snouted monkeys. Theropithecus males have an exceptionally inferior foramen, and in the males of Mandrillus it is located very far to the rear and outside the arc.

Hylobatids and the great apes fall within the general catarrhine trend. The foramen in Hylobates and in Symphalangus is located at the posterior end of the arc.

Some of the New World monkeys (Aotus, Ateles, Cacajao, and Lagothrix) comprise a second arc which is a little superior to that of Old World monkeys. Cephalic indices indicate that these species of platyrrhines have a relatively longer and narrower cranium than other

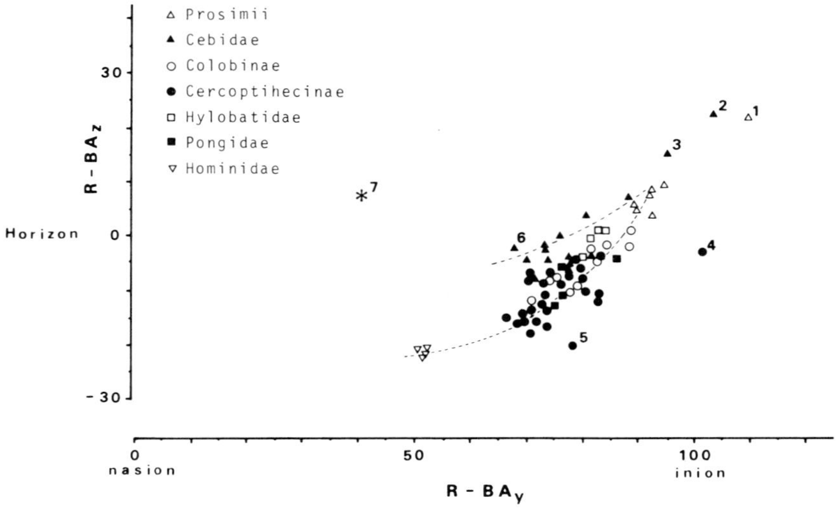

Figure 4. Relative positions of BA. The superior arc indicates the distribution of New World monkeys, the inferior one that of other primates. 1 = Perodicticus, 2 = male Alouatta, 3 = female Alouatta, 4 = male Mandrillus, 5 = male Theropithecus, 6 = male Saimiri (estimated), 7 = the punctum fixum, the center of the basicranial flexion.

species of catarrhines with the same degree of clivus inclination. However, Pithecia and Cebus, whose cephalic indices are relatively low, are near the majority of catarrhines. In Alouatta, because of the upward flexion of the cranial base, the situation of the foramen is exceptionally superior and rearward. Saimiri, which has the lowest cranial equilibrium index (Senyurek, 1938), could not be measured photogrammetrically. However, several male crania of Saimiri sciureus were measured with slide calipers. The mean relative position of the occipital foramen was estimated at 68, -3 and is plotted in Figure 4. The cephalic index of Saimiri is about 66; it thus belongs to the groups of New World monkeys forming the superior arc. The foramen in Saimiri is certainly situated most anteriorly of all nonhuman primates and, on the other hand, is superior to that in catarrhines.

Among prosimians, lemurs occupy the posterior limit of the simian arcs. The foramen magnum in Daubentonia and in Galago are similar in position to that of Lemur. Perodicticus has a superiorly and posteriorly situated foramen, perhaps exceeding even male Alouatta in this respect.

The points representing prosimians, catarrhines, and humans form a quarter of an ellipse. This means that the more anterior the location of the foramen, the shorter the relative length of the clivus becomes.

When a line is drawn forward from the point representing BA for a sample in Figure 2 and following the slope of the clivus, it will generally pass near the point marked with an asterisk in Figure 4. This so-called punctum fixum (coordinate value about 40.5) is the center of the basicranial flexion near the pituitary fossa (Abbie, 1963; Starck, 1965).

DISCUSSION

The coordinate system employed in the present study has several advantages in quantification of the location of the foramen magnum in primates. First, the horizontal is defined by the orbital axis and is equivalent throughout primates. The axis represents a single structure of the skull; it is not influenced by change in cranial shape or orbital size; its direction is always the same, forward at right angles to the pull of gravity. It follows that the positions of the foramen can be quantified independently of other factors.

Secondly, positions were evaluated relative to the neurocranium only. The resulting values are free of influence by the facial part of the skull. Contrary to prior assumptions, head equilibrium is not very meaningful for the understanding of the relative position of the foramen magnum.

Without doubt, the lower the equilibrium index, the better balanced is the skull. Modern humans possess the lowest index; some prosimians and the howler monkey have the highest. Generally, the precondylar part of the skull is much heavier than the postcondylar part, and even in humans total equilibrium has not yet been attained (Senyurek, 1938). However, disproportions between the two parts can easily be compensated for by the development of nuchal musculature, although not by movement of the fulcrum. It does not follow that the occipital foramen or condyles have moved anteriorly to support the skull on the body better. In fact, variety in the location of the foramen in primates does not reflect the degree of deviation from erect posture.

The relative position of the foramen is associated rather with the development of the brain. Although it is difficult to quantify the developmental stages of the brain, these are probably reflected in the basicranial flexture (Matsumoto, 1983). The position and direction of the foramen magnum result from the posterior expansion of the braincase.

From this point of view, the location of the foramen in humans is not necessarily an incomplete solution for balancing the head. The complex of the occipital lobule, cerebellum, and medulla is vertical, and the cervical columns even show lordosis. It can be concluded that the shift of the foramen magnum has reached its maximum in humans, just as bipedalism has. The gap between Homo and other primates is probably bridged by fossil hominids. Although fossils were not considered in the present study, the modern human condition was perhaps already achieved at the Homo sapiens neanderthalensis stage. This process was probably triggered by mutual relationships among upright walking, free use of hands, development of the brain, and so on.

Because of the small sizes of the samples, statistical analysis was of necessity limited in scope. However, significant sexual differences in clivus inclination and the location of the foramen were detected in Lemur, Alouatta, Cebus, Lagothrix, Colobus guereza, Cercopithecus, Pan, and Pongo. Except in Pan, the females of these species have a more inclinated clivus and more anteriorly situated foramen than the males. It is known that, in nonhuman primates, the ontogenetic development of the cranial base is associated with a reduction in the degree. Thus, a more advanced stage of ontogenetic change is achieved by males than by females. In Homo, the development of the cranial base is the converse, i.e., the base becomes more and more flexed. However, there is no significant difference between the sexes in humans, and the adult state is reached by subadults after full eruption of the second molars.

ACKNOWLEDGMENTS

I wish to express my sincere gratitude to Dipl. Biol. H. Zeltner, Department of Anthropology, Goettingen, FRG, for introduction to the photogrammetric system and advice concerning the design of the present study. I am also grateful to Prof. C. Vogel (Department of Anthropology, Goettingen), Prof. H. Felton and Dr. D. Kock (Naturmuseum Senckenberg, Frankfurt am Main), Prof. W. Maier (the Anatomical Institute, Frankfurt, a.M.), and Dr. M. Schultz (the Anatomical Institute, Goettingen) for permitting access to the material in their charge. Parts of the calculations were done with the aid of the Computation Center of Nagoya University, Japan.

REFERENCES

Abbie, A. A.: The cranial center. Z. Morph. Anthrop. 53:6-11. (1963).

Ashton, E. H., and Zuckerman, S.: Some cranial indices of Plesianthropus and other primates. Am. J. Phys. Anthrop. 9:283-296. (1951).

Ashton, E. H., and Zuckerman, S.: Age changes in the position of the occipital condyles in the chimpanzee and gorilla. Am. J. Phys. Anthrop. 10:277-288. (1952).

Ashton, E. H., and Zuckerman. S.: Age changes in the

position of the foraman magnum in hominids. Proc. Zool. Soc. London 126:315-325. (1956).

Bolk, L.: Ueber Lagerung, Verschiebung und Neigung des Foramen magnum am Schaedel der Primaten. Z. Morph. Anthrop. 17:611-692. (1915).

Clark, W. E. Le Gros: New paleontological evidence bearing on the evolution of the Hominoidea. Q. J. Geol. Soc. London 105:225-264. (1950).

Delattre, A., and Fenart, R.: Etude des projections horizontale et verticofrontale du crane au cours de l'hominisation. L'anthropologie 67:525-561. (1963).

Hofer, H.: Die morphologische Analyse des Schaedels des Menschen. In Menschliche Abstammungslehre, G. Heberer Ed. pp. 145-226. Gustav Fischer, Stuttgart. (1965).

Jacobshagen, B.: Grenze konventioneller Techniken und Moeglichkeiten alternativer Ansaetze in der Anthropometrie. Z. Morpho. Anthrop. 71:306-321. (1980).

Jacobshagen, B.: Comparison of morphological factors in the cranial variation of the great apes and man. In Primate Evolutionary Biology. A. B. Chiarelli and R. S. Corruccni Eds. pp. 98-108. Springer, Berlin. (1981).

Lavelle, C. L. B., Schellis, R. P., and Poole, E. F. G.: Evolutionary Changes to the Primate Skull and Dentition. Charles C. Thomas, Springfield. (1977).

Matsumoto, S.: A new coordinate system by the orbital axis and morphological analysis of the primate skulls. Primates 24(4):546-566. (1983).

Schultz, A. H.: Conditions for balancing the head in primates. Am. J. Phys. Anthrop. 29:484-497. (1942).

Schultz, A. H.: The position of the ocipital condyles and of the face relative to the skull base in primates. Am. J. Phys. Anthrop. 13:97-120. (1955).

Senyurek, M. S.: Cranial equilibrium index. Am. J. Phys. Anthrop. 14:23-41. (1938).

Starck, D.: Some remarks on morphological problems of the baboon skull (skull base, development of the skull, variability in homogeneous populations). In: The Baboon in Medical Research. H. Vagborg Ed. pp. 181-190. Univ. Tex Press, Austin. (1965).

13

Morpho-Functional Study of Tarsius's Foot as Compared to the Galagines's: What Does an "Elongate Calcaneus" Mean?

Ch. Berge and F. K. Jouffroy

The tarsal elongation of galagines and tarsiers, as compared to other primates, has been investigated by means of multivariate and bivariate analyses. The reckoning of the calcaneal elongation has been drawn from five species (or subgenera) of galagines (109 specimens in total) and three species of tarsiers (28 specimens in total). A scale of the calcaneal elongation is given for extant strepsirhines and tarsiers in order to permit a reliable interpretation of fossil calcanei. Tarsiers are characterized by the great length of their calcaneus relative to the total length of the tarsus, which is due to the extreme shortness of the cuboid. This characteristic divides the tarsiers from all other primates. Among the galagines there is no relationship between the calcaneal elongation and the size of the animals. Among tarsiers, the calcaneus is slightly more elongate in the lighter species, *T. spectrum* and *T. syrichta* than in the heaviest one, *T. bancanus*. Among the Eocene calcanei, *Notharctus* and the Adapids are devoid of any elongation; in *Teilhardina* and *Hemiacodon*, the relative elongation is close to that in *Cheirogaleus*. In *Necrolemur*, it is similar to the least elongated calcaneus among galagines. None of the Oligocene calcanei from the Fayum are elongated.

INTRODUCTION

In the morphology of the foot, both tarsiers and galagines are characterized among all primates by the great length of their tarsus, compared to the other parts of the foot. This is why both groups were considered during the greater part of the 19th century (after Illiger, 1811) as members of one family, the "Macrotarsi," or "long tarsed," as opposed to the

"Brachytarsi," or "short tarsed," which included all the other prosimians known at that time (e.g., Fitzinger 1861; Brehm, 1868).

A thorough investigation of the tarsus shows that its great relative length, in galagines as well as in tarsiers, is due to the elongation of only two tarsals - the navicular and the calcaneus. Moreover, the elongation of the latter occurs only in that part which is anterior to the talo-calcanear joint. For that reason, the calcaneus provides by itself a reliable index of the tarsal elongation (Schultz, 1963; Hall-Craggs, 1966; Walker, 1967). Except for this common characteristic of the tarsal elongation, tarsiers are completely divided from galagines by several features of their foot morphology (Allen, 1897; Hafferl, 1929; Hall-Craggs, 1966). For instance, the calcaneo-cuboid joint is in galagines, as in most primates, a pivot-joint, with a male tenon in the cuboid articulating with a female mortice in the calcaneus. Such a pivot does not exist in tarsiers; both bones show plane joint facets. Furthermore, the shafts of both calcaneus and navicular are tightly bound in galagines by a calcaneo-navicular joint. Such a joint does not exist in tarsiers; the rod-like shafts of both tarsals are widely separated to the utmost end.

It has been pointed out in a previous study (Jouffroy, et al., 1983) that these differences involve two different grasping mechanisms of the foot, which accord with the phylogenetic heterogeneity of tarsiers and galagines (see also Oxnard, 1983; Jouffroy, et al., 1982). In light of these dissimilarities, the calcaneal elongation is generally regarded as an adaptive feature to the type of vertical leaping that characterizes both galagines and tarsiers in their locomotor behavior. A similar interpretation is generally given for many fossils. Since many elongate calcanei are known from eocene localities, it has been suggested that some form of vertical leaping was probably practiced by most Omomyids (such as <u>Hemiacodon</u>, <u>Teilhardina</u>, <u>Necrolemur</u>) or even by all the members of the family (Gingerich, 1981).

Since the elongation is anterior to the tibio-tarsal joint, it modifies only the load arm of the foot lever and it is only to be found in light animals. Walker (1970) suggested that "there is an allometrical relationship between the lengths of two parts of the calcaneus, the smaller forms being able to afford more elongation of the load arm than the larger ones."

The large number of hypotheses or conclusions that are based upon the elongation of the calcaneus has led us to wonder, what really does an "elongate" calcaneus mean?

In order to go more deeply into the morpho-functional significance of calcaneal elongation, we carried out a biometrical study of the different parts of the foot in five species of galagines and three species of tarsiers, in comparison with other primates.

We firstly tried to ascertain which relative lengths

of the different segments of the foot are the main parameters for discriminating the species or genera.

Secondly, we focused on the calcaneal elongation in order to elucidate the following questions: (1) Is there a one-to-one relationship between the calcaneal elongation and the adaptation to leaping? (2) What about an allometrical phenomenon, as Walker suggested? (3) Where can we locate the Eocene calcanei within the range of the extant elongate calcanei?

METHODS AND MATERIALS

Our sample of skeletons consisted of the following: 28 specimens of *Tarsius*: *T. syrichta* (15); *T. spectrum* (5); *T. bancanus* (8). 109 specimens of *galagines*: *Euoticus elegantulus* (24); *Galago crassicaudatus* (36); *G. alleni* (10); *G. senegalensis* (22); *Galagoides demidovii* (17).

For comparison, we studied at the generic level 148 other strepsirhine Prosimians and 55 Anthropoids (see figure legends). The Eocene fossils, *Notharctus*, *Hemiacodon*, *Teilhardina*, *Necrolemur*, *Adapis*, *Leptadapis* and the Darmstadt Adapid *gen.* and some Oligocene fossils *Apidium*, *Parapithecus* and *Aegyptopithecus* were also considered.

All the specimens of extant Primates belong to the following collections: Museum national d'Histoire naturelle (Paris); Universite Paris 10 Universite de Montpellier; British Museum Natural History (London); Freie Universitat (Berlin); Anatomisches Institut fur Anthropologie (Zurich); Musee Royal de l'Afrique Centrale (Tervuren); Institut Royal des Sciences naturelles de Belgique (Bruxelles); Rijks Museum van de Natuurlijke Historie (Leiden); American Museum of Natural History (New York); Field Museum of Natural History (Chicago); Smithsonian Institution (Washington); SUNY-Stony Brook (Long Island) and private collections.

From a set of measurements of the maximal lengths of the limb bones (Lessertisseur and Jouffroy, 1973; Jouffroy and Lessertisseur, 1978, 1979; Jouffroy, et al., 1982; Jouffroy, et al., 1983) we selected the following data: femur (F); tibia (T); foot (P); tarsus (t); calcaneus (Ct); anterior part of the calcaneus, between the more distal point of the *facies articularis talaris posterior* and the distal end of the bone (Ca); metatarsals I to IV (mI, mII, mIII and mIV); digits I to IV (dI, dII, dIII and dIV).

In as much as many specimens were mounted skeletons, we had to measure Ct and Ca on the lateral side of the calcaneus (Figure 1).

The research has been performed by means of both bivariate and multivariate methods. The multivariate analyses have been carried out by J. F. Ponge according to the "factorial analysis of the correspondences" method (Benzecri, 1973). The details of this method for the

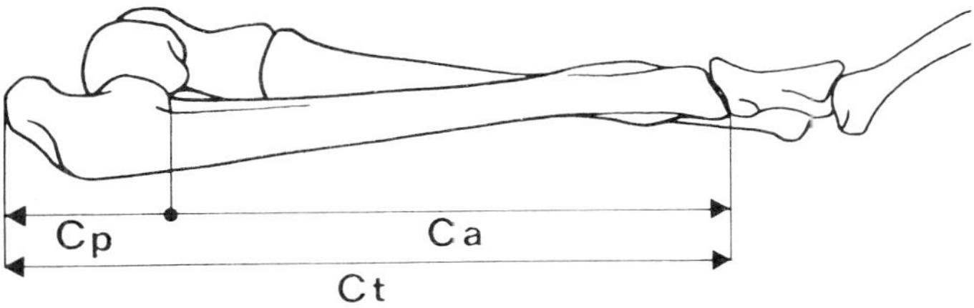

Figure 1. Lateral view of an elongate calcaneus showing the anatomical spots selected to define the calcaneal index.

biometrical study of primate locomotion have been described in previous papers (Berge, 1980; Berge and Ponge, in press). The advantage of this method is in providing a representation of the anatomical parameters as well as of the animals: both kinds of variables are treated simultaneously and represented in the same space (Figure 2).

The bivariate analysis has involved examination of the following ratios: anterior part of the calcaneus relative to the total length of the bone (= calcaneal

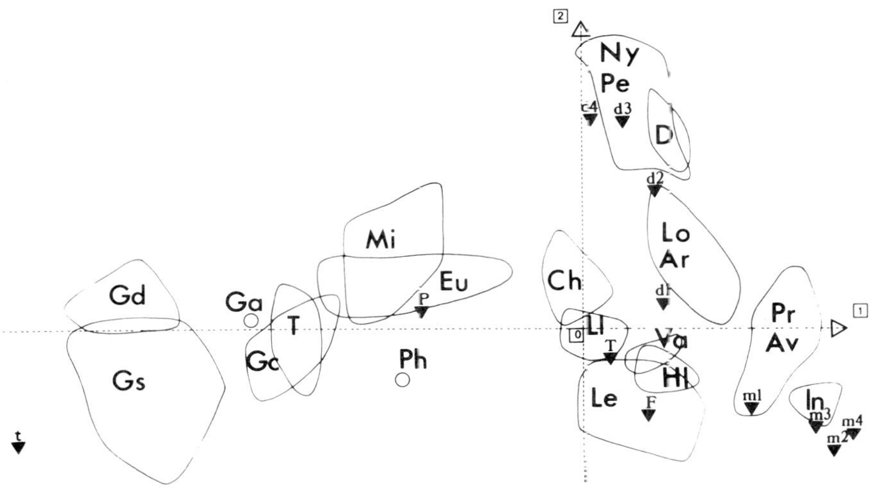

Figure 2. A two dimensional plot (axes 1 and 2) of prosimian genera and subgenera and of anatomical parameters of the hind limb, drawn from a factorial analysis of the correspondences (247 specimens belonging to 21 genera or subgenera, and 12 anatomical parameters). For each genus or subgenus, only the envelope of the clusters of individuals has been represented. The location of the anatomical parameters is noted by black triangles. See the general list of symbols.

index Ca/Ct, Figure 1); length of the calcaneus relative to the length of the tarsus (Ct/t); length of the tarsus relative to the length of the foot (t/P).

RESULTS

Assessment of the anatomical parameters for discriminating the genera (or subgenera).

Figure 2 shows the boundaries of the clusters for each genus (or subgenus) according to axes 1 and 2, drawn from a factorial analysis for 247 prosimians belonging to 21 genera (or subgenera). It is to be noted that the 11 tarsiers have been considered together at the generic level. The 12 anatomical parameters upon which this analysis is based are plotted in the same graph, which allows a morphological interpretation of the discrimination.

Along axis 1, the length of the tarsus (t) and the lengths of the metatarsals (m) 2, 3 and 4 are the main discriminating parameters. They are inversely correlated. On axis 2, the lengths of the third and fourth toes (d) are the main discriminating parameters: both are inversely correlated with the lengths of the metatarsals, of the tarsus, and of the femur (F).

We happen to notice that for both axes, the length of the tibia (T), nearest to the origin of the axes, is the least discriminating as shown in previous studies (Lessertisseur, 1970; Lessertisseur and Jouffroy, 1973; Jouffroy and Lessertisseur, 1979).

Taking into account all parameters, the analysis points out three major trends departing from the central, *Lepilemur*, cluster at the origin (see also Oxnard et al., 1981).

Along axis 1, two of these trends are opposed: a relatively long tarsus and short metatarsus are characteristics of the left oriented trend (including *Microcebus*, *Phaner*, the galagines and the tarsiers). Relatively long metatarsals and a short tarsus characterize the right side trend (indriids). One may notice that both extreme clusters include very good leapers (galagines and indriids, respectively) which have been placed by Napier and Walker (1967) in one locomotor category: the vertical clingers and leapers. Here the opposite location of both groups points out the basic differences in their leaping mechanisms, which are related to the size and the weight of the animals.

The third trend discriminates the species along axis 2 according to the great relative lengths of their third and fourth toes: e.g., *Daubentonia* and the Lorisines. Among the latter, the plump *Perodicticus* and *Nycticebus* are more specialized than the slender *Loris* and *Arctocebus*.

For the purpose of this study, we focus only on the

first trend. Among the animals with a long tarsus, we find successively along axis 1: Microcebus, Phaner and Euoticus; Galago crassicaudatus, G. alleni and Tarsius; Galagoides demidovii and Galago senegalensis, the last one being also the closest to the parameter "tarsus" according to axis 2.

This analysis shows that there is no one-to-one relationship within galagines between the proportions of the hind limb and the weight of the animals: Galago alleni (about 300g in average) and Galago crassicaudatus (1200g) on one hand, and Galagoides demidovii (about 60g) and Galago senegalensis (about 300g) on the other hand, are very close to each other. Tarsius is not apart from Galago crassicaudatus, which means that neither allometric factors nor phylogenetic ones are solely responsible for discriminating the species (or subgenera).

In order to compare the relative lengths of the tarsus and of the calcaneus, we have plotted in Figure 3 the ratios "tarsus/foot" (t/P) along the y axis and "calcaneus/tarsus" (Ct/t) along the x axis. The five species of galagines are located more or less along a line; there is a positive correlation between the elongation of their tarsus related to the foot (between 45 and 55%) and the elongation of their calcaneus related to the tarsus (72-82%). The three species of tarsiers do not show such a correlation; their calcaneus is relatively much longer (up to 92% of the tarsus).

This figure is the biometrical expression of a morphological feature of the feet of tarsiers which divides them from all the other primates - the extreme shortness of the cuboid. Since the length of the tarsus is made up of the length of the calcaneus plus that of the cuboid, the shorter the cuboid, the higher the ratio Ct/t. We have pointed out in a previous study (Jouffroy, et al., 1983) the functional implication of such a morphology and its phylogenetical meaning.

Assessment of the calcaneal elongation.

From a biomechanical viewpoint, one can be misled by taking into consideration only one isolated element of an articular chain (such as the hind limb) or one element of a complex structure (such as the foot). Nevertheless, given the great number of fossil calcanei and the diversity of the locomotor hypotheses drawn from their shapes, we shall focus now upon the questions raised by the elongation of the calcaneus.

We have assessed the calcaneal elongation by means of the calcaneal index Ca/Ct, which allows an easier interpretation than the indices previously described (Schultz, 1963; Hall-Craggs, 1965; Walker, 1967). In most primates (Figure 4), the ratio (in average for each genus) is less than 50%. Among them, anthropoids and hominoids (less than 45%) differ from prosimians (between

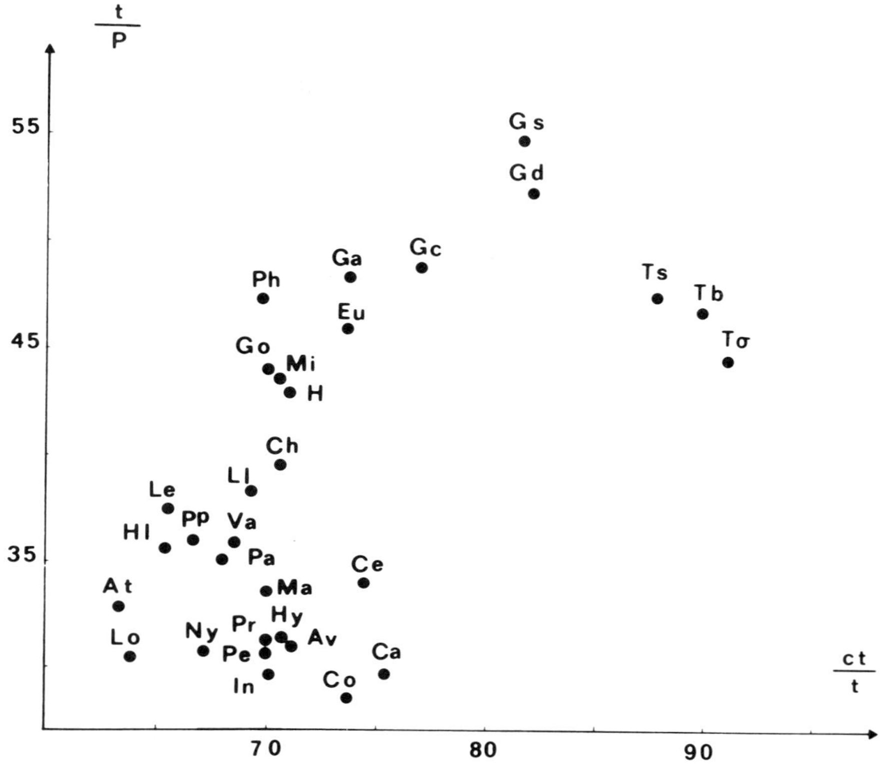

Figure 3. Relative lengths (in percentage) of two segments of the foot, within prosimians: length of the tarsus related to the total length of the foot, along the y axis; relative length of the calcaneus related to the total length of the tarsus along the x axis. See the general list of symbols.

45 and 50%). The ratio is greater than 50% in several prosimians: cheirogaleines (60% in *Microcebus* and *Phaner*), galagines, and tarsiers (up to 78%). It follows from this picture that a calcaneus can be taken as "elongate" when the index Ca/Ct is greater than 50%.

Among the prosimians with elongate calcanei, there is a positive correlation between the calcaneal elongation and the length of the calcaneus related to the length of the whole tarsus. *Galagoides demidovii* is the only one to stand out, because of the relatively greater elongation of the anterior part of its calcaneus.

If we take into consideration only the calcaneal elongation along the x axis, two groups are well defined among galagines and tarsiers. On one hand, *Euoticus elegantulus*, *Galago alleni* and *G. crassicaudatus* (about 65%) and on the other hand, *Galagoides demidovii* and the

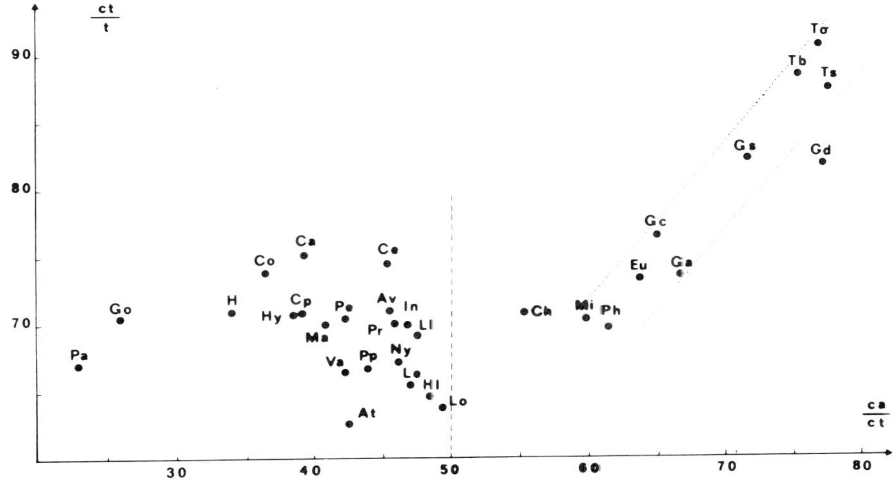

Figure 4. Intrinsic proportions of the tarsus among primates (in percentage). Total length of the calcaneus related to the length of the tarsus along the y axis; anterior part of the calcaneus related to the total length of the bone, along the x axis. See the general list of symbols.

three species of tarsiers (about 75%). *Galago senegalensis* is isolated in between these two groups (about 70%).

If we consider the individual variations in the calcaneal elongation (Figure 5), we observe that the two groups are still clearly divided, but within each one the different species overlap broadly. *Galago senegalensis* is no longer isolated, but it overlaps the two groups.

Turning to the problems posed by Eocene calcanei (Figure 6), one may notice that *Notharctus* and the Adapines are devoid of any elongation. *Teilhardina* and *Hemiacodon* are close to *Cheirogaleus*. In *Necrolemur*, the elongation is similar to the shortest of the galagines and the Oligocene calcanei from Le Fayum are not elongate.

According to the calcaneal index calculated by Walker (1970), the miocene Lorisids from Kenya should be close to *Microcebus* and *Phaner*.

DISCUSSION

Setting out to identify an interspecific allometrical phenomenon, we are led to think that in galagines the calcaneal elongation, as well as the general proportions of the hind limb, are not directly related to the weight of the animals. *Galago*

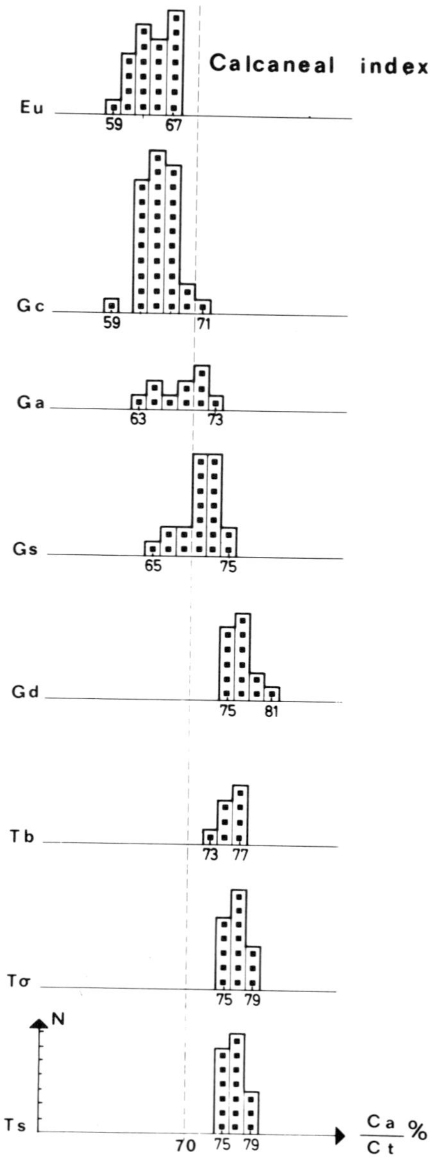

Figure 5. Histograph showing the individual range of variation of the calcaneal elongation (in percentage) within galagines and tarsiers. See the general list of symbols.

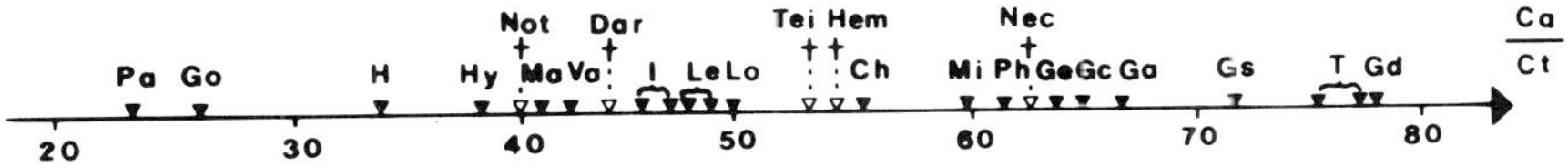

Figure 6. Calcaneal elongation (in percentage) among some extant and extinct primates. See the general list of symbols.

crassicaudatus, the heaviest species (about 1200g) clusters with the medium-sized ones *Euoticus elegantulus* and *Galago alleni* (about 300g/250g), whereas *Galago senegalensis*, which is approximately as heavy as the latter (about 300g), is clearly divided from them and stands between them and the lightest species *Galagoides demidovii* (about 60g).

Among tarsiers, the calcaneus is slightly more elongate in the lighter species *T. spectrum* and *T. syrichta*, than in the heaviest one, *T. bancanus* (Niemitz, 1977).

When comparing the galagines to the tarsiers, it appears that the calcaneal elongation of the latter is similar to that of *Galagoides*, the most elongate one among galagines.

Is there a one-to-one relationship between the calcaneal elongation and the vertical leaping mode of locomotion? It does not seem so.

The calcaneal elongation is not a necessary and sufficient condition for a vertical leaping type of locomotion. On the one hand, *Cheirogaleus*, despite its calcaneal elongation does not leap; on the other hand, indriids, although they are not endowed with elongate calcanei are typical verticle clingers and leapers, according to Napier and Walker's categories (1970).

A comparative study of the limb proportions in primates (Jouffroy and Lessertisseur, 1978) has shown that the hind limbs of *Cheirogaleus* are relatively too short to be able to launch the body (about 110% of the length of the precaudal part of the spine). However, relative to trunk length, indriids' hind limbs are relatively as long as, or even longer, than galagines' (150% of the length of the precaudal spine in *Propithecus* and *Euoticus*, 165% in *Indri* and *Galago senegalensis*, 230% in *Tarsius*). The length of the hind limb related to the size of the body is the main morphofunctional characteristic for predicting the leaping ability. Furthermore, Figure 2 shows that indriids are characterized by their long metatarsals, which suggests a different mechanism of the foot lever, and by their long femur, which suggests a different mechanism of the whole articular chain of the hind limb.

Comparing the elongation of some Eocene calcanei with extant ones, we see that *Teilhardina* and *Hemiacodon* are similar to *Cheirogaleus*, which does not leap. However, by itself, the calcaneal elongation cannot lead us to a reliable interpretation of the locomotor behavior of these fossils, nor can it be taken into account to suggest that the Omomyids, as a whole, practiced a vertical leaping type of locomotion.

These examples should be kept in mind when formulating hypotheses concerning the locomotor behavior of fossils on the basis of the calcaneal elongation.

ACKNOWLEDGMENTS

This research has been supported by the "Centre national de la Recherche scientifique" (CNRS) and the "Museum national d'Historie naturelle" Paris.

We are strongly indebted to J. F. Ponge for defining and processing the multivariate analysis.

This study could not have been carried out without the invaluable assistance of all the Directors and Curators of the Museums and Institutions noted in the Methods section.

All our gratitude goes to J. Fleagle and M. Godinot for allowing us to study their fossil material, to C. Niemitz, R. Thorington, and F. Vincent for discussing and commenting upon our results, and to Mrs. M. C. Dane for help with the translation.

We wish to thank A. Gordon and G. Sanchez for their technical assistance.

LIST OF SYMBOLS

Genera and species:

A.pa	Adapis parigensis(+)	I	Indriids
A.ph	Apidium phiomense(+)	In	Indri
Ag.ze	Aegyptopithecus zeuxis(+)	L	Lemurines
		Le	Lemur
Ar	Arctocebus	Ll	Lepilemur
At	Ateles	Lma	Leptadapis magnus(+)
Av	Avahi	Lo	Loris
Ca	Callithrix	Ma	Macaca
Ce	Cebus	Mi	Microcebus
Ch	Cheirogaleus	Nec	Necrolemur(+)
Co	Colobus	Not	Notarctur(+)
Cp	Cercopithecus	Ny	Nycticebus
D	Daubentonia	Pa	Pan
Dar	Adapid from Darmstadt(+)	Pe	Perodicticus
		Pgr	Parapithecus grangeri(+)
Eu	Euoticus	Ph	Phaner
Ga	Galago alleni	Pp	Papio
Gc	Galago crassicaudatus	Pr	Propithecus
Gd	Galagoides	Se	Semnopithecus
Go	Gorilla	T	Tarsius
Gs	Galago senegalensis	Tb	T. bancanus
H	Homo	Ts	T. spectrum
Hem	Hemiacodon(+)	To	T. syrichta

Hl	Hapalemur	Tei	Teilhardina(+)
Hy	Hylobates	Va	Varecia

Anatomical parameters (lengths):

ca	anterior part of the calcaneus
cp	posterior part of the calcaneus
ct	total length of the calcaneus
d1-d4	toe 1 to 4
F	femur
m1-m4	metatarsal 1 to 4
p	foot
t	tarsus
T	tibia

REFERENCES

Allen, H.: Observations on Tarius fuscus. Proc. Acad. Nat. Sci. pp. 34-55. (1897).

Benzecri, J. P., et al.: L'Analyse des donnees. II. L'analyse des correspondances. Dunod, Paris. (1973).

Berge, Ch.: Biometrie du bassin des Primates. Application aux Primates fossiles de Madagascar et aux Anciens Hominides. These de 3e cycle, Univ. Paris 7. (1980).

Berge, Ch., and Ponge, J. F.: Les caracteristiques du bassin des Australopitheques sont-elles liees a une bipedie de type humain? Bull. Mem. Soc. Anthrop. Paris. (in press).

Brehm, A. E.: La vie des animaux illustree. Trad. Francaise. Balliere, Paris. (1868).

Decker, R. L. and Szalay, F. S.: Origins and function of the pes in the Eocene Adapidae. In: F. A. Jenkins ed. Primate Locomotion. pp. 261-291. Acad. Press, New York. (1974).

Fitzinger, L. J.: Wissenschaftliche populare Naturgeschite der Saugethiere. I. F. K. Hof-und Staatsdruckerei, Wein. (1861).

Gingerich, P. D.: Early Cenozoic Omomyidae and the evolutionary history of Tarsiiform Primates. J. of Human Evolution 10:345-374 (1981).

Hafferl, A.: Bau und Funktion des Affenfusses. Ein Beitrag zur Gelenkund Muskelmecanik. Die Prosimier. Z. Anat. Entw.-Gesch. 90:46-51. (1929).

Hall-Craggs, E. C. B.: An osteometric study of the hind limb of the Galagidae. J. Anat. London 99:119-126. (1965).

Hall-Craggs, E. C. B.: Rotational movements in the foot of Galago senegalensis. Anat. Rec. 154IJ:287-294. (1966).

Illiger, C.: Prodomus systematis Mammalium et Avium. I. Berolini sumptibus g., Salfeld. (1811).

Jouffroy, F. K., Berge, Ch., and Niemitz, C.: Comparative study of the lower extremity in the genus Tarsius.

In: C. Niemitz Ed. Biology of Tarsiers. Gustav Fischer, New York Stuttgart. (1983).

Jouffroy, F. K., and Lessertisseur, J.: Etude ecomorphologique des proportions des membres des Primates et specialement des Prosimiens. Ann. Sc. Nat. Zoologie et Biologie Animale 20:99-128 (1978).

Jouffroy, F. K., and Lessertisseur, J.: Relationships between limb morphology and locomotor adaptations among Prosimians; an osteometric study. In: M. E. Morbeck, H. Preuchoft and N. Gomberg Eds. Environment, Behavior and Morphology. Dynamic Interactions in Primates. G. Fischer, New York, Stuttgart. (1979).

Jouffroy, F. K., Oxnard, C. E., and German, R.: Interpretation phylogenetiques des proportions des membres des Tarsiers. Analyse discriminante. C. R. Acad. Sc. Paris. 295:315-320. (1982).

Lessertisseur, J.: Les proportions du membre posterieur de l'Homme comparees a celles des autres Primates. Leur signification dans l'adaptation a la bipedie erigee. Bull. Mem. Soc. Anthropol. 6:227-241. (1970).

Lessertisseur, J., and Jouffroy, F. K.: Tendances locomotrices des Primates traduites par les proportions du pied. Folia Primatol. 20:125-160. (1973).

Napier, J. R., and Walker, A. C.: Vertical clinging and leaping, a newly recognized category of locomotor behavior among Primates. Folia Primatol. 6:180-203. (1967).

Niemitz, C.: Zur Funktionsmorphologie und Biometrie der Gattung _Tarsius_ Storr 1780. Herleitung von Evolutions-mechanismen bei einen Primaten. Cour. Forsch.-Inst. Seckenberg. 25:1-161. (1977).

Oxnard, C. E.: The place of _Tarsius_ as revealed by multivariate statistical morphometrics. In: C. Niemitz Ed. Biology of Tarsiers. Gustav Fischer, New York, Stuttgart. (1983).

Oxnard, C. E., German, R., Jouffroy, F. K., and Lessertisseur, J.: A morphometric study of limb proportions in leaping prosimians. Am. J. Phys. Anthrop. 54:421-430. (1981).

Schultz, A. H.: Relations between the lengths of the main parts of the foot skeleton in Primates. Folia Primatol. 1:150-171. (1963).

Walker, A. C.: Locomotor adaptations in recent and fossil madagascan lemurs. pp. 1-528. Dissert. Univ. of London Press, London. (1967).

Walker, A. C.: Post-cranial remains of the Miocene Lorisidae of East Africa. Am. J. Phys. Anthrop. 33:249-261 (1970).

14

Computerized Tomographical Biometry of the Distal Humerus: A New Method in Osteological Studies

P. P. LeFloch-Prigent

Biometry by computerized tomography on osseous series has not yet been applied in physical anthropology. Computed tomographical cross-sections were performed on nine distal human humeri, every 3 mm, up to 5 cm on their length. The bi-epicondylar width was the reference level to compare the bones. Several measurements were taken at each level for each bone, and each area was calculated. The results confirmed our previous data on dry bones. Many reproducible and valuable measurements were possible, the accuracy of several ones was increased, and new ones were defined. The morphological analysis and density measurements added new data. The osseous areas were easily integrated. This method which does not alter bones is precise and important in osteological studies in modern and paleontological primatology.

INTRODUCTION

The distal end of the human humerus (Figures 1 and 2) has proved to be an interesting structure both in medical practice (Le Floch, 1978, 1981) and for comparison between primates (McHenry and Corruccini, 1975; Senut, 1978).

Computerized tomography (CT) was created for medicine (Hounsfield, 1973, Ambrose, 1973). It has been used in several other fields such as modern and fossil primatology (Le Floch and Yacoubi, 1978; Jungers and Minns, 1979; Tate and Cann, 1982).

In previous papers (Le Floch, 1982a, 1982b) the distal human humerus was studied on a series of 120 bones by biometrical methods for orthopaedic purposes. Several seriated cross-sections were realized at different levels.

This work is an attempt to explore biometrical

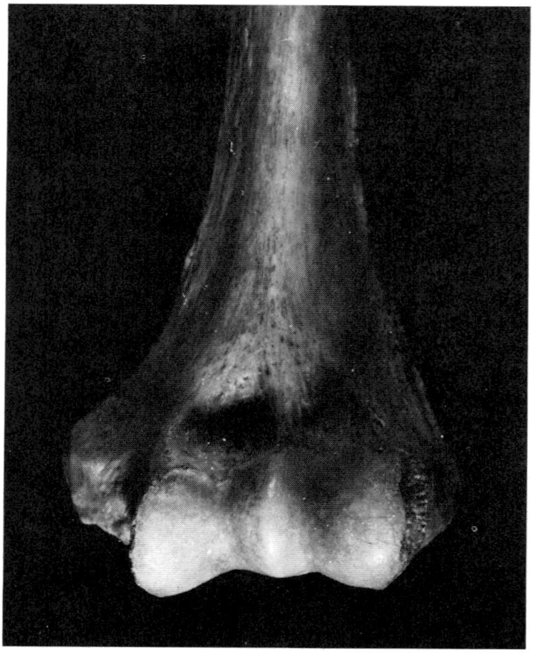

Figure 1. Anterior aspect of a left human distal humerus.

possibilities offered by seriation of osseous horizontal cross-sections performed by CT scanning.

METHODS AND MATERIALS

Nine human humeri from the Paris Medical School were used; the bones were intact, of unknown sex and origin, five from the right side and four from the left. A CT scan, N. D. eight-thousand type from the General Company of Radiology was used in the Department of Neuroradiology of the La Pitie Hospital of Paris (Pr. J. Metzger and Dr. D. Gardeur).

In the method used, the bones were set longitudinally on three rows of three humeri on four layers of foam which were superimposed. The bi-epicondylar width was the reference level. The computed tomograph cross-sectioned the pack in the horizontal plane, every three millimeters, and we obtained 24 levels for each bone (Figure 3).

Each level of cross-section was pictured with a high window level and a window width of 200 H.U. in order to define the osseous outlines with the detailed internal structures.

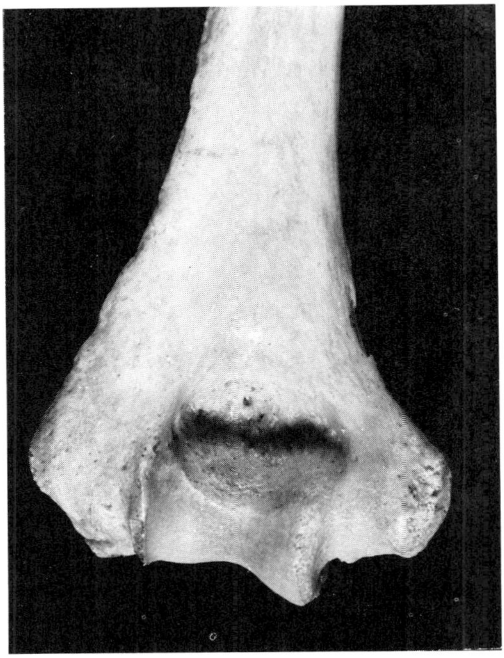

Figure 2. Posterior aspect of a left human distal humerus.

The results were displayed on nine columns, one for each bone. The bi-epicondylar level was readapted, so we obtained only 17 comparable levels which were numbered from the bi-epicondylar line.

The linear measurements were different for the three types of levels: diaphyseal, metaphyseal, and epiphyseal.

For the diaphysis (Figure 4), external measurements were anteroposterior and mediolateral diameters; internal measurements were anteroposterior and mediolateral diameters; and thickness measurements were anterior and posterior.

For the metaphysis (Figure 5), the anteroposterior and mediolateral diameters were measured.

For the epiphysis (Figure 6), we measured the anteroposterior diameter from the olecranon fossa (A-P), the depth of the olecranon fossa (FOD), the mediolateral diameter (M-L), and two special distances from the posteromedial crest of the lateral pillar (PMC LP) to the most salient point of the lateral trochlear crest (LTCS), and to the most salient point of the capitulum (CS).

The area of each level of cross-section was measured for every bone after tracing its outline on a squared paper.

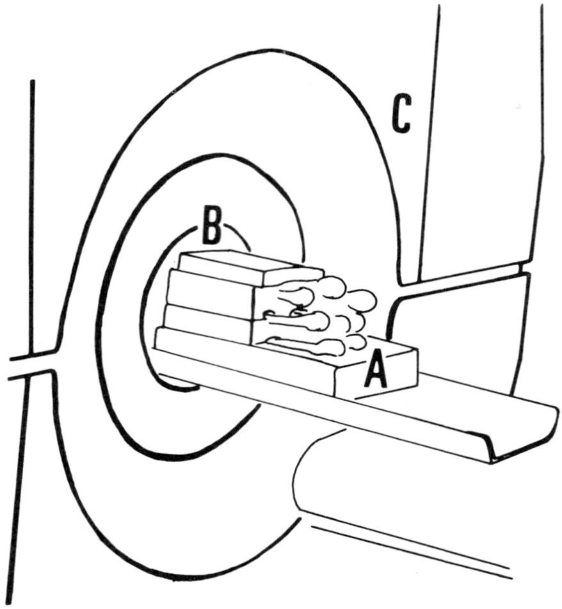

Figure 3. The pack of nine humeri (A) in the gantry (B) of the CT scan (C).

RESULTS

The comparable levels of CT cross-sections are drawn on the anterior, lateral and posterior aspect of a schematic distal human humerus on Figure 7. Six diaphyseal levels (17 to 12) were obtained, six metaphyseal levels (11 to 6), and five epiphyseal levels (5 to 1) (Figure 8).

The complete set of cross-sections for one humerus is shown in Figure 9. The diaphyseal aspects are triangular with a progressive flattening toward the lowest levels. The metaphyseal aspects are characterized by two separate columns: a medial pillar and a lateral pillar separated by the olecranon fossa. The most distal cross-sections were through the trochlea and the medial epicondyle (Figure 10).

The biometrical measurements are shown on Figure 11 and their mean graphically reported. The transverse diameter is regularly increasing from the low diaphysis to the bi-epicondylar line from 2 to 5 cm. The anteroposterior diameter is weakly decreasing in the low diaphysis under 2 cm, unexploitable on the levels of the olecranon fossa, and slightly increasing on the trochlear levels (Figure 12).

The internal diameters of the diaphysis are

DIAPHYSIS

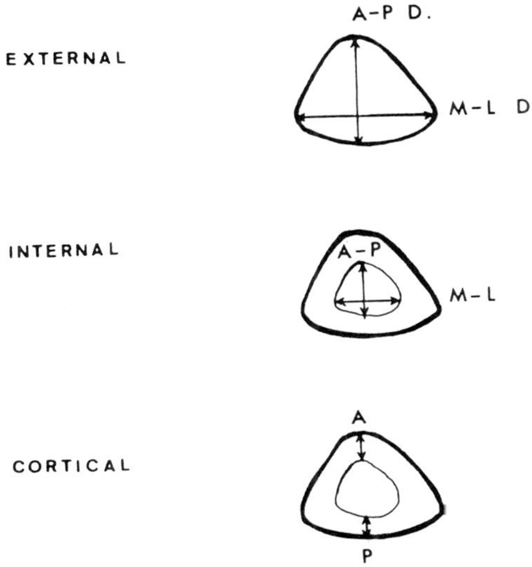

Figure 4. Linear measurements performed on the diaphyseal levels.

METAPHYSIS

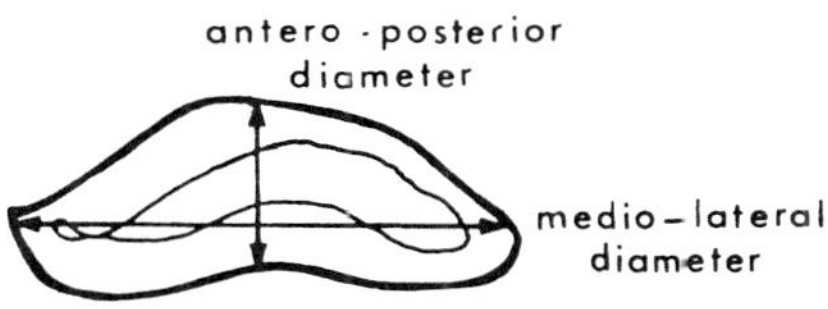

Figure 5. Linear measurements performed on the metaphyseal levels.

EPIPHYSIS

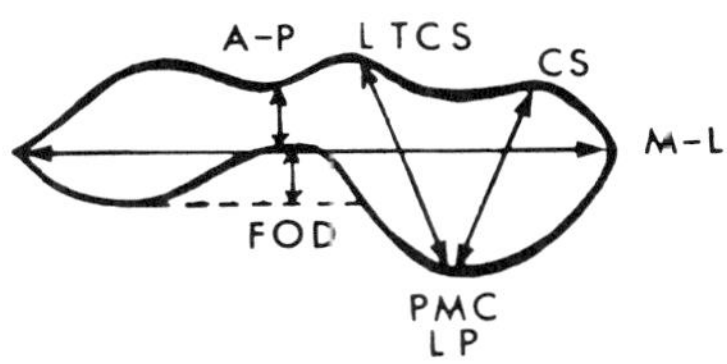

Figure 6. Linear measurements performed on the epiphyseal levels.

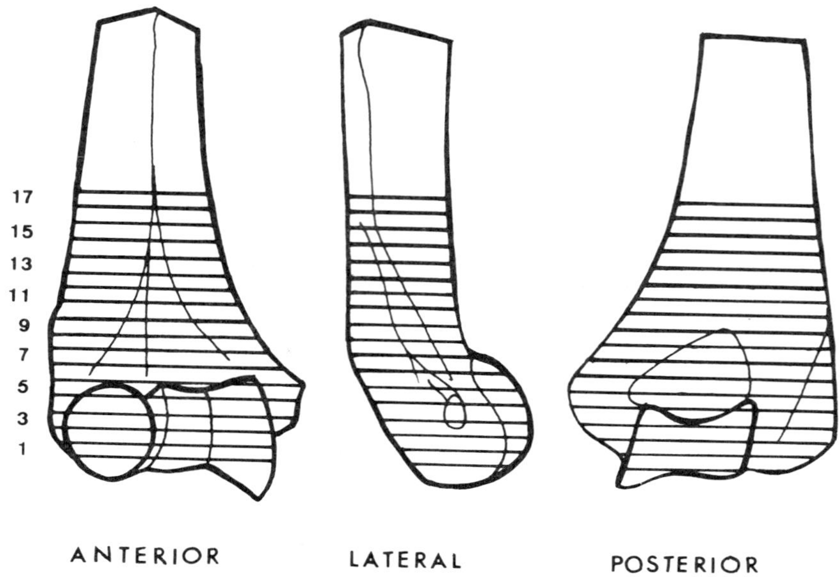

Figure 7. Schematic diagram of the computed tomographical comparable levels: anterior, lateral and posterior aspect.

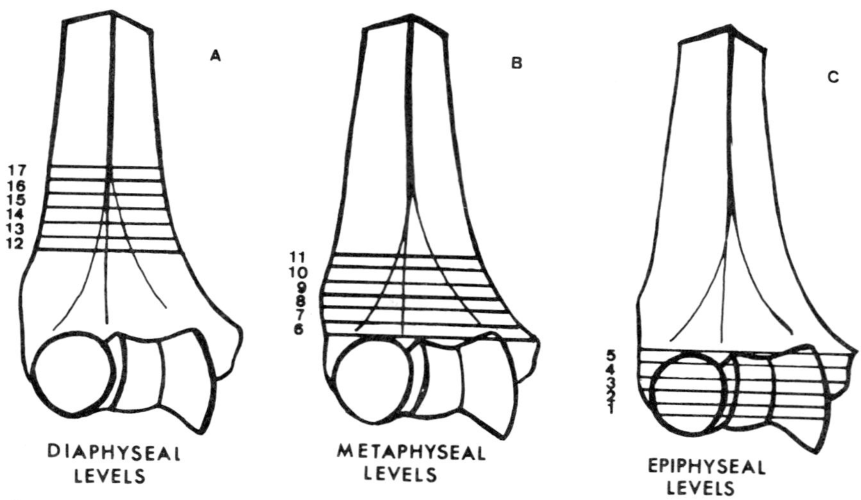

Figure 8. Diaphyseal (A), metaphyseal (B), and epiphyseal (C) levels obtained by CT scanning drawn on the anterior aspect of the distal humerus.

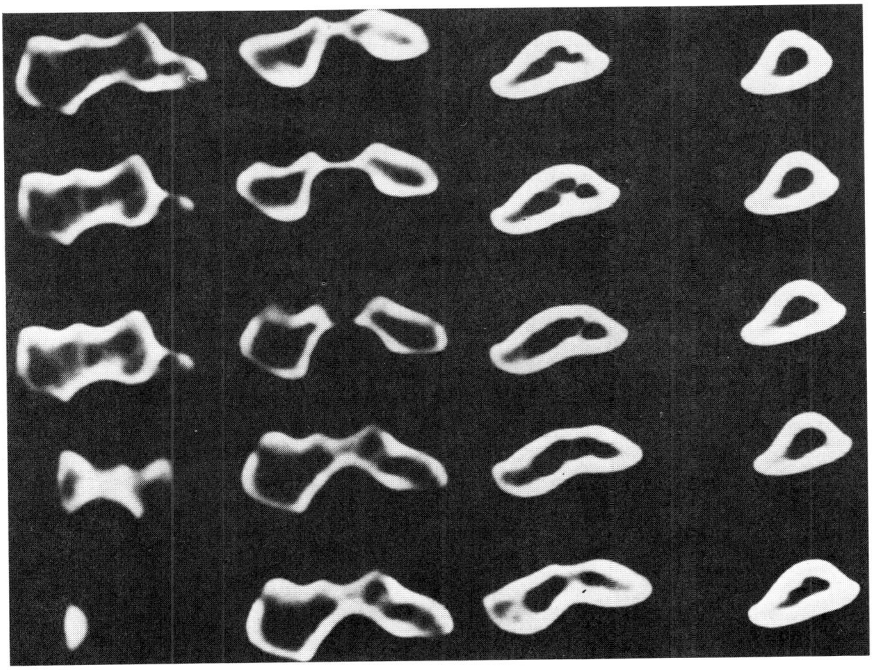

Figure 9. Photographic results cf the computed tomographical cross-sections obtained for one humerus.

regularly increasing distally from 1 to 2 cm in the mediolateral plane and show little variance for the anteroposterior one: less than 1 cm. The cortical thickness was about five mm for the anterior as for the posterior one (Figure 13).

The distal measurements of capitular and trochlear salience were very scant, obtainable only for two levels (Figure 14), the results being approximately 2 cm. The olecranon depth was less than 1 cm with bad standard deviation on the same two levels.

The surface area measurements are reported in Figure 15 and graphically resumed on Figure 16 for their mean. The total area increases slightly from the diaphyseal levels to the metaphyseal ones and more rapidly for the epiphyseal ones with the exception of the lowest level which is decreasing.

DISCUSSION

This study proves the computed tomographic scan to be a valuable method for measuring complex osseous structures. The possibility of performing cross-sections

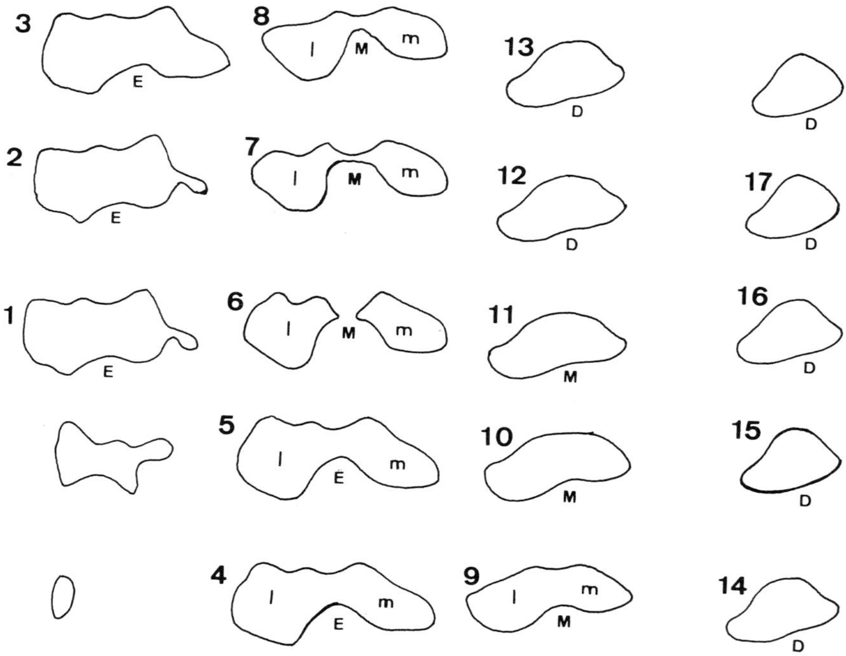

Figure 10. Drawing of Figure 9. D = Diaphyseal levels, M = metaphyseal levels, E = Epiphyseal levels, m = medial pillar, and l = lateral pillar.

without any alteration and analyzing their shape and structure is unique for paleoanthropological and primatological studies.

However, some difficulties have appeared. The main one is alignment of the bones from one reference level and comparison of equal cross-sections at the same intervals for bones of different sizes. The second is that it is not always an easy method to use since CT scans exist only in hospitals and are still expensive.

A lot of new information can be obtained with this instrument in order to complement previous works.

We cannot discuss here the density measurements that we found at about 900 to 980 Hounsfield's Units for cortical bone on the diaphysis levels and from 300 to 500 for cancellous bone in the epiphysis.

Thus, CT scanning applied to systematic cross-sections of the distal human humerus confirms the data obtained by sectioning dry bones, allows comparison of several samples on an identical level, and presents new and numerous perspectives in biometry and density analysis. Surgical and primatological applications are useful.

Level	Transverse Diameter	S-D.	Ant.-Post. Diameter	S-D.	Internal Transverse Diameter	S-D.	Internal Ant.-Post. Diameter	S-D.
17	21	2	15	1	8	3	5	1
16	22	3	15	1	9	4	5	1
15	23	4	15	1	10	5	5	1
14	23	3	15	1	11	5	5	1
13	25	3	14	1	13	4	5	1
12	27	3	14	1	14	1	5	1
11	26	10	14	1				
10	31	3	14	1				
9	33	3						
8	37	4						
7	40	4						
6	41	14						
5	48	5	4	2				
4	49	3	7	4				
3	48	2	11	2				
2	47	2	12	3				
1	46	2	13	1				

Figure 11. Values of the linear measurements expressed in millimeters.

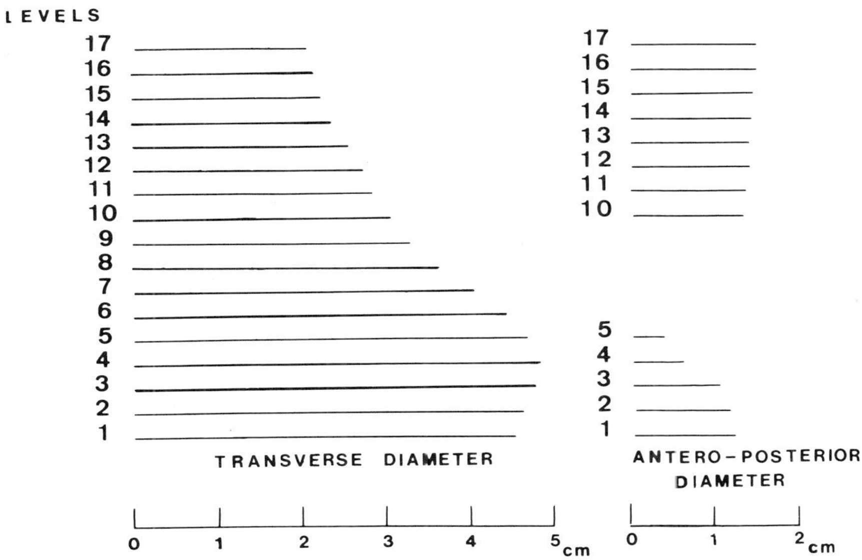

Figure 12. Diagrams of the transverse and anteroposterior diameters.

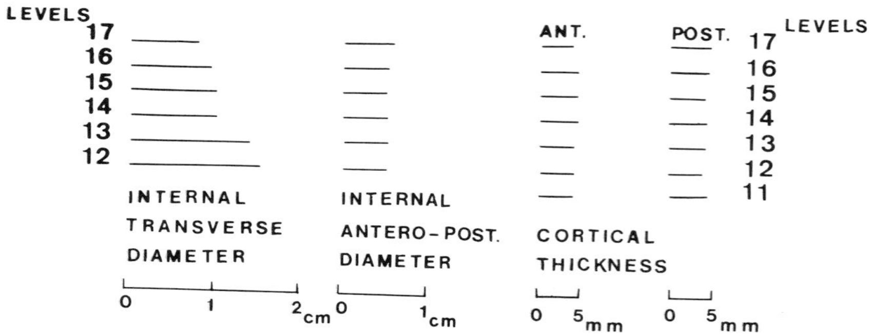

Figure 13. Internal measurements of the diaphysis and thickness of the cortical bone.

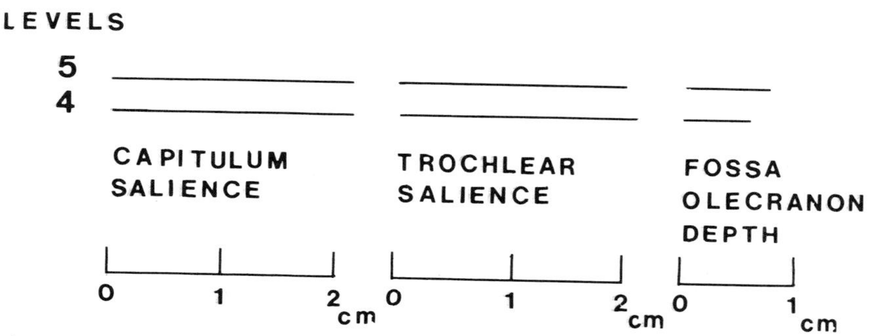

Figure 14. Measurements of the capitulum and trochlear salience anad of the fossa olecranon depth.

BONE N° / LEVEL	1	2	3	4	5	6	7	8	9	mean	standard-deviation
17	418	311	399	249	265	265	339	339	244	314	64
16	486	335	397	315	283	263	359	354	248	338	73
15	505	317	407	315	292	273	384	373	260	347	78
14	502	337	425	343	321	301	386	413	288	368	69
13	533	359	476	371	346	319	395	418	298	391	75
12	559	380	552	406	384	344	423	457	347	428	80
11	592	473	582	402	384	341	456	471	382	454	88
10	636	438	579	471	438	403	450	522	372	479	85
9	670	487	641	520	467	409	497	575	393	518	95
8	695	516	649	598	573	483	508	602	452	564	80
7	782	555	705	618	553	449	553	569	489	586	103
6	839	540	821	620	591	505	591	687	534	636	122
5	1041	524	989	825	637	733	551	877	503	742	202
4	1023	684	1163	924	895	933	633	881	520	851	202
3	1170	833	1165	985	967	1004	805	964	756	961	146
2	1119	881	1070	973	980	995	1006	885	790	967	101
1	1061	773	1028	893	969	939	1008	894	756	925	107

Figure 15. Values of the area's measurements expressed in millimeters per square.

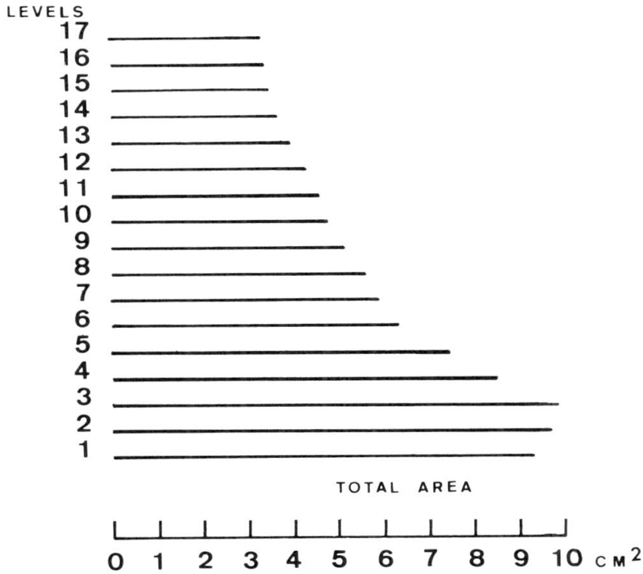

Figure 16. Diagram of the areas' values.

ACKNOWLEDGMENTS

We thank Pr. G. Hidden for a grant from the D. G. R. S. T. and Pr. F. A. King. Linear measurements were reviewed with B. Senut.

REFERENCES

Ambrose, J.: Computerized transverse axial scanning tomography. Part 2: Clinical application. Brit. J. Radiol. 46:1023-1047. (1973).

Hounsfield, G. N.: Computerized transverse axial scanning tomography. Part 1: Description of system. Brit. J. Radiol. 46:1016-1022. (1973).

Jungers, W. L., and Minns, R. J.: Computed tomography and biomechanical analysis of fossil long bones. Amer. J. Phys. Anthrop. 50:285-290. (1979).

McHenry, H. M., and Corruccini, R. S.: Distal humerus in hominoid evolution. Folia Primat. 23:227-244. (1975).

Le Floch, P.: Les piliers de la palette humerale. Memoire du Laboratoire d'Anatomie, Paris. (1978).

Le Floch, P.: Les deux colonnes de l'extremite inferieure de l'humerus. Bull. Ass. Anat. 65:125-133. (1981).

Le Floch, P.: The lower end of the humerus and its two columns. Folia Morphol. 30:136-140. (1982).

Le Floch, P.: The distal humerus: a structure with two pillars. Anat. Clin. 4:235-244. (1982).

Le Floch, P., and Yacoubi, A.: Asymetrie de la palette humerale. IV: Confrontation anatomo-densitome-trique dans le plan horizontal. Bull. Soc. Anat. Paris. 3:79. (1978).

Senut, B.: Contribution a l'etude de l'humerus et de ses articulations chez les hominides du Plio-pleistocene. These de Doctorat de 3eme Cycle de l'Universite Pierre et Marie Curie, Paris. (1978).

Tate, J. R., and Cann, C. E.: High resolution computed tomography for the comparative study of fossil and extant bone. Amer. J. Phys. Anthrop. 58:67-73. (1982).

15

Computed Tomography of Distal Humerus in Some Living Hominoid Primates

B. Senut

This paper presents the main results obtained from a computed tomography scan for comparative study of three genera of modern hominoid primates (*Pan troglodytes*, *Pongo pygmaeus*, and *Hylobates lar*). Contiguous 3 mm sections were taken every 3 mm on distal humeral ends and permitted the differentiation of the primates anatomically as well as biometrically. In particular, on diaphyseal levels, the shaft is triangular in *Pan* and *Pongo* (with the anterior border medially salient in *Pan* and laterally in *Pongo*) and rounded in *Hylobates*. On epiphyseal levels, the anterior lateral lip of the trochlea humeri is salient in the three genera (as salient as the capitulum humeri in *Pan*, less salient in *Pongo*, and more salient in *Hylobates*). Some newly described measurements (impossible to estimate directly on dry bone) were taken, including salience of the lateral crest to the fossa olecrani or projection of the lateral anterior lateral lip to the trochlea humeri. Finally, at diaphyseal levels the proportions of cortical bone are variable in *Pan*, *Pongo*, and *Hylobates*, and the densities of bone are equivalent in these genera.

INTRODUCTION

Since its first description in 1973 by Hounsfield, computerized tomography (CT) has been mainly used for medical purposes and has not often been applied to primatology and anthropology (see, however, Jungers and Minns, 1981; Tate and Cann, 1982; Pahl, 1981; Wong, 1981; Fleagle and Simons, 1982). Le Floch performed systematic cross-sections of limbs by tomodensitometry in modern humans and proved the interest of the method for studying post-cranial anatomy (Le Floch and Yacoubi, 1978; Le Floch et al., 1978, 1979; Le Floch, 1982). Moreover, the distal humerus appeared to be useful for taxonomic differentiations in some living primates as well as in Plio-Pleistocene hominids (Senut, 1978a, 1980; Le Floch and Senut, 1980; McHenry and Corruccini, 1975; Feldesman, 1982). It was thus quite interesting to study this anatomical complex in nonhuman primates by this new method of investigation in an attempt to isolate anatomical differences of the genera. This work is limited to Pan, Pongo, and Hylobates.

METHODS AND MATERIALS

Prof. J. Metzger and Dr. D. Gardeur from the Department of Neuroradiology of the Hospital La Pitie in Paris allowed access to the CT scanner, type N.D. 8000 from C.G.R. (Metzger et al., 1978). Bones are completely coffered in foam sheets, which have been cut away to hold the individual specimens (Figure 1). The posterior face of the humerus, the most flattened one, is set horizontally and the biepicondylar width is used as reference for a horizontal line. Several pieces of foam are then superimposed following alignment at the biepicondylar line. This line is superimposed with the laser-ray of the CT scan and bones are introduced in it. Contiguous sections every 3 mm are taken on the most distal end and every 6 mm on the diaphysis. The thickness of each section is 3 mm distally but reaches 6 or 9 mm proximally. Eighteen to 21 sections are thus obtained for each bone. Window-levels (40) and window-widths (1130) are equivalent to ease the comparisons between the different genera. Radiographs are photographed and mounted for study and illustration.

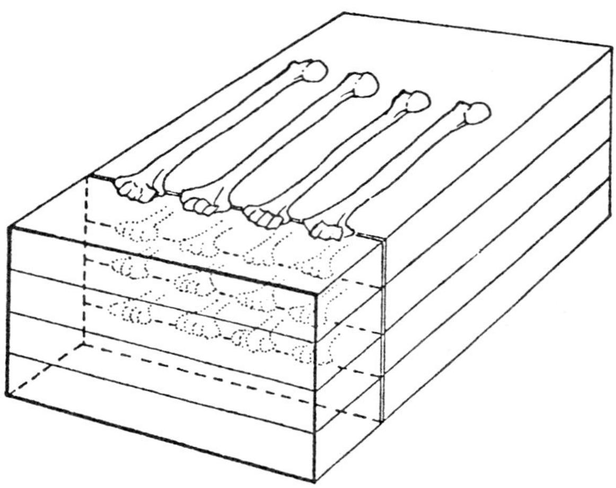

Figure 1. Method for clamping bones.

Four Pongo pygmaeus were loaned by Prof. C. Smeenk from the Zoological Department of the Rijksmuseum van Natuurlijke Historie in Leiden. Prof. J. Anthony and Prof. Y. Coppens permitted study of four Pan troglodytes and four Hylobates lar from the collections in their care at the Laboratory of Comparative Anatomy and Laboratory of Anthropology in the Museum National d'Histoire Naturelle, Paris. All the specimens are wild and adult, including males and females.

RESULTS

Morphology

After tracing and projection on a horizontal line (e.g., parallel to the biepicondylar one; Figure 2), it can be noted that the lateral and medial borders converge superiorly in gibbons and orangutans. In chimpanzees, the lateral border is straighter and as a result the bone appears asymetrical.

At diaphyseal levels (Figure 3), cut sections are roughly triangular in chimpanzees and orangutans, but rounded or circular in gibbons. In chimpanzees, the lateral border is flattened and the medial one rounded; in orangutans both borders are sharp and in gibbons both are rounded. Distally, the anterior border is medially

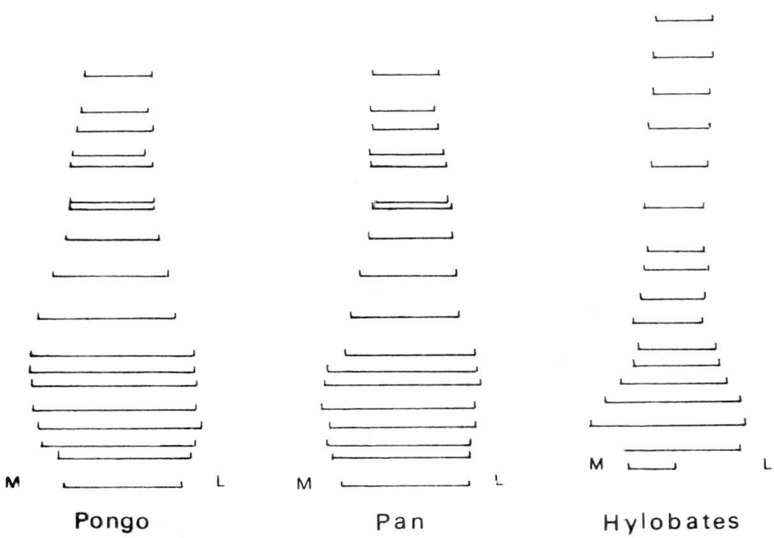

Figure 2. Projection on a horizontal line parallel to the biepicondylar one of the maximal width per level of section. One bone is chosen for each genus (right humeri).

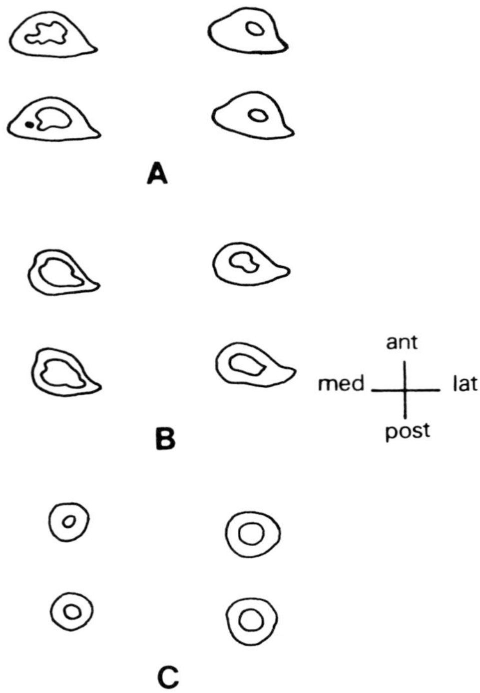

Figure 3. Sections on two contiguous diaphyseal levels. Proximal views of right humeri. For each genus, two specimens are represented. A = Pongo; B = Pan; C = Hylobates.

salient in Pan and more laterally in Pongo. The proportion of cortical bone is variable in each genus and not linked with sex as far as we can determine from this small sample.

At metaphyseal levels (Figure 4), we can see evidence of medial and lateral humeral pillars. Considering the size of the different bones, the pillars originate at the same level (18 mm above the biepicondylar bone in Pongo, 15 mm in Pan, and 10 mm in Hylobates). The sections seem much flattened in Pongo and less in the two other genera.

At epiphyseal levels (Figure 5) the lateral and medial pillars are triangular in the three genera studied. The posterolateral border of the lateral pillar is clearly oblique, and the medial one more anteroposteriorly flattened. The medial border is slightly flattened in Pan and Hylobates and sharp in Pongo. The fossa olecrani is roughly rectangular and limited laterally by a sharp crest in Pan and Pongo and by a smoother one in Hylobates; it is strongly salient in

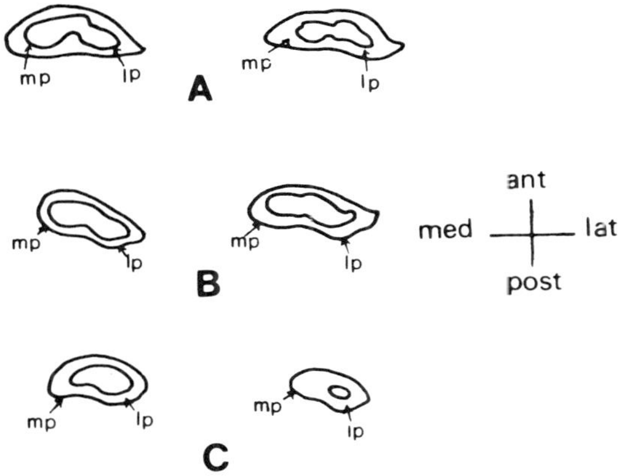

Figure 4. Sections on one metaphyseal level. Proximal views of right humeri. A = *Pongo*; B = *Pan*; C = *Hylobates*. For each genus, two specimens are represented. mp = medial pillar; lp = lateral pillar.

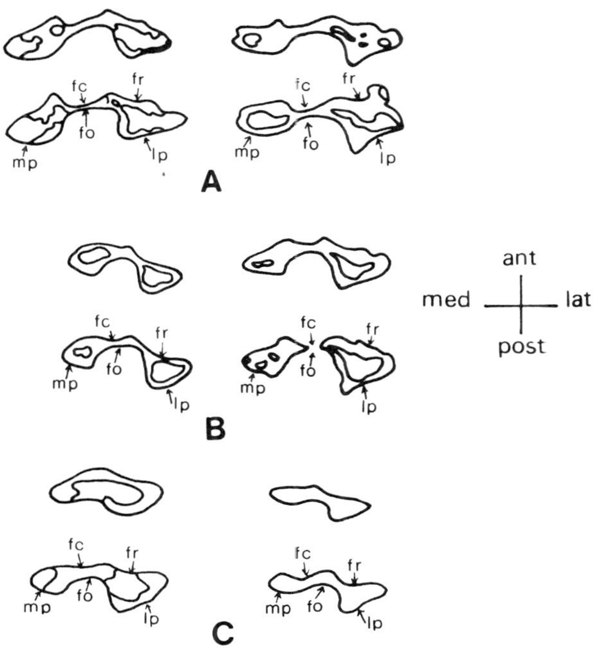

Figure 5. Sections on two epiphyseal levels. Proximal views of right humeri. For each genus, two specimens are represented. A = *Pongo*; B = *Pan*; C = *Hylobates*. mp = medial pillar; lp = lateral pillar; fc = fossa coronoidea; fr = fossa radialis; fo = fossa olecrani.

Pan. The salience of the anterior lateral trochlear lip and the spool-shaped trochlear humeri are clear and more marked in chimpanzees and gibbons.

Biometry

Seriation of the bone permits more precise biometry and definition of some measurements which cannot be taken directly on the bone (Figure 6). After tracing the pictures, 10 linear measurements were taken with a set of calipers:

1. Total mediolateral width line: joining the most salient points on a horizontal line (taken on all levels).
2. Mediolateral width of the medullar cavity: line joining the most salient points of the medullar cavity, taken parallel to 1 (only on diaphyseal and metaphyseal levels).
3. Total anteroposterior width: line joining the most projected points on a line perpendicular to 3 (only on diaphyseal and epiphyseal levels).
4. Anterior to posterior lateral crest distance: line joining the most anteriorly salient point of the lateral trochlear lip and the most

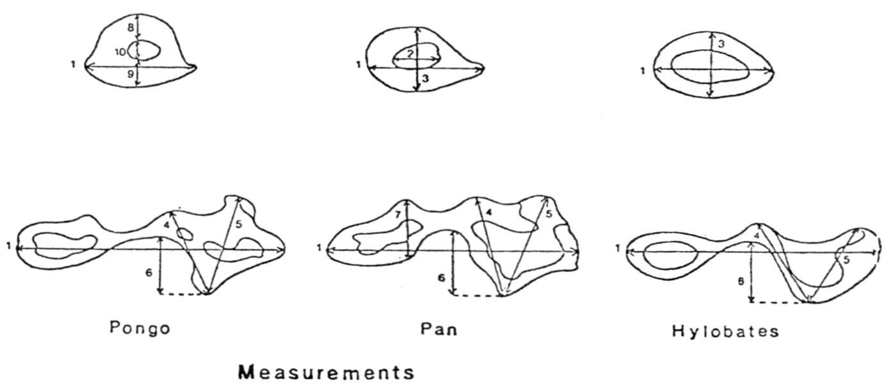

Figure 6. Linear measurements. 1 = total mediolateral width; 2 = mediolateral width of the medullar cavity; 3 = total anteroposterior width; 4 = anterior to posterior lateral crest distance; 5 = capitulum humeri to posterior lateral crest distance; 6 = salience of the lateral crest to the fossa olecrani; 7 = anterior to posterior medial trochlear lip distance; 8 = anterior cortical thickness; 9 = posterior cortical thickness; 10 = anteroposterior width of the medullar cavity.

posteriorly salient point of the lateral crest to the fossa olecrani (epiphyseal levels only).

5. Capitulum humeri to posterior lateral crest distance: line joining the most anteriorly salient point of the capitulum humeri (on a line parallel to the biepicondylar one) and the most salient point of the lateral posterior crest to the fossa olecrani (epiphyseal levels only).
6. Salience of the lateral crest of the fossa olecrani: line joining orthogonally the deepest point of the fossa and the parallel to the biepicondylar line issued from the most salient point of the lateral crest to the fossa olecrani (epiphyseal levels only).
7. Anterior to posterior medial trochlear lip distance: line joining the most projected point anteriorly and posteriorly of the medial crest to the fossa olecrani (epiphyseal levels only).
8. Anterior cortical thickness: thickness of the cortical bone of the anterior border taken on 3 (diaphyseal and metaphyseal levels).
9. Posterior cortical thickness: thickness of the cortical bone of the posterior border taken on 3 (diaphyseal and metaphyseal levels).
10. Anteroposterior width of the medullar cavity: line joining the most projected points taken on 3 (diaphyseal and metaphyseal levels).

Some ratios were then calculated:

1. 4/5 times 100 compares the salience of the lateral anterior crest of the trochlea humeri with that of the capitulum humeri (Table 1). In Pan, the lateral anterior crest of the trochlea humeri is as salient as the capitulum humeri: the ratio approximates 100. In Hylobates, the lateral anterior crest of the trochlea humeri seems more salient: the ratio is superior to 100. In Pongo, the reverse is observed - the capitulum humeri is more anteriorly projected: the ratio is inferior to 90.
2. 6/1 times 100 expresses the different development of the lateral crest to the fossa olecrani (Table 1). It is strongly salient in Pan (ratio close to 28), less salient in Hylobates (ratio close to 25), and much less salient in Pongo (ratio close to 21).
3. 10/3 times 100 and 2/1 times 100 shed some light on the variation of cortical thickness in each genus (Table 1). Some areas of bone were measured at equivalent diaphyseal level, confirming the variation previously mentioned (Table 2). Measures of osseous densities were attempted on diaphyseal levels: values approximate 997 HU in each genus.

TABLE 1. Ratios calculated in hominoids Studied*

	4/5 x 100	6/1 x 100	2/1 x 100	10/3 x 100
Hylobates				
1961.287	106.1	26.7	49.5	81.5
1880.1169	107.1	24.2	31.0	85.8
A.3856	105.4	26.5	46.3	85.5
1948.95	113.8	24.8	64.5	72.3
Pongo				
10495	88.7	20.9	48.5	80.7
288	90.0	23.0	28.8	91.5
5018	83.5	20.8	48.2	79.4
y			64.1	73.5
Pan				
26079	96.0	30.7	70.4	64.5
A	99.2	27.3	45.6	83.8
A.12761	99.4	29.2	33.3	78.6
1947.17	98.7	26.7	58.9	44.0

* 4/5 x 100 taken at equivalent epiphyseal level;
6/1 x 100 taken at biepicondylar level;
2/1 x 100 taken at equivalent diaphyseal level;
10/3 x 100 taken at equivalent diaphyseal level.

DISCUSSION

The 3 mm cut sections taken every 3 mm are sufficient to obtain precise morphological details for large bones. However, it would be necessary and more precise to get 1 mm cut sections every 1 mm for small bones such as those from gibbons (some levels are difficult to read because too many details are superimposed). Compared with the method of tracings employed previously (Senut, 1978a, 1978b), computerized tomography is slightly more precise for morphology as the bone does not move at all during the procedure. The main advantages are the lack of alteration of the bone and the possibility for exploring external as well as internal morphology.

Previous results on the external morphology of sections are confirmed. The lateral flattening of the shaft observed in chimpanzees is due to the strong brachioradialis flange which does not occur in the asiatic apes. At epiphyseal and metaphyseal levels, the posterior face of the bone appears to be more anteroposteriorly compressed in Pongo than in the other apes. This method permits evaluation of the differences in the development of the components of the condylus humeri (e.g., trochlea humeri and capitulum humeri). The strong anterior projection of the capitulum humeri in Pongo is striking. In the nonhuman hominoids examined in this study, the lateral pillar exhibits a triangular shape partly due to the position of capitulum humeri

TABLE 2. Measures of areas of cortical bone taken at equivalent diaphyseal levels

Hylobates	1961.287	1880.1169	A.3856	1948.95
	81.5	85.8	85.5	72.3
Pongo	10495	288	5018	y
	80.7	91.5	79.4	73.5
Pan	26079	A	A.12761	1947.17
	64.5	83.8	78.6	62.1

which is set lower than the epicondylis lateralis. As soon as we cross the capitulum humeri (more distally), the shape of the pillar changes and it becomes quadrangular and more human-like (cf., Le Floch, 1982). However, Le Floch (1982) has shown that the lateral pillar remains quadrangular on several levels above the biepicondylar level in modern *Homo* who seems to be unique among the hominoids.

Although the biometrical results are from a small sample (four specimens of each genus), it seemed more interesting to obtain data for different genera instead of more data of one genus only. Thus, our results are limited but give a basis for future work. We note the strongest projection of the lateral crest to the fossa olecrani in *Pan*. As far as we know, this crest is linked with stabilization of the elbow joint during quadrupedal gait. Many terrestrial quadrupedal primates and most nonprimate mammals exhibit this feature. In chimpanzees, it seems to be linked with knuckle-walking. Cortical thickness is highly variable in each genus, but more data are needed to explain these differences. It does not seem to be linked with sex since in two male *Pongo* (10495 and 288, cf. Table 2) 80.7 and 91.5 were obtained for areas of cortical bone. Densities on diaphyseal levels do not distinguish the individual genera and are equivalent to that found in modern humans (Le Floch, 1982).

Seriation of bone by computerized tomography provides additional criteria for distinguishing some modern hominoid primates anatomically and biometrically.

The most important diagnostic features are: morphology of the shaft on diaphyseal levels (especially the anterior, medial, and lateral borders), shape of the pillars, development of the lateral crest to the fossa olecrani, and salience of the anterior lateral lip to the trochlea humeri. In addition, the variation in cortical proportions and equivalence of bone densities are noted.

ACKNOWLEDGMENTS

I wish to thank Prof. J. Metzger and Dr. D. Gardeur for permitting me to make the CT scans in their department of Neuroradiology (Hospital La Pitie, Paris) and for assistance. Prof. J. Anthony, Prof. Y. Coppens and Prof. C. Smeenk kindly provided modern hominoid primates in their care. I am grateful to Dr. P. Le Floch for his advice and the anonymous reviewer for helpful comments. This work was supported by the L.A.49 (C.N.R.S.). I am indebted to Dr. M. Lathrop for his help.

REFERENCES

Feldesman, M. R.: Morphometric analysis of the distal humerus of some cenozoic catarrhines: the late divergence hypothesis revisited. Am. J. Phys. Anthrop. 59:73-95. (1982).

Fleagle, J. G., and Simons, E. L.: The humerus of Aegyptopithecus zeuxis: A primitive anthropoid. Am. J. Phys. Anthrop. 59:175-193. (1982).

Hounsfield, G. N.: Computerized transverse axial scanning (tomography). Part 1. Description of system. Br. J. Radiol. 46:1016-1022. (1973).

Jungers, W. L., and Minns, R. J.: Computed tomography and biomechanical analysis of fossil long bones. Am. J. Phys. Anthrop. 50:285-290. (1981).

Le Floch, P.: Tomodensitometrie osseuse seriee fine: exemple de l'extremite distale humerale chez l'homme moderne. C. R. Acad. Sci. Paris. 294:609-612. (1982).

Le Floch, P., Merland, J., Hidden, G., and Chiras, J.: Coupes horizontales seriees des membres: anatomie et densitometrie. I. Bras et avant-bras. Bull. Soc. Anat. Paris. 3:84. (1978).

Le Floch, P., Merland, J., Hidden, G., and Chiras, J.: Coupes horizontales seriees des membres tous les 9 mm. Anatomie et tomodensitometrie. Bull Soc. Anat. Paris. 4:10-14. (1979).

Le Floch, P., and Senut, B.: Sections osseuses seriees fines: approche morphologique, comparative et paleontologique. Exemple de l'extremite distale de l'humerus chez les primates hominoides. C. R. Acad. Sci. Paris. 292:157-162. (1981).

Le Floch, P., and Yacoubi, A.: Asymetrie de la palette humerale. IV: Confrontation anatomo-densitometrique dans le plan horizontal. Bull. Soc. Anat. Paris. 3:79. (1978).

McHenry, H. M., and Corruccini, R. S.: Distal humerus in hominoid evolution. Folia Primatol. 23:227-244. (1975).

Metzger, J., Gardeur, D., and Millard, J. C.: Tomodensitometrie Cephalique. Doin, Paris. (1978).

Pahl, W. M.: La tomographie par ordinateur appliquee aux momies egyptiennes. Apercu de l'etat actuel des recherches. Bull. Mem. Soc. Anthrop. Paris. 8:343-346. (1981).

Senut, B.: Etude comparative des piliers de la palette humerale. Premiere partie. Cah. Anthrop. 3:1-8. (1978a).

Senut, B.: Etude comparative des piliers de la palette humerale. Deuxieme partie. Cah. Anthrop. 4:1-20. (1978b).

Senut, B.: Nouvelles donnees sur l'humerus et ses articulations chez les Hominides plio-pleistocenes. L'Anthropologie, Paris. 84:125-131. (1980).

Tate, J. R., and Cann, C. E.: High resolution computed tomography for the comparative study of fossil and extant bones. Am. J. Phys. Anthrop. 58:67-73. (1982).

Wong, P. A.: Computed tomography in paleopathology: technique and case study. Am. J. Phys. Anthrop. 55:101-110. (1981).

16

The Knee Joint in Three Hominoid Primates: Application to Plio-Pleistocene Hominids and Evolutionary Implications

C. Tardieu

The general osteology of the knee in chimpanzees, gorillas, and humans is compared. The differences observed between humans and the two great apes are applied to fossil hominids and lead to the distinction of two groups: one, the fully modern *Homo* and the other more primitive *Australopithecus*. Two different types of insertion of the external meniscus on the tibia are found, one characteristic of humans and the other observed in chimpanzees and gorillas. On the Plio-Pleistocene hominids, these two types are represented, confirming the distinction of the two groups. This feature allows the interpretation of a very controversial fossil, AL 333 4, x 26 (Afar, Ethiopia). To determine the functional meaning of these morphological differences, measurements were made of knee rotation in a human and a chimpanzee. The results, confirmed by cineradiographic analysis, show that the amplitude of rotation is much greater in the chimpanzee. These results suggest that in the Plio-Pleistocene, all the hominids were able to practice a bipedal locomotion. Some of them showed a fully modern adaptation to bipedalism; others, more primitive, were probably not exclusively biped and presented a different bipedal behavior.

INTRODUCTION

In previous works (Tardieu, 1979, 1981, 1983) it has been shown that the general osteology of the knee differentiates between pongids and humans by the following femoral characteristics: (1) Degree of obliquity of the femoral diaphysis; (2) Shape of the femoral trochlea; (3) Lateral profile of the external condyle; (4) Proportions of the femoral distal epiphysis; (5) Symmetry of the epiphysis about the parasagittal plane passing through the middle of the trochlea. Shape of the intercondylar notch.

The first three human characteristics are present in all the hominid fossils (Lovejoy et al., 1973), probably associated with the practice of a kind of bipedalism. The other characteristics divide them into two groups. In one group is KNMER 1481 and modern *Homo sapiens* with all the characteristics of human bipedalism (Figure 1). In the other group the morphology is more primitive (*Australopithecus*), with traits that are observed in extant pongids and are probably primitive hominoid features. The fossil AL 129 from Ethiopia is the best representative of this group (Figure 1).

One key feature differentiates these two groups: the fit of the tibial spines in the intercondylar notch. In the modern *Homo* group including KNMER 1481, the tibial spine fits tightly within the intercondylar notch of the

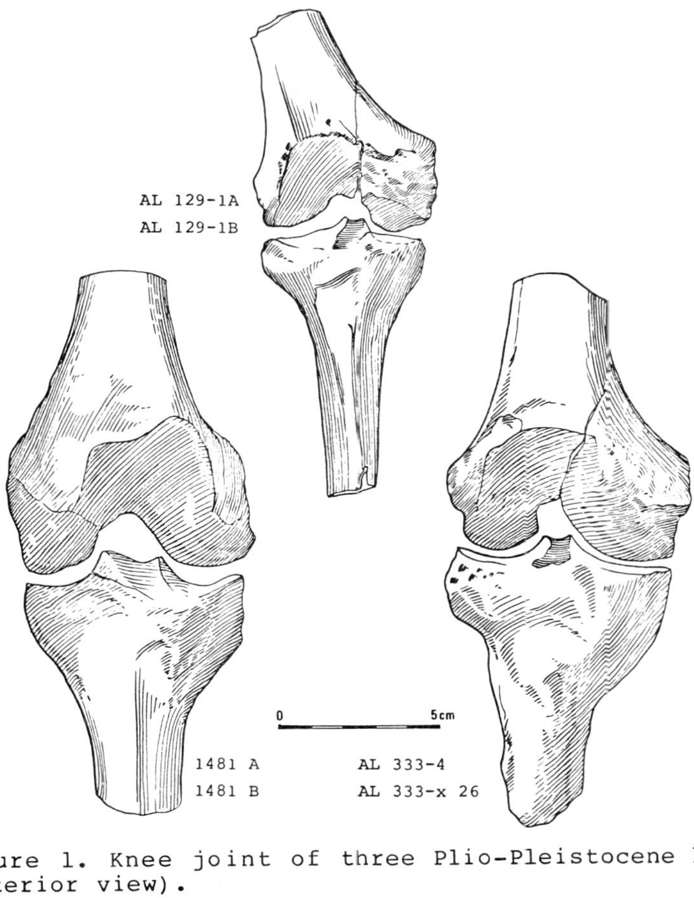

Figure 1. Knee joint of three Plio-Pleistocene hominids (anterior view).

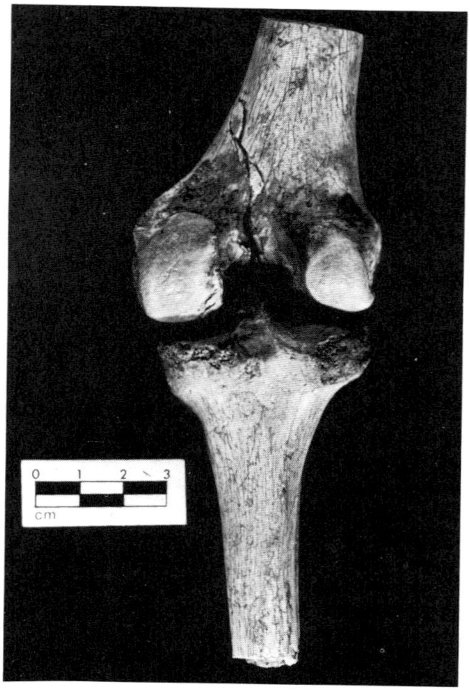

Figure 2. Posterior view of the fossil AL 129 1A 1B. Note the loose fitting of the tibial spines in the intercondylar femoral notch. (Photograph courtesy of Dr. D. C. Johanson, Cleveland.)

femur (Figure 2), but in the primitive group (Australopithecus and modern pongids), the two are loose fitting.

In this paper these observations are supplemented, particularly the last mentioned feature by the dissection of the knee meniscus of two men, two gorillas, and one chimpanzee. The functional meaning of these differences are assessed by angular knee rotation measurements and cineradiographic analysis in a man, a gorilla, and a chimpanzee.

METHODS AND MATERIALS

Angular rotation measurements

These measurements were made on primates whose hip joints were completely preserved, as well as the knee muscles, ligaments, and meniscus.

Knee dissection was performed in 2 gorillas (Vallois

+ 1)*, 1 chimpanzee (Vallois + 3), and 2 humans (Vallois + 10). Associated femurs and tibias were obtained from 30 gorillas, 30 chimpanzees, and 95 humans. Knee rotation was measured in 1 chimpanzee and 1 human. Cineradiographic analysis of the knee was performed on 1 gorilla, 1 chimpanzee, and 1 human.

The fossil material (Figures 1 and 2) used was the complete knee joints of KNMER 1481 1 and B (East Turkana, Kenya), AL 129 1a, 1b (Afar, Ethiopia), AL 288 1AP, 1AQ (Afar, Ethiopia), and AL 333 4, X 26 (Afar, Ethiopia).

The tibia was cut at mid-height and three rods were implanted as reference lines (see Figure 3) as follows:

1. A transcondylar femoral rod, which determines the flexion-extension axis and corresponds to the condylar insertion of the internal lateral ligament. This rod remains fixed during the whole experiment. The variation of the flexion-extension angle was measured by a goniometer centered on this transcondylar rod;
2. A tibial rod, implanted on the tibial tuberosity perpendicular to the longitudinal axis of the tibia and perpendicular to the first rod in position of rotation zero (no rotation);
3. A control rod, transversely implanted on the tibial diaphysis parallel to the femoral rod in position of rotation zero.

The two tibial rods are mobile about the femoral rod in internal and external knee rotation. The experiment consists of performing a maximum external rotation movement of the tibia and then a maximum internal rotation movement of the tibia with the knee flexed 20°. A picture is taken at each 20° of flexion in the position of rotation zero and in each extreme external and internal position. This records the angular variations of the femoral and tibial rods one about the other. The angles are then measured on pictures.

Cineradiographic analysis

The knee joint was examined by cineradiography in lateral and frontal planes during flexion-extension and external-internal rotation movements. A Massiot-Phillips Medio 20 X-Ray generator was employed. The image is produced on a small screen which fluoresces when subjected to x-ray radiation and is retained on movie film. Three views were obtained using Kodak 16 mm Plus X

----

* The number in brackets represents the number of primates Vallois dissected in his thesis (1914), pointing out on each primate the same meniscal observations as in this study, thus extending the present sample.

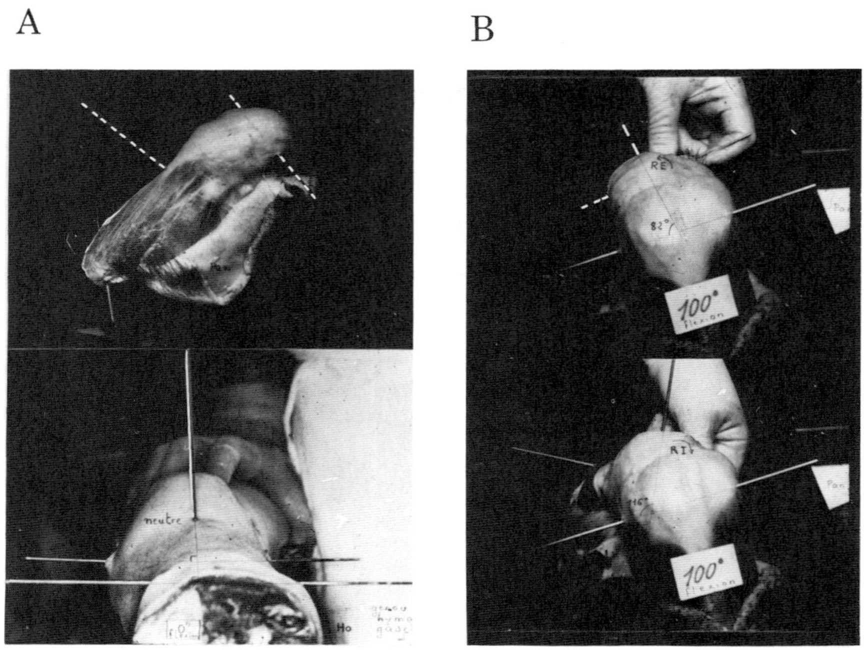

Figure 3. Method of knee rotation measurements: (A) Superior: lateral view of the flexed knee joint of a chimpanzee, with the three reference rods. Inferior: distal view of the extended knee joint of a human in position of rotation zero. (B) Proximal view of the flexed left knee joint (100°) of a chimpanzee. Superior: tibia in external rotation; inferior: tibia in internal rotation.

reversible film, one for each primate. The diameter of the image intensifier was 15 cm. These films were studied with a Spekto Motion Analyser.

RESULTS

During dissection it was observed that the shape and tibial insertions of the internal meniscus was very similar in human, chimpanzee, and gorilla (Figure 4). In contrast, the external meniscus was strikingly different. In the humans studied, it was crescent shaped and displayed two tibial insertions, one anterior and the other posterior to the external spine. This last one is exclusive of humans and unique among primates. These two meniscal insertions are short, narrow, and thick. In the

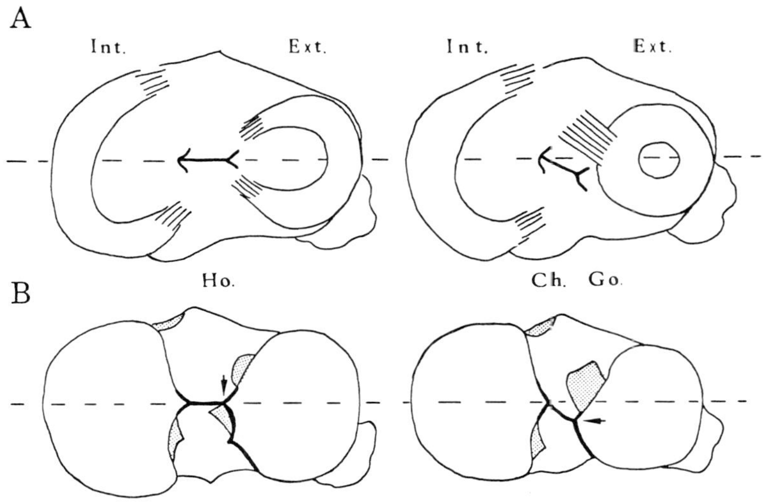

Figure 4. (A) Compared shape and tibial insertions of the two knee menisci in human, chimpanzee, and gorilla (right tibia). (B) Compared areas of insertion of the menisci on the tibia in a human, chimpanzee, and gorilla (right tibia). The arrow points out the summit of the external tibial spine. Note the different morphology of the posterior medial border of the external plateau.

chimpanzee and gorilla, the external meniscus was ring shaped and displayed one single tibial insertion anterior to the external spine. This insertion is wide, long, and made of very supple fibers.

On the dry tibia the areas of insertion of the external meniscus were different. In the humans there were two areas of insertion which set free the external spine: the tibial spines are set at the level of the middle of the tibial plateau, and the posterior border of the external plateau is long, discontinuous, and notched by the posterior insertion of the meniscus. In the chimpanzee and gorilla there was one area of insertion of the external meniscus. The tibial spines were displaced backward on the tibial plateau. The posterior border of the external plateau was short, continuous, and abrupt.

With these observations in mind, the following points can be made about the fossils (Figure 5).

1. On the right tibia of KNMER 1481B, the tibial spines are set at the level of the middle of the tibial plateaus and the posterior border of the external plateau is long, discontinuous, and notched by the posterior insertion of the external meniscus. The external meniscus of KNMER 1481 is crecent-shaped with two insertions as in modern humans.

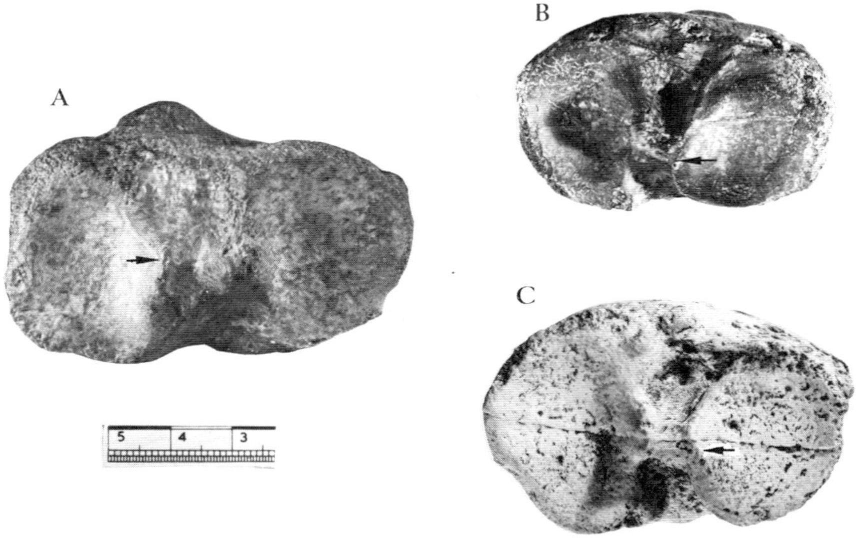

Figure 5. Proximal view of three Plio-Pleistocene hominid tibiae: (A) KNMER 1481 B (left tibia); (B) AL 129 1B (right tibia); (C) AL 333 x 26 (right tibia). The arrow points out the summit of the external KNMER 1481 and the more primitive morphology of AL 129 and AL 333.

2. On the left tibia of Al 129 lb (and Al 288 1 AQ), the tibial spines are displaced backward on the tibial plateaus and the posterior border of the external plateau is short, continuous and abrupt.
3. The external meniscus of Al 129 presented one single tibial insertion, anterior to the external spine. Notice the strong proximity of the two tibial spines.
4. On the left tibia Al 333 x 26 the posterior border of the external plateau is short, continuous, and abrupt. As in Al 129, there was no posterior insertion of the external meniscus on the tibia Al 333.

Thus, these meniscal features divide the hominid fossils into two groups, the same groups that had been distinguished from the osteological features, with a fully modern group *Homo* and the more primitive *Australopithecus*.

So in the hominid evolution, one can suggest that the knee joint has evolved from a single insertion to a double insertion of the external meniscus. The

Plio-Pleistocene hominids give some very representative specimens of each type.

Cineradiographic analysis

The graph (Figure 6) displays the results of angular rotation measurements on one chimpanzee and one man. Note that there is stronger rotation, particularly internal rotation, in the chimpanzee. Internal rotation is stronger than external rotation in humans and particularly in chimpanzees. In extension, no rotation occurs in humans, but strong rotation occurs in the chimpanzee. Thus, we observe a stronger knee stability in humans, especially in extension, and a stronger knee laxity in the chimpanzee.

Cineradiographic analysis confirms the previous angular measurements but cannot be presented here. The presentation of the complete comparative analysis made on the knees of four primates (*Homo*, *Pan*, *Gorilla*, *Cercopithecus*) needs further investigation.

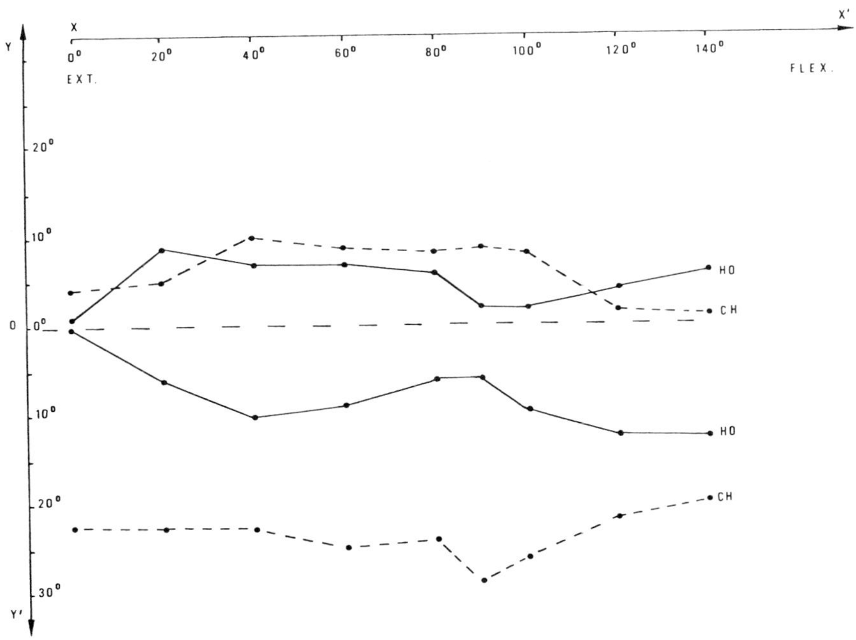

Figure 6. Compared knee rotation measurements in human and chimpanzee: xx' shows extension-flexion axis; yy' shows rotation axis. oy = external rotation; oy' = internal rotation.

DISCUSSION

Functional link between the different results

Results from osteology, dissection, and measurements are closely linked in a functional point of view. In humans, the close fitting of the joint and the double insertion of the external meniscus limit the amplitude of rotation movement. In chimpanzees the looser fitting of the joint and the single external meniscal insertion permit a greater amplitude of rotation. In fact, the mode of insertion of the meniscus determines its mobility, which thus determines the amplitude of rotation.

Functional hypothesis concerning the evolution of hominid locomotion

Numerous knee rotation measurements have been performed on humans with different methods (Bugnion, 1892; Dempster, 1956; Langa, 1963; Kapandji, 1970) but none has been performed on primates. More precise results are obtained by means of an articulograph. Nevertheless, in this study the important point is the use of the same method on humans and chimpanzees which permits comparative functional results that can be applied to fossil hominids. I have shown that in hominid evolution the knee joint has probably evolved from a single to a double insertion of the external meniscus, with representative specimens of each type among the Plio-Pleistocene hominids. These functional results suggest that during the Plio-Pleistocene some bipedal hominids had a knee joint in which stability was similar to modern humans with limited rotation. These hominids had bipedal behavior that was fully modern. Other bipedal hominids had a femorotibial joint with significantly stronger rotation. This consequent laxity of the knee means that the leg and the foot could be placed on the substrate in a much freer fashion than in humans, and that the mobility and prehensibility of the foot were greatly complemented, particularly in Lucy (Al 288). Lucy was probably not exclusively biped and presented a different bipedal behavior (cf., also Jungers, 1982).

Provisional interpretation of the fossil Al 333 x 26

Al 333 x 26 is a difficult and controversial fossil. Johanson and White (1979) allocated this fossil to the species _Australopithecus afarensis_ together with the specimens Al 288 and Al 129. Others think that the differences are too strong between these fossils to put them in the same species.

By the common presence of a single insertion of the external meniscus, Al 333 is close to Al 129.

Nevertheless, we know that the single insertion of the ring-shaped external meniscus is not only a hominid primitive feature but an hominoid and even catarhine primitive feature (Vallois, 1914). So the common presence of this characteristic in Al 333 and Al 129 cannot constitute an argument to place these two fossils in the same species because it is a primitive common feature and not a derived common feature. From this meniscal feature, we can only conclude that Al 129 and Al 133 are both more primitive hominids than KNMER 1481.

ACKNOWLEDGMENTS

I would like to thank R. E. F. Leakey (Nairobi) and Dr. D. C. Johanson (Cleveland) for their kind permission to study the original fossils in their care; Prof. Anthony and Prof. Y. Coppens (Paris) for access to their primate and human collections and for practice of primate dissections; Dr. P. Le Floch (Paris) and Dr. Ph. Orengo (Pontoise), orthopedic surgeons, specialists of the knee joint, for their collaboration in human and primate dissection and for their excellent scientific advice.

REFERENCES

Bugnion, E.: Le mecanisme du genou. These de Medecine. pp. 333-374. Rec. Inaugural Univ., Lausanne. (1892).

Dempster, W. T.: The range of motion of cadaver joints: the lower limb. Bull. Medical, Univ. of Michigan. 22:364-379. (1956).

Johanson, D. C., and White, T. D.: A systematic assessment of Early African hominids. Science. 203:321-330. (1979).

Jungers, W. L.: Lucy's limbs: skeletal allometry and locomotion in Australopithecus afarensis. Nature. 297:676-678. (1982).

Kapandji, I. A.: Physiologie articulaire. Membre inferieur. Fasc. II. Maloine, Paris. (1970).

Langa, G. S.: Experimental observations and interpretations of the relationship between the morphology and function of the human knee-joint. Acta Anatomica. 55:12. (1963).

Lovejoy, C. O., Heiple, K. G., and Burstein, A. H.: The gait of Australopithecus. Am. J. Phys. Anthrop. 38:757-780. (1973).

Tardieu, C.: Aspects biomecaniques de l'articulation du genou chez les Primates. Bull. Soc. Anat. Paris. 4:66-86. (1979).

Tardieu, C.: Morpho-functional analysis of the articular surfaces of the knee-joint in Primates. In Primate Evolutionary Biology. A. B. Chiarelli, and R. S. Corruccini. Springer-Verlag, Berlin, Heidelberg. (1981).

Tardieu, C.: Analyse morpho-fonctionnelle de l'articulation du genou chez les primates. Application aux hominides fossiles. Cahiers de Paleoanthropologie. These Doctorat, Universite Paris VI. C.N.R.S., Paris. (1983).

Vallois, H. V.: Etude anatomique de l'articulation du genou chez les primates. Imprim. Coop. ouvriere "L'Abeille", Montpellier. (1914).

17

Morphological Characteristics of Upper Molars of Squirrel Monkeys (*Saimiri*) and Their Individual Variation

M. Natori

The relationships between the trigon and the hypocone in Saimiri are interpreted as a series composed of five morphological categories. In type A, one extremity of the series, the entoflexus is deep and long transversely. This pattern is interpreted as a possibly primitive character among platyrrhines. In type E, at the opposite pole, the lingual part of the postprotocrista and the prehypocrista form a single structure. The buccal part of the postprotocrista (entometacrista) crest does not connect to this structure. This type is regarded as the most derived character state. Type B is defined by the prehypocrista's connection to the trigon below the pronounced postprotocrista. On DP^4 and M^1, the frequency of type B is greater than 90%; on M^2 type B also occurs most often but other variations are common. Among the three teeth examined here, DP^4, M^{1,2}, M^2 are most variable.

The individual variations observed in Alouatta, Aotus, Callimico, and Cebus correspond to one or more of the five categories seen in Saimiri. This suggests that the Saimiri pattern is indeed a good model for ancestral platyrrhine molar morphology, as others have suggested.

INTRODUCTION

Thomas (1913) pointed out that there are gradations in the size of the hypocone in New World monkeys, bridging the morphological gap between three- and four-cusped molars. For example, Callithrix has no trace of the hypocone and its presence is relatively rare in Saguinus. Callimico has a slight rise on the lingual cingulum, and in Saimiri, a fully developed hypocone cusp is present. Callicebus and other living platyrrhines have complicated molars with a large hypocone connecting to the trigon.

It has been suggested that the upper molars in *Saimiri* possess many primitive characters among South American monkeys (Orlosky and Swindler, 1975; Rosenberger, 1977; Maier, 1977; Orlosky, 1980; Setoguchi, in press), particularly the condition of the hypocone and its relation to the protocone. However, data on its morphological variability are scant. Kinzey (1973) noted that the hypocone was always present on M^1 and M^2 of *Saimiri*, while on M^3 it was generally absent. Swindler (1976) demonstrated that this cusp is absent on 35% of the total specimens. Although these studies establish the presence of an $M^{1,2}$ hypocone as a universal *Saimiri* feature, they provide no other information regarding its anatomical variation. The purpose of this paper is to examine the relationships between the trigon and the hypocone in the context of the occlusal crest system of mammalian molars.

METHODS AND MATERIALS

Observations were made on 101 specimens of *Saimiri sciureus* belonging to the Primate Research Institute, Kyoto University, Japan. These specimens were introduced from Bolivia and are now part of the study collection of the Primate Research Institute. The total sample of specimens included 51 DP^4, 100 M^1, and 59 M^2. Eleven individuals of *Alouatta* and ten of *Aotus* are also examined. All specimens of *Alouatta* and the six skulls of *Aotus* come from the collection of the National University of Colombia, and four specimens of *Aotus* belong to the Primate Research Institute. Only specimens in which crowns are not excessively worn were selected for study so that structure could be observed clearly.

The terminology of molar crowns (Figure 1) largely follows Rosenberger and Kinzey (1976). The term prehypocrista (Kay, 1980) is used for the ridge running

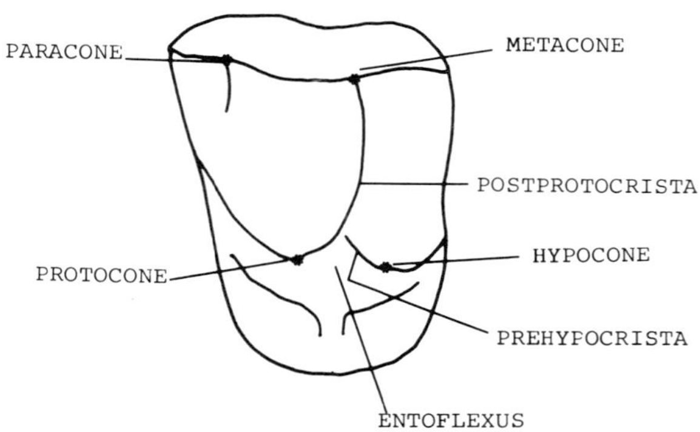

Figure 1. Hypothetical upper molar of *Saimiri* illustrating the nomenclature of cusps and ridges.

mesiobuccally from the hypocone, and entometacrista (Maier, 1980) is used for the buccal part of the postprotocrista attaching to the metacone.

The relationships between the trigon and the hypocone are divided into the following five morphological categories (Figure 2):

Type A: The protocone is connected with the metacone by the postprotocrista. The entoflexus is long along the distal side of the postprotocrista. The prehypocrista is not present.

Type B: The protocone is connected with the hypocone (and the metacone) by the postprotocrista. The low and short prehypocrista connects the hypocone to the trigon below the postprotocrista at about one-third the distance between protocone and metacone.

Type C: As in type B, but with a better developed hypocone. The prehypocrista is more developed than in type B and becomes as clear as the postprotocrista. Thus, the entoflexus is confined to the lingual side of the postprotocrista.

Type D: The protocone is connected with the hypocone by the lingual part of the postprotocrista and the prehypocrista. The low crest running lingually from the metacone, the entometacrista, connects to this ridge at the midpoint between the protocone and the hypocone.

Type E: As in type D, but the entometacrista disappears before connecting to the postprotocrista-prehypocrista.

RESULTS

On DP^4, M^1, and M^2 the hypocone is always present. The frequency of types A to E on DP^4, M^1, and M^2 is shown in Figure 3. It is clear that the left and right tooth rows show the same tendency. On DP^4 and M^1, almost all specimens (90% or so) show type B. On M^2, type B occurs most often but other variations occur with greater incidence than on DP^4 or M^1. On the whole, type B predominates and may be taken as the typical relationship between the trigon and the hypocone in *Saimiri*.

The relationships between the trigon and the hypocone in *Saimiri*

The observed structural variations are relatively minor and seem to reflect changes in two regions of the crown. The differences between types A, B, and C are related to the relative size of the hypocone. Types C and D are distinguished by the strength of the postprotocrista. The five occlusal patterns may be interpreted as a series of which one extremity is type A and the opposite extremity is type E. This series

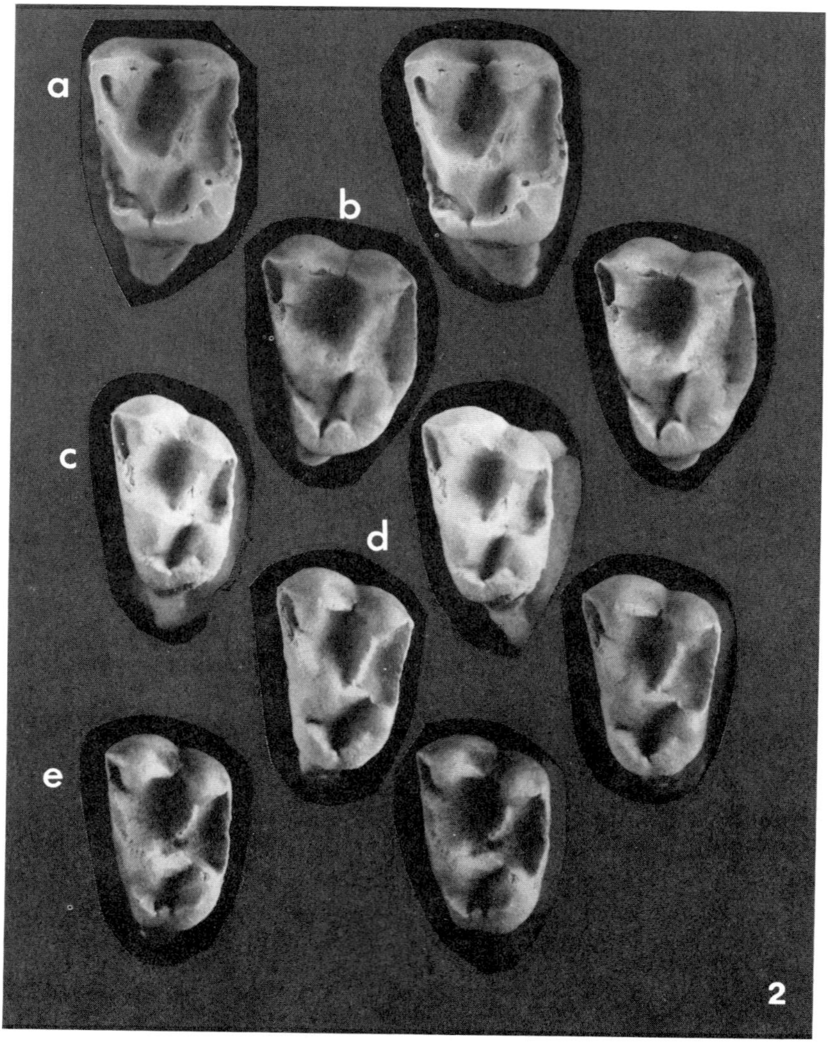

Figure 2. Diagrams of the five types. The diagrams indicate the five morphological categories discussed in the text. PR = protocone, HY = hypocone, ME = metacone, BPP = the buccal part of the postprotocrista, LPP = the lingual part of the postprotocrista, and PHC = prehypocrista.

includes no abrupt transitions. The entometacrista seen in type D and E corresponds to the buccal part of the postprotocrista. Hence, in some Saimiri, the prehypocrista and the lingual part of the postprotocrista form a single structure, which connects the protocone with the hypocone.

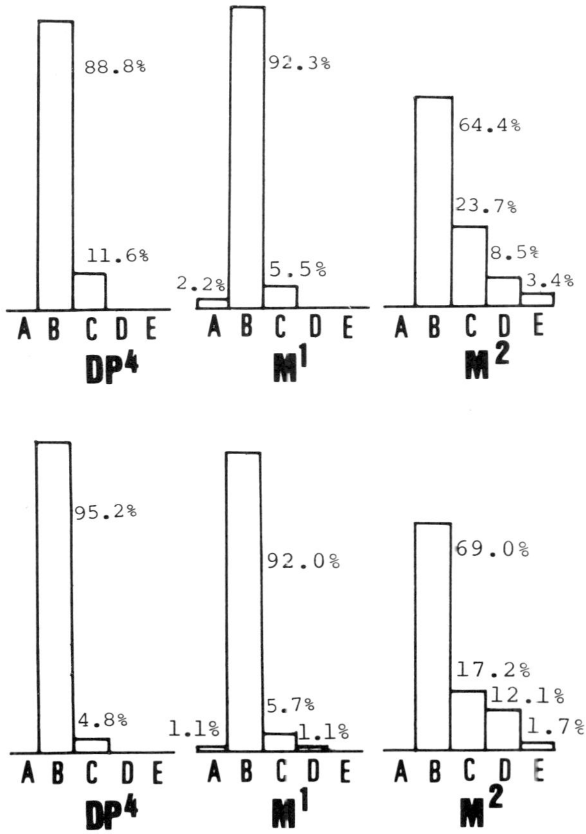

Figure 3. Frequency (%) of occurrence of each pattern type. The upper histograms are for left tooth row and the lower ones for right. Sample size: left DP^4, 43; left M^1, 91; left M^2, 59; right DP^4, 42; right M^1, 87; right M^2, 58.

Branisella is the oldest known monkey in South America. The primitive platyrrhine upper molar morphology is best represented by *Branisella* (e.g., Rosenberger, 1977, 1981; Orlosky, 1980). Here, the hypocone is present, the postprotocrista connects the metacone with the protocone, and the entoflexus is deep and long transversely. This resembles type A as described here for *Saimiri*, suggesting that the primitive character of platyrrhines is not represented in type B, which is most frequent, but by type A.

If type A shows the primitive character of platyrrhines, type E is probably the most derived character in the series from type A to type E. The frequency of type B is higher on DP^4 and M^1 than M^2. But type C, D, and E occur more often on M^2 than on DP^4 and M^1. This evidence also suggests that M^2 has a more derived morphology than DP^4 and M^1 in *Saimiri*. This hypothesis is consistent with Kinzey's (1973) findings that the lingual cingulum in *Saimiri* is rarely reduced on M^1, while it is relatively reduced on M^2. However, the complete lingual cingulum was the primitive pattern of platyrrhines. This shows that M^2 has more derived characters than M^1 as far as the lingual cingulum of *Saimiri* is concerned. Thus, M^2 may generally be more derived morphologically than M^1 among upper molars in *Saimiri*.

The relationships between the trigon and the hypocone in some platyrrhines

The M^1 of *Callimico* is distinguished from four-cusped platyrrhines by having a small hypocone widely separated from the trigon. A prehypocrista is essentially nonexistent. This pattern closely resembles type A as seen in *Saimiri* in low frequency.

On M^1 and M^2 in *Alouatta*, a ridge connects the hypocone with the protocone, and the extremity of the crest running lingually from the metacone disappears before connecting to it. Thus, the relationships between the trigon and the hypocone in *Alouatta* is similar to the type E in *Saimiri*. If the type A-like structure is primitive in platyrrhines, the morphology in *Alouatta* must have been derived from the type A. As a result, the ridge connecting the protocone with the hypocone corresponds to the prehypocrista posteriorly and the lingual part of the postprotocrista anteriorly, and the crest running lingually from the metacone corresponds to the buccal part of the postprotocrista. Therefore, the series of individual variation in *Saimiri* suggests the following homology between the primitive character of platyrrhines and the morphology in *Alouatta*. On upper molars in *Alouatta*, the ridge connecting the protocone with the hypocone corresponds to the lingual part of the postprotocrista anteriorly, the prehypocrista posteriorly, and the entometacrista crest running lingually from the metacone corresponds to the buccal part of the postprotocrista.

On M^1 and M^2 in *Alouatta*, in general, the relationships between the trigon and the hypocone exhibits those of type E. On one of the 11 specimens examined, the postprotocrista connects the protocone with the metacone and the prehypocrista meets the postprotocrista. This structure is similar to type C of *Saimiri*. On M^1 and M^2 in *Aotus*, in general, the postprotocrista connects the protocone with the metacone, and the short and clear prehyposcrista connects to the distal side of the postprotocrista at a point one-third the distance from the protocone to the metacone. This relationship between the trigon and the hypocone is similar to the type C in Saimiri. Some specimens show the relationships like type B and D in Saimiri, however. Thus individual variations in Alouatta and Aotus are also found in five morphological categories in Saimiri.

On M^1 and M^2 of Cebus, the molar crown tends to be divided into mesial and distal parts, giving the appearance of an incipiently bilophodont condition (Hill, 1960; Rosenberger and Kinzey, 1976; Swindler, 1976). A transverse crest connects the metacone with the hypocone. According to Swindler (1976), this structure appears on over 50% of Cebus apella. This transverse crest corresponds to the prehypocrista and the buccal part of the postprotocrista. In general, this pattern is similar to type B or type C of Saimiri, with the addition of a much deeper entoflexus.

This study of individual variation in Saimiri demonstrates that, as several other authors have suggested, this genus may be regarded as a good model of the ancestral platyrrhine molar morphology. Several of the alternative molar designs, as we find in Aotus, Alouatta, Cebus, and Callimico, are all seen as individual variants in Saimiri, showing that selection for relatively simple modifications of a Saimiri-like morphotype can be transformed thusly. In conclusion, one can see how the analysis of intrageneric variation can contribute to the understanding of supraspecific evolution.

ACKNOWLEDGMENTS

I would like to express appreciation to the following persons for their guidance and encouragement: Prof. Akiyoshi Ehara, Prof. Yasuo Nogami, Dr. Mitsuru Aimi, Dr. Takeshi Setoguchi, Mr. Toshio Mouri, Mr. Shin Matsumoto, Mr. Minoru Kinoshita, Primate Research Institute, Kyoto University, Dr. Nobuo Shigehara, Dokkyo University School of Medicine, Japan. I am also grateful to Prof. Alberto Cadena, the National University of Colombia, Colombia, for permission to examine the specimens in his care. Also to the assistance of Dr. T. Setoguchi and Dr. A. L. Rosenberger, University of Illinois at Chicago, through their jointly held U. S. - Japan Cooperative Sciences Program grant (INT 8111229).

REFERENCES

Hershkovitz, P.: Basic crown pattern and cusp homology and evolution. In Dahlber. A. A. (ed.). Dental Morphology and Evolution. Univ. Chicago Press, Chicago, pp. 95-150. (1971).

Herschkovitz, P.: Living New World Monkeys (Platyrrhini), with an Introduction to Primates, Vol. 1. Univ. Chicago Press, Chicago. (1977).

Hill,W. C. O.: Primates: Comparative Anatomy and

Taxonomy, Vol. IV, Platyrrhini, Cebidae, Part A. Univ. Edinburgh Press, Edinburgh. (1960).

Kay, R. F.: Platyrrhine origins: A reappraisal of the dental evidence. In: Evolutionary Biology of the New World Monkeys and Continental Drift. Cìochon, R. L. & Chiarelli, A. B. (eds.). Plenum Press, New York, London: pp. 159-188. (1980).

Kinzey, W. G.: Reduction of the cingulum in Ceboidea. In: Craniofacial Biology of Primates. Zingeser, M. R. (ed.). Karger, Basel: pp. 101-127. (1973).

Maier, W.: Die bilopholodonten Moalren der Indridae (Primates) - ein evolutionsmorphologischer Modellfall. Z. Morphol. Anthropol. 68:307-344. (1977).

Maier, W.: Konstruktionsmorphologische Untersuchungen am GrebiB der rezenten Prosimae (Primates). Abh. Seneckenb. Naturforsch. Ges. (1980).

Orlosky, F. J.: Dental evolutionary trends of relevance to the origin and dispersion of the platyrrhine monkeys. In: Evolutionary Biology of the New World Monkeys and Continental Drift. Ciochon, R. L. and Chiarelli, A. B. (eds.). Plenum Press, New York, London: pp. 189-200. (1980).

Orlosky, F. J., and Swindler, D. R.: Origins of New World monkeys. J. Hum. Evol. 4:77-83. (1975).

Rosenberger, A. L.: <u>Xenothrix</u> and ceboid phylogeny. J. Hum. Evol. 6:461-481. (1977).

Rosenberger, A. L.: A mandible of <u>Branisella boliviana</u> (Platyrrhini, Primates) from the Oligocene of South America. Int. J. Primatol. 2:1-7. (1981).

Rosenberger, A. L., and Kinzey, W. G.: Functional patterns of molar occlusion in platyrrhine primates. Am. J. Phys. Anthrop. 45:281-298. (1976).

Setoguchi, T.: On the individual variations of upper molars of howlers and the phyletic relations among South American monkeys based on molar structure. J. Anthropol. Soc. Nippon. (in press).

Setoguchi, T., Watanabe, T., and Kondo, S.: Do South American monkeys have a pseudohypocone? J. Anthropol. Soc. Nippon 89:7-26 (in Japanese). (1981).

Swindler, D. R.: Dentition of Living Primates. Academic Press, New York. (1976).

Thomas, O.: On some rare Amazonian mammals from the collection of the Para Museum. Ann. Mag. Nat. Hist. 11:130-136. (1913).

18

Relation Between Morphology and Function of the Dentition in the *Stirtonia-Alouatta* Lineage (Ceboidea)

T. Setoguchi

The upper premolars and molars of the middle Miocene *Stirtonia* are described for the first time. They were found in the La Venta badlands of Colombia, South America. The general morphology of these teeth is similar to that of living howler monkeys, *Alouatta*. It is concluded that *Stirtonia* is a direct ancestor of *Alouatta*. The premolars of *Stirtonia* are clearly wider transversely than in *Alouatta*. This warrants the separation of *Stirtonia* from *Alouatta* on a generic level. On the upper molars of *Stirtonia*, no direct connection between the postero-internal cusp and the protocone is observable. The postprotocrista extends postero-buccally from the protocone towards the posterior side of the metacone and the entoflexus separating the postero-internal cusp from the protocone is deep and long transversely. The postero-internal cusp connects with the posterior cingulum posteriorly. Thus, the postero-internal cusp of *Stirtonia* is a hypocone, not a pseudohypocone. The entoflexus occludes with the entoconid, and their major function is grinding or crushing food material. On extant *Alouatta*, the postprotocrista shifts its position lingua-posteriorly and connects with the prehypocrista forming a single structure. In occlusion, the postprotocrista-prehypocrista alternates with the buccal face of the entoconid, and thus alternation or shear is the major function between them. The shearing function is more emphasized in *Alouatta* than in *Stirtonia*. The dietary specialization of *Alouatta* as exclusive leaf-eaters may be closely related to the functional adaptation of its dentition.

INTRODUCTION

In 1979, the joint Paleontological Expedition of Kyoto University, Japan, and the Instituto Nacional de

Investigaciones Geologico-Mineras, Colombia, discovered the upper premolars and molars of *Stirtonia tatacoensis* in the Miocene La Venta badlands of Colombia, South America. As the geological work has not yet been completed, a report of the geology of the area will be published separately later. The discovery was made in the same general area as the University of California field parties worked during the 1940s (Stirton, 1951). The vertebrate fauna from the La Venta badlands is correlated with faunas of Friasian Age in Argentina (Hirschfeld and Marshall, 1976) and therefore is dated as approximately middle Miocene (Marshall et al., 1977).

The University of California field party discovered a primate mandible (UCMP-38989) from the "Monkey unit" in the La Venta badlands. The material was described originally as *Homunculus tatacoensis* by Stirton (1951) and redescribed as a new genus, *Stirtonia tatacoensis* by Hershkovitz (1970). *S. tatacoensis* has long been known only from the lower dentition. During the 1979 field season, the joint expedition between Japan and Colombia succeeded in discovering the upper dentition (P^{2-4}, M^{1-2}) which has not previously been described. This is described below. Stirton (1951) also described another specimen (M_1, UCMP-39206) of "*Homunculus*" found in the "Fish Bed" which overlies the "Monkey unit." Stirton considered that it probably represented a different species from *Stirtonia tatacoensis*. Hershkovitz (1970, p. 16), however, felt that the morphological difference between the specimen and the holotype of *S. tatacoensis* probably lies within the limits of individual variation. By comparing it with the holotype, new material treated here, and several undescribed specimens obtained in the 1982 field season, it is now clear that Stirton's lower molar from the "Fish Bed" does not belong to *S. tatacoensis* though it shows a close tie with that species.

With the discovery of the upper dentition of *Stirtonia*, the criteria that can be used to trace the phylogenetic history of the postero-internal cusp of upper molars of extant howler monkeys (*Alouatta*) are presented. *Stirtonia* is considered to be a direct ancestor to *Alouatta* as discussed below. Based on the general similarity seen in the upper molar configuration between North American Eocene *Notharctus* and South American extant *Alouatta*, Gregory (1920) wrote that the postero-internal cusp of the upper molars of *Alouatta* is apparently a pseudohypocone as in *Notharctus*, because in *Alouatta* the postero-internal cusp is connected with the protocone by a ridge. On the newly discovered upper molars of *Stirtonia*, no direct connection between the postero-internal cusp and protocone is observable. Instead, the postprotocrista runs postero-buccally from the protocone towards the posterior side of the metacone and the entoflexus separating the postero-internal cusp from the protocone is deep and long transversely. The postero-internal cusp connects with the posterior cingulum posteriorly. Thus, the postero-internal cusp of Stirtonia is a hypocone or euhypocone (Remane, 1960), not a pseudohypocone or pseudypocone (Stehlin, 1916).

The structural differences in the molars of Stirtonia and Alouatta result in functional differences in the molars between these two taxa. These are discussed below.

METHODS AND MATERIALS

The systematic account used for Stirtonia is as follows: Class: Mammalia; Order: Primates; Superfamily: Ceboidea; Family: Cebidae (Bonaparte, 1831); Genus: Stirtonia (Hershkovitz, 1970); Species: tatacoensis (Stirton, 1951). The following abbreviations are used: IGM = Instituto Nacional de Investigaciones Geologico-Mineras, Colombia; KU = Kyoto University Primate Research Institute, Japan; and UCMP = Museum of Paleontology of the University of California, Berkeley, USA.

The fossil materials examined, all belonging to a single individual, (IGM-KU-III-1), were P^2, P^3, P^4, M^1, M^2, and M_1 (see Figures 1-7). This specimen was located near River Tatacoa, northeast of the town of Villa Vieja, Huila Department, Colombia, South America, and is dated at middle Miocene.

P^2 is somewhat wider transversely than it is long. As in P^3 and P^4, in P^2 the paracone is the only prominent cusp. The metacone is not formed. The paracone is conical but a little compressed transversely. The paracone is situated a little anterior to the longitudinal midpoint of the tooth. Two acute ridges run from the tip of the paracone: one runs anteriorly but a little buccally, and the other posteriorly but a little lingually. One broad ridge descends antero-lingually from the tip of the paracone. The lingual cingulum is very broad and occupies the lingual two-thirds of the tooth. The edge of the lingual cingulum forms a continuous acute ridge. There are no small cusps or conules on the lingual cingulum. The anterior half of the ridge of the lingual cingulum runs almost horizontally around the antero-lingual base of the paracone. The posterior half of the ridge descends

TABLE 1. Dimensions of teeth of Stirtonia tatacoensis, IGM-KU-III-1 (in mm)

	P^2	P^3	P^4	M^1	M^2	M_1	
Length	4.1	4.5	4.5	6.3	6.6	6.9	
Width	6.0	6.8	7.2	7.2	8.4	4.0	(trigonid)
						4.9	(talonid)

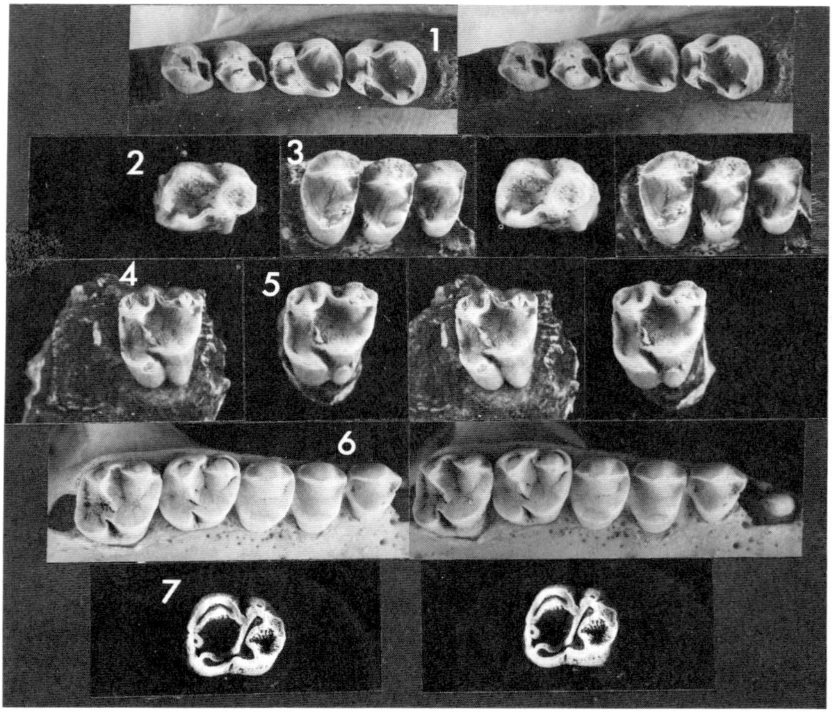

Figure 1. Cast of the holotype of Stirtonia tatacoensis, UCMP-38989. Right P_{3-4}, M_{1-2}. x2.

Figures 2-5. Stirtonia tatacoensis, IGM-KU-III-1. (2) Left M_1, (3) Right P^{2-4}, (4) Right M^2, (5) Right M^1. x2.

Figure 6. Alouatta seniculus. Right P^{2-4}, M^{1-2}. x2.

Figure 7. "Fish Bed" molar, UCMP-39206. Left M_1. x2.

postero-buccally from the lingual corner of the cingulum. The ridge continues to run horizontally and ascends again to join the acute posterior ridge from the tip of the paracone.

The morphology of P^3 is almost identical to that of P^2. P^3 is wider transversely and more robust than P^2. The base of the paracone on P^3 is broader than that of P^2. The parcone on P^3 is situated a little more anteriorly than on P^2. The anterior ridge from the paracone descends antero-buccally and terminates buccally to the posterior end of the posterior ridge from the paracone of P^2. The posterior ridge from the paracone runs almost posteriorly. The lingual cingulum occupies the lingual two-thirds of the tooth as in P^2. The ridge forming the edge of the cingulum is more robust than in P^2. At the lingual corner of the cingulum, a small and low conule is formed (tentatively, this conule is called conule B). A somewhat broad ridge extends anteriorly from conule B and the ridge connects with the lingual base of a conspicuous conule (tentatively called conule A). Conule A is situated just anterior to the tip of the paracone and elongated bucco-lingually on the lingual cingulum. The tip of conule A is clearly higher than that of conule B and much lower than the paracone. Buccal to conule A the lingual cingulum runs almost horizontally. Posterior to conule B, the lingual cingulum descends a little postero-buccally forming the postero-lingual edge of the lingual cingulum.

P^4 is wider and more slender transversely than in P^3. The paracone is conical and situated just on the longitudinal midpoint of the tooth. The antero-lingual ridge from the paracone is not prominent, whereas in P^2 and P^3 this broad ridge is conspicuous making the lingual configuration of the paracone asymmetrical. Because the tooth is transversely elongated lingually, the lingual cingulum occupies almost the lingual three-fourths of the tooth. Just posterior to the lingual corner of the lingual cingulum, the small but conspicuous conule B is formed. Conule B on P^4 is larger than that on P^3. Anterior to conule B, the tooth structure is almost identical to that of P^3. Posterior to conule B, the lingual cingulum runs lingually and a little posteriorly. The ridge descends only slightly making this part of the tooth higher crowned than in P^2 and P^3. Any trace of the hypocone is not found.

M^1 is square shaped. The paracone is larger and taller than the metacone. The paracone is situated very close to the anterior border of the tooth so that the parastyle is not well defined. The paracone is elongated antero-posteriorly. One acute ridge descends from the tip of the paracone posteriorly but slightly buccally and continues to run postero-buccally. This ridge connects with the antero-lingual corner of the mesostyle which is somewhat compressed antero-posteriorly. The mesostyle is situated posterior to the longitudinal midpoint of the tooth. The stylar shelf between the paracone and the mesostyle is concave buccally.

The metacone is V-shaped with the apex facing lingually. The antero-lingual and the postero-lingual walls of the metacone are almost vertical. Two acute ridges run from the apex of the metacone: one descends antero-buccally and runs horizontally connecting with the postero-lingual corner of the mesostyle. Thus, the mesostyle connects with both the paracone and the metacone as seen in the dilambdodont dentition. The mesostyle is clearly lower than these two cusps.

The other ridge from the metacone descends postero-buccally and connects with the postero-lingual corner of the obliquely elongated metastyle. The metastyle forms the postero-buccal corner of the tooth. The metastyle is elongated antero-buccally so that the concave stylar shelf between the metastyle and mesostyle becomes narrower buccally.

The protocone is situated lingually to the midpoint between the paracone and the mesostyle. The protocone is somewhat elongated antero-posteriorly with the apex pointing upward. The protocone is as high as the metacone. The preprotocrista gradually descends antero-buccally from the apex of the protocone and then gradually ascends almost buccally to connect with the anterior extremity of the paracone. The buccal wall of the protocone forms an almost flat plane. Posterior to the protocone, a ridge descends postero-buccally, then runs straight postero-buccally and disappears at the point between the metacone and the postero-internal cusp. This ridge is completely separated from the latter cusp by the deep and transversely long entoflexus, and is identified as the postprotocrista. Thus, no direct connection between the protocone and the postero-internal cusp is seen.

Antero-buccally to the point where the postprotocrista disappears, a remnant of a small ridge runs transversely and its buccal extremity connects with the lingual base of the metacone. However, the lingual extremity of this ridge is separated from the postprotocrista by a small valley. The small ridge is situated very low in comparison with the metacone. The homology of this structure is uncertain. At this stage, it is not possible to decide whether this structure can be identified as the metaconule or as the remnant of the buccal half of the postprotocrista. Tentatively, this ridge is called the medial posterior transverse crest.

The postero-internal cusp is situated posteriorly and slightly lingual to the protocone; it is lower than the protocone. Between these two cusps, the V-shaped entoflexus runs obliquely postero-buccally. From the apex of the postero-internal cusp, one broad ridge descends towards the postero-lingual corner of the metastyle. This broad ridge forms the posterior cingulum. The prehypocrista is not present. As the posterior cingulum and the postero-internal cusp form a continuous single structure, and the Nannopithex-fold is not present, the postero-internal cusp must be identified as the hypocone, not as a pseudohypocone which is developed on the Nannopithex-fold or is formed by the splitting off from the protocone.

The morphology of M^2 closely agrees with that of M^1. M^2 is a little more elongated transversely than in M^1. The buccal side of the paracone and the buccal half of the mesostyle are more developed than in M^1. The small ridge between the postprotocrista and the metacone is more developed and well defined than in M^1. On the middle of this ridge, there is a small cusp. The entoflexus is narrower than in M^1, and forms an acute V-shaped valley. The crown base lingual to the protocone and the hypocone extends lingually forming a W-shaped ledge. This structure is more developed in M^2 than in M^1.

The morphology of M_1 in this sample is almost identical to that of the corresponding ooth of the holotype of *Stirtonia tatacoensis*. The only obvious difference between them is in the morphology of the postero-lingual corner of the tooth. On the present specimen, the entoconid is clearly separated from the hyopconulid by a small but distinct notch, whereas in the holotype, these two cusps are united by a low ridge. The rest of the morphology is identical in both specimens.

DISCUSSION

Because of the presence of the conspicuous hypocone, *Stirtonia* is not phylogenetically related to the Callithricidae nor to *Callimico*. The upper molar of *Stirtonia* is characterized by the prominence of the mesostyle. Most of the Cebidae lack a conspicuous mesostyle except for *Brachyteles* and *Alouatta*. In *Brachyteles*, the paracone and the metacone are elongated antero-posteriorly and situated closer together than in *Stirtonia*. The antero-lingual part of M^1 in the former is truncated so that the molar is not squared as it is in the latter. Phylogenetically, *Stirtonia* does not have any direct relationship with *Brachyteles*. Only *Alouatta* shares similar morphological characters of molars with *Stirtonia*. In fact, based on the morphological characteristics of *Stirtonia* described above, there are no persuasive reasons to significantly set this Miocene fossil apart from extant *Alouatta*.

The obvious differences in dental morphology between *Stirtonia* and *Alouatta* are seen in: (1) the premolar structure, (2) degree of robustness of the major cusps, and, (3) the relation between the protocone and the hypocone. The latter two points are discussed later.

The premolars of *Stirtonia* are slender and clearly wider transversely than those of any species of *Alouatta*, and especially in *A. seniculus*, *A. caraja* and *A. palliata*. In *Stirtonia*, P^2 is clearly wider than long and on P^4 the width is almost twice the length. The difference in premolar structure warrants the separation of *Stirtonia* from *Alouatta* at the generic level. The degree of development of the conules A and B on the lingual cingulum varies among the species of *Alouatta*. Conule A is usually more prominent and situated a little more lingually than in *Stirtonia*. Most species of *Alouatta* have the tiny but distinct conule A even on P^2. *A. seniculus*, *A. caraja* and *A. palliata* lack conule B or possess a tiny one. In *A. palliata* conule A is situated near the lingual border of the tooth and separated from the posterior cingulum by a distinct notch. In *A. guariba* conule A is robust but conule B is low and elongated antero-posteriorly. These two conules are separated by a notch. Compared with the molar structure, this kind of development of conule B and the posterior cingulum on premolars gives us the impression of the incipient development of the hypoconal shelf. If this is the case, the conule A of *Stirtonia* must be regarded as an incipient protocone.

The holotype of *Homunculus patagonicus*, the genotype of *Homunculus*, is a lower jaw fragment with heavily worn M_1 and M_2. Judging from the outline of the crown base of lower molars, the width of the trigonid is almost the same as that of the talonid on both these teeth. In "*Homunculus*" *tatacoensis*, the talonid is much wider transversely than the trigonid and this warrants the separation of "*Homunculus*" *tatacoensis* from *Homunculus* at a generic level as Hershkovitz thought.

UCMP 39206 (Figure 7) is a lower molar found in the "Fish Bed" which overlies the "Monkey unit." Although the tooth is moderately worn, the lophids or ridges can be clearly seen. The basal coronal outline is characterized by its squared-shape, whereas the outline of lower molars of *Stirtonia* is more elongated antero-posteriorly with transversely wider talonid. In the "Fish Bed" molar, the metastylid connecting the metaconid with the entoconid is shorter and the crista obliqua running from the hypoconid towards the posterior base of the protoconid extends more transversely than in *Stirtonia*. Consequently, the entoconid and the hypoconid are situated closer to the metaconid and the protoconid, respectively, than in *Stirtonia*. On the posterior margin of the teeth, the hypoconulid lies postero-buccally to the entoconid making the talonid basin broader antero-posteriorly in *Stirtonia*. In the "Fish Bed" molar, the hypoconulid is seen on the buccal side of the entoconid so that the talonid basin is compressed antero-posteriorly. The "Fish Bed" lower molar is characterized by its shortened talonid and this cannot be produced by attrition alone, as Hershkovitz (1970) believed. The shortness of the talonid cannot be regarded as being within the limits of individual variation of *Stirtonia tatacoensis*. The present molar surely represents a distinct species, but it is inadvisable to propose a new name on the basis of a single worn specimen.

Protocone-Hypocone relations

In *Stirtonia*, no direct connection between the protocone and the hypocone is seen. The deep and long entoflexus separates them. In *Alouatta*, by contrast, a conspicuous ridge connects the protocone with the hypocone, and the small entoflexus is confined lingually to this ridge.

The concept of a pseudohypocone for the postero-internal cusp of upper molars which represents a posterior budding from the protocone was introduced by Stehlin (1916) who originally designated it as a pseudypocone. The concept was applied to the North American Eocene *Pelycodus*-*Notharctus* lineage and its validity has been established for this lineage. Because of the general morphological similarities between

Notharctus and extant South American platyrrhines, Gregory (1920) believed that the postero-internal cusp of the upper molars of platyrrhines is "apparently a pseudohypocone." Thus, the question of whether or not the platyrrhine ceboids have a pseudohypocone on their upper molars was introduced by Gregory. Since Gregory, arguments for or against his opinion have been based only on living materials as paleontological specimens were lacking.

No upper dentition ancestral to extant Alouatta was known until recently. Now, however, the first paleontological evidence to apply to these controversies is available. In Stirtonia, the one ridge extends posteriorly from the protocone and continues between the metacone and the postero-internal cusp. The ridge is clearly separated from the postero-internal cusp by a deep entoflexus and the latter cusp is developed on the posterior cingulum. Thus the ridge must be identified as the postprotocrista. The Nannopithex-fold is not present on the upper molars of Stirtonia. Since the postero-internal cusp is raised on the posterior cingulum rather than on the Nannopithex-fold, the cusp in Stirtonia is not regarded as a pseudohypocone but as a hypocone or euhypocone (Remane, 1960). Hence the same cusp of the upper molars of Alouatta must also be regarded as a hypocone.

In Alouatta, the one ridge extends posteriorly from the protocone as in Stirtonia. This ridge runs postero-buccally and acutely changes direction lingually making an acute V-shaped ridge with the apex buccally. The ridge surrounds the remnant of the entoflexus buccally. However in Alouatta, the posterior portion of the ridge posterior to the apex must be regarded as the prehypocrista. The anterior portion of the ridge anterior to the apex is the postprotocrista. In Stirtonia, the postprotocrista runs towards the posterior base of the metacone. During the course of evolution from the middle Miocene Stirtonia to recent Alouatta, the postprotocrista may have migrated lingually, and simultaneously the prehypocrista, undeveloped in Stirtonia, may have begun to develop buccally to the hypocone. Finally, these two ridges are united forming a continuous single structure. Thus, the connection between the protocone and the hypocone in Alouatta is surely secondary and the Notharctus-like characters of the molars of Alouatta developed by homoplastic evolution.

Dental function in the Stirtonia-Alouatta lineage

The obvious difference in the dental morphology between Stirtonia and Alouatta is also seen in the degree of robustness of the major cusps. In Alouatta, the major cusps are more robust than in Stirtonia; the major cusps of upper and lower molars of Alouatta are taller and have a broader base than in Stirtonia.

In *Stirtonia*, the hypocone is conical and the prehypocrista is unformed so that the entoflexus is deep and rather wide. The anterior and posterior walls of the entoflexus support wear surfaces 6 and 8, respectively, where they wear against the entoconid which is a slender cusp with a narrow base. In occlusion, their major function is grinding and partly crushing food material because no shearing blades are formed there except for the weak metastylid which occludes with the lingual face of the protocone. On the contrary, in *Alouatta*, because the prehypocrista is well developed and the connecting point with the postprotocrista is a little below the level of the tip of the hypocone, the prehypocrista and the postprotocrista form a high wall above the entoflexus. In relation to this, the hypocone is elongated antero-posteriorly to form a rather flat ledge pointing lingually. The entoconid has a broader base and is taller with a pointed tip than that in *Stirtonia*. The metastylid, supporting wear facet 6, and the postcristid running postero-buccally from the entoconid, supporting wear facet 8, form a sharp edge, and they occlude with the postprotocrista and the prehypocrista, respectively. These ridges form sharp cutting blades so that the major function in their occlusion is shearing food material. Because the tip of the entoconid occludes into the embrasure between the protocone and the hypocone and hence with the small entoflexus, a grinding function is still retained though as a minor part of the function.

In the previous section, the structural differences of the protocone and the hypocone between *Stirtonia* and *Alouatta* were stressed. The differences in the molars result in a functional transformation from grinding or crushing food material in *Stirtonia* to shearing in *Alouatta*. The same sort of the structure-function relationship is seen in the other parts of molars in this lineage.

On the lingual part of teeth, the metaconid is conical and taller than the protoconid, but the protolophid connecting these two cusps is not well

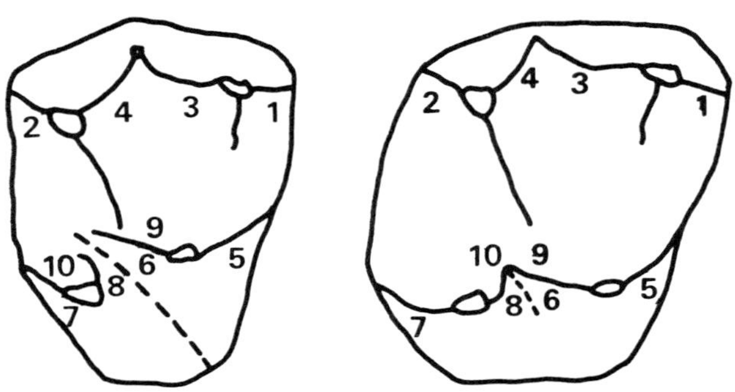

Figure 8. Right M^1 of *Stirtonia* (left) and *Alouatta* (right) to indicate the positions of wear facets.

developed and borders the posterior margin of the trigonid in Stirtonia. In Alouatta, the metaconid is more robust and taller with a pointed tip. The base of this cusp is also broad supporting the protolophid buccally and the metastylid posteriorly. The protolophid, supporting wear facet 5, is a sharp edge. Because the trigonid is highly elevated above the talonid, the protolophid forms an acute cutting blade. In occlusion, the metaconid itself occludes into the embrasure between the protocone posteriorly, and the hypocone or the posterior cingulum of the preceding tooth anteriorly. The antero-lingual face of the preprotocrista running antero-buccally from the protocone and the posterior face of the protolophid alternate to function as shearing blades. Here, too, the shearing function is more exaggerated in Alouatta than in Stirtonia.

On the buccal part of teeth, the exaggeration of the shearing function in Alouatta is clearly observed. In Stirtonia, the paracone is lower than the metacone; both cusps are conical but slender, and not as tall as the cusps of Alouatta. The postparacrista, supporting wear facet 3, and the premetacrista, supporting wear facet 4 are low and weak. On the lower molars, the crista obliqua, supporting wear facet 3, is also low, and the postcristid running postero-lingually from the hypoconid, supporting wear facet 4, is again low and not sharp. At the beginning of wear, they may function as shearing blades, but with wear these ridges are soon flattened so that shearing is not a major part of dental function there. Because the trigon basin is broad and shallow, and occludes with the broad hypoconid, grinding or crushing must be their major function in Stirtonia. In Alouatta, the paracone has a broader base which supports the acute postparacrista posteriorly. The metacone is a very large cusp having a broad and triangular base. Antero-buccally from the cusp, the sharp premetacrista extends to the mesostyle supporting wear facet 4. These ridges wear

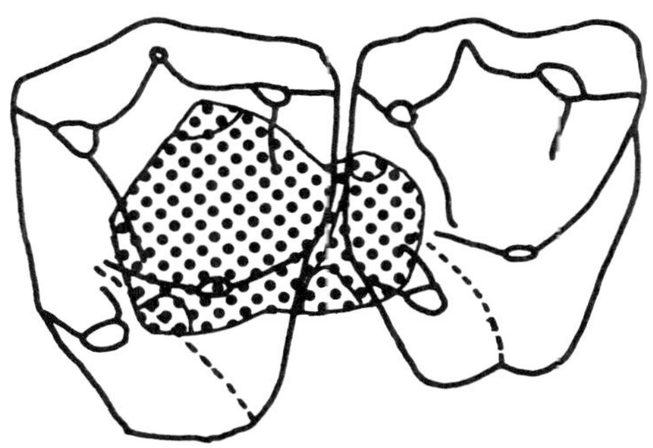

Figure 9. Occlusal relation between upper and lower molars of Stirtonia. Right M^{1-2} and M_2 are shown. Shaded is a lower molar.

against the crista obliqua and the postcristid, respectively. Here, also, a shearing function is more emphasized in *Alouatta*.

Another notable structural difference between *Stirtonia* and *Alouatta* is observable on the lingual side of the metacone. In *Stirtonia*, the small and short mesial posterior transverse crest runs lingually from the tip of the metacone. In *Alouatta*, this crest is well developed and forms an acute ridge on the broad lingual slope of the metacone. Evidently, the posterior face of the medial posterior transverse crest supports another wear face (a part of 2a of Kay, 1977: Figure 2). This crest wears against the paralophid running antero-lingually from the protoconid of the succeeding lower molars. In occlusion, the posterior face of the transverse crest on the upper molars alternates with the antero-buccal face of the paralophid of the succeeding lower molars in *Alouatta*. The function there is surely shearing, and the function has been attained through the course of evolution from *Stirtonia* to *Alouatta*, because alternation in these teeth has not been seen in *Stirtonia*.

To summarize, on the molars of *Alouatta*, the wear facets 3, 4, and part of 2a on the buccal side, and the facets 5, 6, 8 and 7 on the lingual side are more exaggerated than in *Stirtonia*. These parts of morphology of the dentition of *Alouatta* function to shear more herbaceous food material. The dietary specialization of *Alouatta* as exclusively leaf-eaters may be closely related to the functional adaptation of its dentition.

ACKNOWLEDGMENTS

This study was supported by a Grant-in-Aid for Overseas Scientific Survey from the Ministry of Education, Science and Culture, Japan, Nos. 404135 for 1979 and 504331 for 1980. I am grateful to the following people for their kind help in various ways; Dr. Shiro Kondo, Professor Emeritus and leader of our expedition, Dr. Tsuyoshi Watanabe and Mr. Toshio Mouri of the Primate Research Institute, Kyoto University, Japan: Dr. Michel Hermelin, former Director, Dr. Alfonso Lopez, Director, Dr. Francisco Zambrano, Dr. Herman Duque of the Instituto Nacional de Investigaciones Geologico-Mineras, and Dr. Alberto Cadena of Universidad Nacional, Colombia: Dr. Jose F. Bonaparte of the Museo Argentino de Ciencias Naturales "B. Rivadavia," and Mr. Jamie E. Powell of the Miguel Lillo Fundacion, Argentina.

REFERENCES

Gregory, W. K.: On the structure and relations of *Notharctus*, an American Eocene Primate. Mem. Amer. Mus. Nat. Hist. N.S. 3:49-243. (1920).

Hershkovitz, P.: Notes on Tertiary platyrrhine monkeys

and description of a new genus from the late Miocene of Colombia. Folia Primat. 12:1-37. (1970).

Hirschfeld, S. E., and Marshall, L. G.: Revised faunal list of the La Venta fauna, Friasian-Miocene, of Colombia, South America. Journ. Paleont. 50:433-436. (1976).

Kay, R. F.: The evolution of molar occlusion in the Cercopithecidae and early catarrhines. Amer. J. Phys. Anthrop. 46:327-352. (1977).

Marshall, L. G., Pascual, R., Curtis, G. H., and Drake, R. E.: South American geochronology: radiometric time scale for middle to late Tertiary mammal-bearing horizons in Patagonia. Science. 195:1325-1328. (1977).

Remane, A.: Zahne und Gebiss. In H. Hofer, A. H. Schultz, and D. Starck. Primatologia. 3:637-846. Karger, Basel. (1960).

Stehlin, H. G.: Die Saugetiere des schweizerischen Eocaens. Critischer Catalog der Materialien. Teil VII, Halfte II: _Caenopithecus_ - _Necrolemur_ - _Microchoerus_ - _Nannopithex_ - _Anchomomys_ - _Periconodon_ - _Amphichiromys_ - _Heterochiromys_ - Nachtrage zu _Adapis_ - Schlussbetrach - tungen zu den Primaten. Abh. schweiz. palaont. Ges. 41:1299-1552. (1916).

Stirton, R. A.: Ceboid monkeys from the Miocene of Colombia. Univ. Calif. Publ. Bull. Dept. Geol. Sci. 28:315-356. (1951).

19

Inbreeding in the Maternal and Paternal Lines of Four Captive Groups of Rhesus Monkeys (*Macaca mulatta*)

D. G. Smith

Paternity exclusion analysis was employed to identify the fathers of 129 offspring born in 4 captive multimale groups of rhesus monkeys and to test the hypothesis of random mating with respect to kinship. Calculations of the relative reproductive success of each of the sexually mature males in each group were used to estimate the expected number of inbred offspring under this hypothesis. With the possible exception of mother-son matings, no evidence of inbreeding avoidance was detected. Both the observed incidence of inbreeding and that expected, based upon relative reproductive success of the males, exceeded theoretical expectations based upon assumptions of completely random mating because variance in reproductive success among adult males is high. Under conditions of high variance in male reproductive success and random mating with respect to kinship, the deleterious consequences of inbreeding will accumulate rapidly, with substantial losses in colony productivity. Colony management strategies should be designed, therefore, to reduce the opportunity for inbreeding.

INTRODUCTION

Theory of population genetics predicts that there will be increases in homozygosity as the average level of inbreeding increases in a population. That level is traditionally measured by the average value of $\underline{F}$, Wright's classic coefficient of inbreeding, $\bar{F}$. Concomitantly, a loss of fitness is experienced due to overdominant loci, where heterozygotes enjoy a selective advantage over all homozygotes. The proportion of that population lost due to high mortality and morbidity is linearly correlated with $\bar{F}$ and further losses result from lower levels of fertility of inbred members of the population. Ralls and Ballou (1982), for example, found that inbred members of 15 of 16 different primate taxa exhibit higher rates of infant mortality than their noninbred cage mates.

Empirical studies of human populations suggest increases in prereproductive mortality of about 3% of all everborn offspring for every 10% of inbreeding (Schull and Neel, 1965). Litter size for several species of domestic animals has been shown to decline by about 3% for every 10% of inbreeding in the offspring and by about 2.5% for every 10% of inbreeding in their mothers (Dickerson, 1954; Falconer, 1960). Thus, the total loss in productivity of surviving offspring of same-generation parents is at least 8% for every 10% of inbreeding.

Although inbreeding involving maternal half-sibs has been reported in free-ranging groups (Loy, 1971), it has long been argued that close inbreeding within the maternal lines of macaque groups is usually avoided (Sade, 1968; Missakian, 1973). The low incidence of maternal inbreeding has usually been attributed to the facts that (1) kinsmen in the maternal line are easily recognized by mutual association with the mother who maintains long term ties with all her offspring and (2) natal, subadult males usually emigrate and join a new group prior to reproductive age. The degree of intergroup admixture provided by the emigration of all subadult males from their natal groups has usually been considered to be far in excess of that sufficient to prevent deleterious effects of inbreeding.

Some have argued that paternal kin, especially fathers, are also recognized. While some empirical evidence has been reported which is consistent with this hypothesis (Wu et al., 1980), this evidence remains controversial. It is doubtful that inbreeding can in fact be either consciously or indirectly avoided unless kin recognition does occur. Therefore, the avoidance of inbreeding in the paternal line would both obviate the necessity of emigration of all males to prevent deleterious effects of inbreeding and provide circumstantial evidence for paternal kin recognition.

In response to embargos recently imposed on the importation of rhesus macaques from India, efforts have intensified to establish domestic captive breeding colonies of macaques. The success of these colonies is crucial to the continuation of biomedical research efforts in this country. A popular and economical strategy has been to breed 30-40 adult females to a much smaller number of males in a large outdoor enclosure. In such captive groups, where emigration of males is restricted, or, more often, prevented altogether, even completely random mating leads to an increase in the value of $\overline{F}$ (Smith, 1980a, 1982b). In a completely random mating group with effective population size N_e, the average value of $\overline{F}$ at the tth generation of inbreeding will be $\overline{F}_t = 1-(1- \Delta F)^t$ where $\Delta F = 1/(2N_e)$,

$N_e=(4N_mN_f)/N_m+N_f$ and N_m and N_f are the numbers of breeding males and females, respectively, in that group. If mating is nonrandom due to variation in reproductive success, N_e is further reduced by the factor

$$\frac{1}{1 + \frac{V - \bar{k}}{\frac{k}{\bar{k}^2}}}$$

but accurate estimates of $\bar{k}$ and V_k, the average number of surviving offspring produced per adult animal and variance in that number, have not been made. If inbreeding avoidance is practiced, F_t will increase much more slowly.

For the reasons stated above, knowledge of the extent and pattern of inbreeding in groups of macaques is of both practical and theoretical importance. We have reported preliminary evidence elsewhere that close inbreeding in the paternal line is not in fact avoided in captive groups of rhesus monkeys (Smith, 1982a). In this paper we have updated that study with a larger sample, some members of which are potentially inbred in the maternal line, compared results to theoretical expectations and estimated expected losses to the colony resulting from inbreeding.

METHODS AND MATERIALS

Four captive groups of rhesus monkeys, housed since 1976 in outdoor, half-acre corrals at the California Primate Research Center (CPRC), comprised the groups studied here. Each group contained between 25 and 35 adult females and between 2 and 4 adult males in 1976. Since that time colony management practices have been followed which limit the number of reproducing females to between 35 and 40 and the number of adult and subadult males to fewer than 10. No more than 2 of the adult males in a given cage, however, have produced the majority of all offspring born during any given year (Smith, 1981). While variance in male reproductive success will lead to overestimates of effective population size and can only be estimated from long term studies, the preceding formula provides a maximum estimate for N_e in these four groups of about 25. Thus the theoretical value of ΔF is about 0.020.

Since no members of the parental generation in any corral were known to be related, $\bar{F}_1=0$ and all inbreeding which has occurred has resulted from parent-offspring, full-sib or half-sib matings. Offspring of such matings carry inbreeding coefficients of 0.25, 0.25 and 0.125, respectively. An offspring was regarded as potentially inbred if its grandfather or uncle was sexually mature and living in the corral during the breeding season preceding its birth. A male was regarded as sexually mature if it had reached the age of 4 during that breeding season. The number of offspring expected to be inbred based upon the null hypothesis of random mating with respect to kinship was computed by multiplying the relative reproductive success of each sexually mature male during that year by the number of offspring born during that year who are related to him, then summing over all sexually mature males.

Phenotypes for more than 30 electrophoretically and serologically defined polymorphisms were employed for paternity exclusion analysis to determine the fathers of all offspring born in the 4 corrals during years when potentially inbred matings occurred. These markers and the phenotypes they identify have been described elsewhere (Smith, 1980, 1981, 1982b; Sullivan et al., 1977) and have provided for the successful paternity identification of about 90% of all rhesus monkeys born in most of the outdoor, multimale corrals at the CPRC since 1976. When all but a single sexually mature male could not be excluded from paternity with respect to a given offspring, relative reproductive success of those males was calculated by assigning equal portions of that offspring to all of the remaining possible fathers.

The observed and expected numbers of offspring inbred and not inbred in the paternal lines were compared using the chi-square statistic. The resulting value of $\bar{F}$ for the first inbred generation was compared both with its expected value, based upon observed reproductive success of each male under the assumption of random mating with respect to kinship, and with its theoretical value of 0.02, based upon the population structure under the assumption of completely random mating.

RESULTS

A total of 129 offspring were born during the years in which inbreeding might have occurred (Table 1). During the single year of potential inbreeding for corral 1 (NC 1), a total of 12 offspring were produced by 3 different

TABLE 1. Number of possible fathers (n), offspring produced (N), and number of fathers excluded from paternity

Corral	1980					1981					1982				
NC	n	N	1	2	3	n	N	1	2	3	n	N	1	2	3
1	-	-	-	-	-	3	12	12	0	0	-	-	-	-	-
2	5	15	10	4	1	7	15	13	2	0	6	23	20	3	0
3	-	-	-	-	-	8	21	17	3	1	9	22	17	4	1
7	-	-	-	-	-	-	-	-	-	-	6	21	17	4	0

fathers. NC 2 contained 5 sexually mature males during 1980, 7 in 1981 and 6 in 1982 and they produced 15, 15 and 23 offspring, respectively, during these years. NC 3, which was not potentially inbred until 1981, contained 8 sexually mature males during 1981 and 9 in 1982 who produced 21 and 22 offspring,respectively. Finally, potentially inbred offspring first appeared in NC 7 during 1982 and were fathered by 6 different males who produced a total of 21 offspring that year.

While one subadult male in NC 1 reached the age of 4 in 1980 and was therefore judged to be sexually mature, he has never produced any offspring and is excluded from Table 1. Similarly, three sexually mature subadult males in NC 2 and seven in NC 3 produced none of the offspring born in 1981. Two of these subadult males in NC 2 and six of these seven in NC 3 were removed from their respective corrals following the 1981 birth season. Two subadult males in NC 2 and one in NC 7 who reached age 4 during 1981 fathered none of the offspring born in NC 2 during 1982 and are also excluded from Table 1. Finally, one adult male each in NC 1 and NC 3 and two in NC 2 produced no offspring during the years covered by this study and are excluded from Table 1. Although these males who produced no offspring during the years indicated might be viewed as possible mates who were avoided by certain animals, they were rarely, if ever, observed to engage in sexual behavior with any female (Stern and Smith, 1984). Moreover, the exclusion of these reproductively unsuccessful males cannot effect the estimation of the expected number of inbred matings since this number is a function of relative reproductive success.

Only three of the 15 offspring born during 1980 in NC 2 were potentially inbred. All three were potentially inbred in the paternal line and one of these three offspring was potentially inbred in both the paternal and maternal lines. In 1981, NC 1 produced three potential paternally inbred offspring and NC 2 and NC 3 produced five and six, respectively. One additional offspring born in NC 2 and three in NC 3 were potential issues of mother-son matings while one additional offspring born in NC 2 could have been fathered by his mother's maternal half-brother. In addition, two potentially (paternally) inbred stillbirths, one each in NC 3 and NC 7, occurred that year (a blood sample could be recovered in suitable condition for paternity exclusion analysis only from the NC 3 stillbirth).

Prior to the 1982 breeding season, NC 1 was disbanded. NC 2, NC 3 and NC 7 produced seven, eight and 11 offspring, respectively, during that year who were potentially inbred only in the paternal line. Potential maternally inbred matings during that breeding season included one potential mother-son mating each in NC 2 and NC 7 and two such matings in NC 3. Only two potentially inbred maternal half-sib matings occurred during 1982, both in NC 3. One of these two matings was also potentially inbred in the paternal line. Finally, one potential paternally inbred stillbirth each occurred in NC 2, NC 3 and NC 7 during 1982 but a suitable blood sample could be recovered only from those occurring in NC 3 and NC 7.

Thus, over the 3 year period, blood samples were drawn from 57 potentially inbred offspring; 10 who were potentially inbred in the maternal line, 45 who were potentially inbred in the paternal line and 2 who were potentially inbred in both the paternal and maternal lines. The proportion of all offspring born during this period who were potentially inbred in either the maternal or paternal line increased from 20% to about 40% between 1980 and 1981 and to 56% in 1982.

The fathers of 106 of the 129 offspring described above could be identified by paternity exclusion analysis. In 20 and 3 cases, respectively, the father was proven to be either one of two or one of three of the sexually mature males. In all cases where all but one of the males could not be excluded by paternity exclusion analysis, the remaining possible fathers were all either related or were unrelated to the offspring under consideration. This is, of course, due to the fact that relatives share more genes in common than do unrelated individuals. Thus, all of the total of 57 offspring who were potentially inbred in either the paternal or maternal line (excepting the 2 stillbirths from whom a suitable blood sample could not be drawn) could be unequivocally proven to be either inbred or not inbred.

Table 2 shows the relative reproductive success (A) and expected number of paternally inbred offspring (C) for each of the sexually mature males (those described in Table 1, and based upon the null hypothesis of random mating with respect to kinship) and the observed number of inbred matings (D). While a total of 18.01 of the 47 possible paternally inbred matings were expected by chance alone to be inbred, 15 were in fact discovered. Of these at least four (and perhaps as many as 6) represented matings between a father and his daughter (two inbred offspring born in NC 2 during 1982 were fathered either by 7820, their mother's father, or 17274, their mother's paternal half-brother).

Table 3 gives the observed and expected numbers of maternally inbred offspring. Of the total of 12 potential maternally inbred matings, 1.41 (i.e., 0.98 mother-son matings and 0.43 maternal half-sib matings) were expected to produce inbred offspring under the null hypothesis. One such mating (a maternal half-sib mating) actually occurred, which was, in fact, a full sib mating between the son and daughter of the alpha male and the alpha female in NC 2. While this might appear coincidental, exactly 0.29 of the 0.43 maternal half-sib matings were expected, by chance alone, to be, in fact, full sib matings, since most females who shared a mother with a sexually mature cage-born male also shared a father with that male.

Fewer cases of paternal inbreeding occurred than

TABLE 2. Relative reproductive success (A), number of potentially inbred offspring (B), random probability of inbred matings (AxB=C) and observed number of paternally inbred matings (D) for each sexually mature male by corral and year of offspring's birth

	1980				1981				1982				Total	
Corral 1														
Fathers	A	B[1]	C	D	A	B[1]	C	D	A	B[1]	C	D	C	D
7835	-	-	-	-	.167	2	.334	0	-	-	-	-	.334	0
16254	-	-	-	-	.167	0	.000	0	-	-	-	-	.000	0
16255	-	-	-	-	.667	1	.667	1	-	-	-	-	.667	1
Total	-	-	-	-	1.001	3	1.001	1	-	-	-	-	1.001	1
Corral 2														
7820	.189	3	.567	1	.266	5	1.330	1	.109	6	.654	1	2.551	3[2]
8173	.055	3	.165	0	-	-	-	-	-	-	-	-	0.165	0
8175	.300	3	.900	1	.167	5	.835	1	.087	6	.522	0	2.257	2
8241	.456	0	.000	0	.134	0	.000	0	.000	1	.000	0	0.000	0
16684	.000	0	.000	0	.134	0	.000	0	.043	0	.000	-	0.000	0
17178	-	-	-	-	.033	5	.165	0	-	-	-	-	0.165	0[2]
17274	-	-	-	-	.199	5	.995	1	.565	6	3.390	3	4.385	4[2]
17280	-	-	-	-	-	-	-	-	.196	1	.196	0	0.196	0
17298	-	-	-	-	.067	0	.000	0	-	-	-	-	0.000	0
Total	1.000	3	1.632	2	1.000	5	3.325	3	1.000	7	4.762	4	9.719	9
Corral 3														
4749	-	-	-	-	.100	5	.500	0	.182	8	1.456	1	1.956	1
7051	-	-	-	-	.075	1	.075	0	-	-	-	-	0.075	0
7059	-	-	-	-	.350	1	.350	0	.091	1	.091	0	0.441	0
16953	-	-	-	-	.075	1	.075	0	.136	1	.136	0	0.211	0
17150	-	-	-	-	-	-	-	-	.091	8	.728	0	0.728	0
17155	-	-	-	-	.100	5	.500	1	.046	8	.368	0	0.868	1
17200	-	-	-	-	.050	5	.250	1	.046	8	.368	0	0.618	1
17204	-	-	-	-	.150	1	.150	1	.363	1	.363	1	0.513	2
17235	-	-	-	-	.100	0	.000	0	.000	1	-	0	0.000	0
17638	-	-	-	-	-	-	-	-	.045	8	.360	0	0.360	0
Total	-	-	-	-	1.000	7	1.900	3	1.000	10	3.874	2	5.770	5
Corral 7														
11	-	-	-	-	-	-	-	-	.048	5	.240	0	.240	0
16888	-	-	-	-	-	-	-	-	.428	1	.428	0	.428	0
16898	-	-	-	-	-	-	-	-	.048	0	.000	0	.000	0
17179	-	-	-	-	-	-	-	-	.238	1	.238	0	.238	0
17466	-	-	-	-	-	-	-	-	.095	5	.475	0	.475	0
17502	-	-	-	-	-	-	-	-	.143	1	.143	0	.143	
Total	-	-	-	-	-	-	-	-	1.000	12	1.524	0	1.524	0
Total													18.01	15

1 The number given for "Total" in this column is not the sum of all entries in this column, because some of the fathers are related, but is rather the total number of offspring born that year in that corral who are related to any adult male who is their possible father.

2 The father of two of the 7 offspring produced by 7820 and his son, 17274, is not known. Here, and for estimating F, one of these offspring has been arbitrarily assigned to each father.

TABLE 3. Possible fathers and expected number of maternally inbred offspring based upon the null hypothesis of random mating

Corral Number	Year of Birth	Possible father who is related to offspring	Relation to offspring's mother	Probability of inbreeding[1]
2	1981	17274	half-sib[2]	0.199[3]
3	1982	17150	half-sib[2]	0.091
3	1982	16953	half-sib	0.136
2	1980	16673	son	0.000
2	1981	17298	son	0.067
2	1982	17280	son	0.196
3	1981	17200	son	0.050
3	1981	17155	son	0.100
3	1981	17204	son	0.150
3	1982	17200	son	0.046
3	1982	16953	son	0.136
7	1982	17179	son	0.238
Total				1.409

[1] The sum of these probabilities is the expected number of maternally inbred offspring under the assumption of random mating with respect to kinship.

[2] Maternal half-sibs also share the same father.

[3] Maternally inbred matings.

were expected (based upon random mating with respect to kinship) but this difference was not statistically significant (chi-square = 0.81, $p > 0.05$). The small number of maternally inbred offspring which were expected by chance alone did not permit statistical testing.

DISCUSSION

No evidence of avoidance of inbreeding could be clearly identified in these four groups of rhesus monkeys. Although the random probability of a mother-son mating was about twice that of maternal half-sib mating, the single case of maternal inbreeding was a sib, rather than a mother-son, mating. Thus, with the possible exception of avoidance of mother-son matings, the parents of the 57 potentially inbred offspring in this study mated randomly with respect to kinship. The observed and expected values of $\overline{F}$ for members of the first generation of potential paternally inbred offspring studied here were about 0.053 and 0.066, respectively. Maternal inbreeding contributes little extra inbreeding to these values. Since almost all $\overline{F}_2$ offspring will potentially inbred in the paternal line within a few years time, 0.06 is probably a good estimate of $\overline{F}_1$.

Under the ideal conditions of completely random mating, the theoretical value of $\overline{F}$ for any generation of potentially inbred offspring is determined by ΔF, which in these 4 groups is about 0.020. Thus, the level of inbreeding in these 4 groups is expected to, and in fact does, considerably exceed the theoretical expectations. This is due both to generational overlap in mating and variation among males in their reproductive success.

Since $\overline{F}_1 = \Delta F$, then 0.06, should provide an empirical estimate of ΔF which does not require the unrealistic assumption of low variance in reproductive success of adult males. Since $\Delta F = 1/(2N_e)$, the true effective population size of each of the 4 groups, say N_e', is probably no greater than 10 ($N_e' = 1/(2x0.06)$ or between one-fourth and one-fifth the size of the breeding population, N, as a result of variation in reproductive success of (primarily) adult males. This finding has important practical and theoretical implications.

Since the first inbred offspring was born in 1980, the first members of the second potentially inbred, or F_3, generation (i.e., those with F_2 generation parents) could be born during the spring of 1984 and their average inbreeding coefficient, assuming $\Delta F = 0.06$, would be almost 0.12. Approximately 7 or 8% of this F_3 generation can be expected to be lost due to excess mortality and their numbers further will be reduced by 1 or 2% due to reduced fertility of inbred F_2 generation parents. Higher proportions will be lost from future generations as $\overline{F}_t$ increases. Well before the year 2000 two randomly chosen F_6 parents would be more closely related than are half-sibs and their numbers will have been reduced by more than 20% by inbreeding depression. It is, therefore, paramount that colony management practices be employed which prevent the cumulation of inbreeding in the paternal lines of captive groups (Smith, 1982b).

The reproductive success of the older males in all four groups declined dramatically after 1980 as their sons reached sexual maturity. Due to the high variance in male reproductive success, the highest values of $\overline{F}$ will occur in groups in which young males succeed their fathers as the most successful breeders in the group. For example, of the 15 females who have produced potential paternally inbred offspring in NC 2, 14 are daughters of male 7820. His son, 17274, who produced more than half of all offspring born in NC 2 during 1982, is therefore a paternal half-brother to almost all F_1 females. Therefore, the actual rate at which inbreeding is expected to accumulate in a given group will depend upon the history of population dynamics unique to that group and management strategies should be formulated accordingly.

Previous estimates of N_e for numerous species have shown that N_e is approximately 0.75 N, where N is the number of breeding adults in the population (Cavalli-Sforza and Bodmer, 1971). Values of N_e estimated for free-ranging groups of rhesus have been in close agreement with 0.75 N (Olivier et al., 1974) but have not taken variance in reproductive success into account. It is reasonable to assume that the value of $\bar{k}$ for free-ranging groups cannot have varied appreciably from the value 2.0 for any length of time without initiating severe overpopulation or complete extinction of the group. The value of

$$\frac{1}{1 + \frac{V_k - \bar{k}}{\bar{k}^2}}$$

for such a group can be estimated by equating it with

$$N_e' \quad \frac{4N_m N_f}{N_m + N_f}$$ or approximately 10/25 which is 0.40.

In such cases, the value of V_k is 8.0. If, as in human populations, family size for mothers, whose mean value is k, follows a Poissan distribution when $\bar{k} = 2.0$, V_k^f for females alone, $V_k^f/\bar{k}^f$, is also 2.0. $V_k = mV_k^m + (1-m) V_k^f + m(1-m)(\bar{k}^m - \bar{k}^f)^2$ where m is the proportion of adult animals who are male, $\bar{k}^m$ and $\bar{k}^f$ are the values of $\bar{k}$ for males alone and females alone and V_k^m and V_k^f are the variances in k^m and k^f (Crow and Kimura, 1970). If, free-ranging groups, $N_m = N_f$, the $\bar{k}^m = \bar{k}^f$ and $V_k^m = 14.0$. Thus, for a group in which $N = 40$, $N_e = 16$ and $\Delta F = 0.03$. This value of ΔF is two-and-one-half fold that estimated when variation in reproductive success of males is not considered. If, as seems more likely, $N_m < N_f$, then V_k^m and ΔF will be even higher.

By ignoring the high variance in reproductive success, the true value of N_e and the rate of genetically effective migration, whose estimate is based upon N_e, may have been considerably overestimated and underestimated, respectively, for free-ranging groups. Under such conditions of population structure and mating behavior, rates of male emigration which have previously been considered to be quite in excess of those sufficient to minimize deleterious effects of inbreeding might be expected. Specifically, the emigration of all members of one sex might be necessary to prevent eventual extinction of many free-ranging groups which undoubtedly experience frequent population bottleneck effects.

ACKNOWLEDGMENTS

This research was supported by U.S.P.H.S. grants RR00169 and N01-HD-0-2828. The author is indebted to the skilled laboratory assistance of Ms. Becky Rolfs and to the entire colony management staff of the California Primate Research Center for providing animal records and cooperating in the collection of blood samples and other data essential to the success of this study.

REFERENCES

Cavalli-Sforza, L. L., and Bodmer, W. F.: Genetics of Human Populations. W. H. Freeman, San Francisco. (1971).

Crow, J. F., and Kimura, M.: An Introduction to Population Genetics Theory. Harper and Row, New York. (1970).

Dickerson, G. E.: Evaluation of selection in developing inbred lines of swine. Res. Bull. Mo. Agricul. Exp. Sta. No. 551. (1954).

Falconer, D. S.: Introduction to Quantitative Genetics. Ronald Press, New York. (1960).

Loy, J. D.: Estrous behavior of free-ranging rhesus monkeys. Primates. 12:1-13. (1971).

Missakian, E. A.: Genealogical mating activity in free-ranging groups of rhesus monkeys (Macaca mulatta) on Cayo Santiago. Behaviour. 45:224-241. (1973).

Olivier, T. J., Buettner-Janusch, J., and Buettner-Janusch, V.: Carbonic anydrase isoenzymes in nine troops of Kenya baboons, Papio cynocephalus (Linnaeus 1766). Am. J. Phys. Anthrop. 41:175-190. (1974).

Ralls, K., and Ballou, J.: Effects of inbreeding on infant mortality in captive primates. Int. J. Primatol. 3:491-505. (1982).

Sade, D. S.: Inhibition of son-mother mating among free-ranging rhesus monkeys. Scie. Psychoanal. 12:18-38. (1968).

Schull, W. J., and Neel, J.: The Effects of Inbreeding on Japanese Children. Harper and Row, New York. (1965).

Smith, D. G.: Potential for cumulative inbreeding and its effects upon survival in captive groups of nonhuman primates. Primates. 21:430-436. (1980a).

Smith, D. G.: Paternity exclusion in six captive groups of rhesus monkeys (Macaca mulatta). Am. J. Phys. Anthrop. 53:243-249. (1980b).

Smith, D. G.: The association between rank and reproductive success of male rhesus monkeys. Am. J. Primatol. 1:83-90. (1981).

Smith, D. G.: Inbreeding in three captive groups of rhesus monkeys. Am. J. Phys. Anthrop. 58:447-451. (1982a).

Smith, D. G.: Use of genetic markers in the colony management of non-human primates: A review. Lab. Anim. Sci. 32:540-546. (1982b).

Stern, B., and Smith, D. G.: Sexual activity and paternity in three captive groups of rhesus monkeys (*Macaca mulatta*). Anim. Behav. 32(1):1-15. (1984).

Sullivan, P. T., Blystad, C., and Stone, W. H.: Immunogenetic studies on the rhesus monkeys (*Macaca mulatta*): XI Use of blood groups in problems of parentage. Lab. Anim. Sci. 27:348-351. (1977).

Wu, H. M. H., Holmes, W. G., Medina, S. R., and Sackett, G. P.: Kin preference in infant *Macaca nemestrina*. Nature. 285:225-227. (1980).

20

Evolutionary Cytogenetics of the Siabon (Gibbon-Siamang) Hybrid Apes

D. A. Shafer

INTRODUCTION

The first successful matings of siamang and gibbon produced 2 female offspring born August 11, 1975 and August 20, 1976. These hybrids, designated "siabon" by Rumbaugh et al. (1976) occurred in unplanned matings in the Atlanta Zoo between a male gray gibbon (Hylobates muelleri abotti) and a female siamang (Symphalangus syndactylus). Both offspring were healthy and morphologically normal, and they were intermediate in gibbon and siamang physical characteristics. The second hybrid died from an infection following a compound fracture of the fibula of unknown cause. At 3 months of age, the first siabon was moved to the Georgia State University Primate Behavior Laboratory and in 1981 she was moved to the Yerkes Primate Research Center and remains in good health. (Figure 1).

The history, morphological development, vocal behavior, and initial cytogenetic studies of the siabons have been reported previously (Rumbaugh et al., 1976; Shafer and Myers, 1977; Wolkin, 1977; Myers and Shafer, 1979; Wolkin and Myers, 1980; Shafer et al., 1982). In general, the hybrids exhibit characteristics of both species. However, siamang features are somewhat predominate although there is no throat sac or alter-developing ischial callosities. The surviving siabon now has the black coat, white beard, and syndactyly of the second and third toes typical of siamangs, but it has the larger eyes, greater hair density, and shorter forelimbs more characteristic of gibbons. Vocalization patterns from sonogram studies indicate a developmental shift from the short gibbon-like "hoop" calls to the longer and more elaborate great call patterns characteristic of siamangs (despite the lack of an expandable throat sac). All parameters of this

Figure 1. Siabon hybrid at 3 years, first progeny.

developmental history suggest that these siabons are as physically normal as other common hybrids. Since gibbon and siamang have chromosomes of dissimilar number and morphology (as in the horse-donkey cross), a cytogenetic study of siabon hybrids might yield important information about the genetic mechanisms of species divergence and the phylogenetic evolution of these nonhuman primates.

METHODS AND RESULTS

Cytogenetics

Chromosome banding studies of the hybrid and its gibbon and siamang parents demonstrated cytogenetic differences of a greater magnitude than expected based on previous evidence. Prior to the development of chromosome banding techniques, cytogenetic investigations had concluded that all lesser apes studied had 44 chromosomes with similar morphology except that the siamang had 50 chromosomes and the concolor had 52 chromosomes (Chiarelli, 1971). Based on the initial H. lar chromosome banding descriptions of Warburton et al. (1975) and Tantravahi et al. (1975), this predominant 44 chromosome pattern will be referred to as the lar gibbon pattern. We found that the gray gibbon father of the siabon also had 44 chromosomes and the chromosome morphology and banding patterns typical of the lar. Most of the chromosomes of the gibbon parent were submetacentric or metacentric. There were 6 pairs of acrocentric chromosomes and a unique pair of submetacentric marker chromosomes identified by a negative staining nucleolus organizer region. While all 44 chromosome lar gibbon forms appear to share this chromosome morphology, comparison of the banding patterns of H. lar and H. moloch revealed a possible banding polymorphism since one of the number 7 chromosomes of the single moloch studied had a clear banding inversion. (Normal polymorphic banding inversions occur in the orangutan; Seuanez et al., 1976.) Despite this potential variation, present evidence suggests that the 44 chromosome lar gibbon forms may all share essentially identical banding patterns in the same manner as is evident among the macaques, the baboons, and domestic dogs.

In contrast, the siamang mother of the siabon, with 50 chromosomes, exhibited the typical siamang chromosomme morphology -- all of them were submetacentric or metacentric except for the telocentric pair of marker chromosomes which showed a nucleolus organizer region. Our previous chromosome banding analysis of the first siabon hybrid and its gibbon and siamang parents (Shafer and Myers, 1976; Myers and Shafer, 1979) demonstrated that the difference in chromosome number and morphology between these species were greatly accentuated by additional major differences in G-banding patterns. Essentially, there were no identical patterns between the gibbon and siamang chromosomes. This was true of the parental chromosomes and of the corresponding chromosomes in the siabon hybrid. (The siabons exhibited 47 chromosomes per cell with 22 of them matching the 22 pairs of the gibbon parent and 25 matching the 25 pairs of the siamang parent.) Based upon G-banding, only 2 pairs of chromosomes were comparable. There was overall similarity but not identity between the patterns of gibbon chromosome pair 5 and siamang chromosome pair 5, and the marker chromosomes of the 2 species seem to represent the same chromosome but with an obvious internal rearrangement of one segment. Otherwise, homologous chromosomes could not be demonstrated between these species.

These observations contrast with both: 1) the near identity between the banding patterns of many morphologically distinct species and genera of old world monkeys (Cambefort, et al., 1976a, 1976b; Brumback, 1975; deVries et al., 1975) and 2) the close similarity in banding patterns between the great apes and with humans (Turleau, et al., 1972; Warburton et al., 1973; Lin et al., 1973; Miller, 1977; Seuanez, 1979). Comparisons of human, gorilla, chimpanzee, and orangutan chromosomes reveal a limited number of distinct and traceable G-band and R-band rearrangements (deGrouchy et al., 1973; Turleau et al., 1975; Dutrillaux et al., 1975a, 1975b; Paris Conference Supplement Report, 1975; Seuanez et al., 1976). Even comparisons between genera with very different karyotypes, such as the African green monkey with 60 chromosomes and the rhesus macaque with 42 chromosomes, show underlying G-band identity obscured by numerous rearrangements (Stock and Hsu, 1973). The cytogenetic differences between these genera apparently did not result from the loss or gain of euchromatic G-band regions. Indeed, with R-banding (largely the inverse of G-band patterns) a pattern continuity has been traced from the lemurs, through the old world monkeys, to the great apes and humans (Dutrillaux, 1979; Dutrillaux and Rumpler, 1980). In this broad picture of cytogenetic conservativism, the contrast between the gibbon and siamang G-band patterns seems paradoxical - particularly since viable siabon hybrids could be produced between these species. It is possible that further banding studies now underway using extended prometaphase chromosomes could reveal specific regions of banding similarity between the lar gibbon and siamang patterns. This level of analysis was not, however, required for the various primate cytogenetic comparisons discussed above. Clearly then, there is a significant difference between the gibbon and siamang chromosomes which, at the least, represents a series of multiple chromosome rearrangements in one or both species.

With C-banding comparisons, a more distinct contrast between the gibbon and siamang chromosomes was found (Figure 2). In most mammalian species, C-banding stains exclusively a singular chromosome region at or adjacent to the centromere. Most C-band variation is based on relative differences in the size of such regions or their presence or absence. The gibbon parent chromosomes and the gibbon chromosomes of the siabon hybrid showed the typical mammalian C-band patterning with all of the 44 gibbon chromosomes showing a single centromeric C-band. On the other hand, the siamang parent chromosomes and the siamang chromosomes of the siabon hybrid produced a

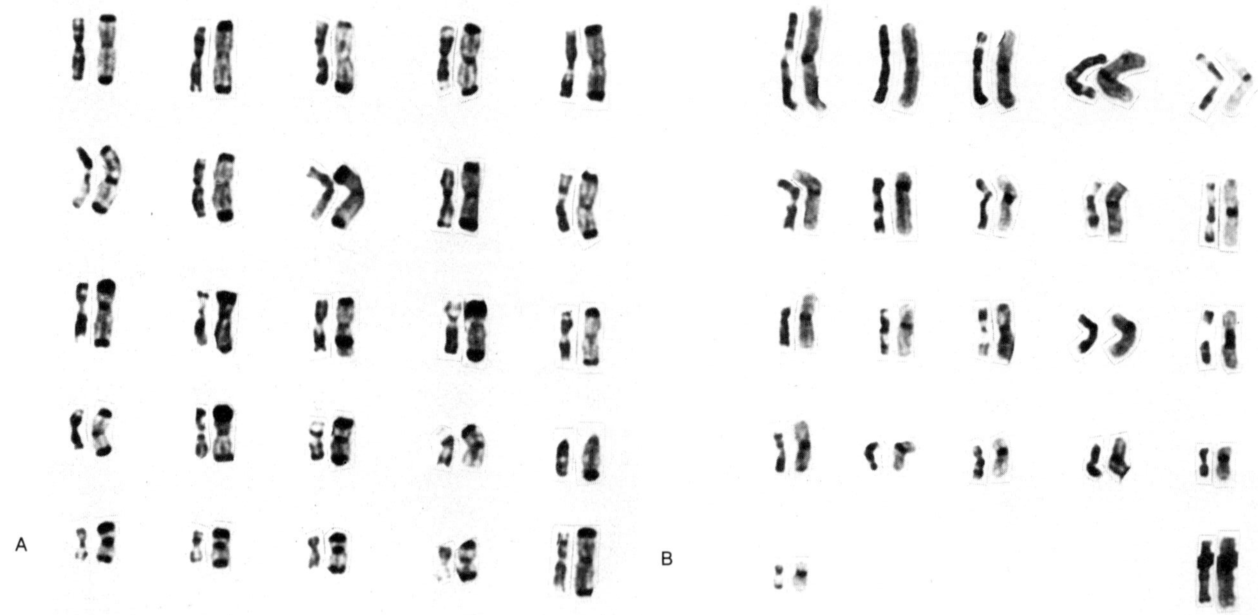

Figure 2. Sequential G-banding and C-banding of siabon hybrid chromosomes from a single metaphase cell. The chromosomes are first G-banded, photographed and destained, and then C-banded and rephotographed. They have been divided into the siamang-type chromosomes (a) and the gibbon-type chromosomes (b) and arranged according to prior karyotypes of these species with the X chromosomes in the lower right position. The siabon's siamang chromosomes all have one or two terminal C-bands, while in the siabon's gibbon chromosomes, the C-bands are singular and located adjacent to the centromere.

unique pattern of terminal C-bands usually consisting of very prominent dark bands at one or both ends of the chromosomes. In the 50 siamang chromosomes, 3 pairs including the marker chromosome pair 20 have a single terminal C-band while the rest have terminal C-bands at both ends. In addition, siamang chromosome pair 13 and the comparable siabon chromosome has a unique large C-band in the middle of the long arm while the other arm has a terminal C-band. This exceptional C-band location is presumed to reflect a recent rearrangement and is identifiable by G-banding as a large dark band in chromosome pair 13. Aside from the general pattern of prominent terminal C-bands in the siamang chromosomes or the chromosomes of siamang origin of the siabon, some C-band preparations also showed smaller centromeric C-bands in 6 pairs of chromosomes. This finding supports the hypothesis that the terminal C-bands do not represent a rearrangement and subsequent enlargement of former centromeric C-bands. Rather, it suggests that the siamang-type terminal C-bands were formed _de novo_ with some functional relation to their terminal location. Biochemical studies of gibbon and siamang DNAs further support this hypothesis with the finding that siamang DNAs contain a unique repetitive DNA fraction of relatively recent origin (Donehower and Gillespie, 1979). While the function of C-bands is not yet known, it is established that these chromosome regions are rich in satellite DNAs, that is, highly repetitive, simple sequence DNA (Arrighi et al., 1971). Therefore, the siamang terminal C-bands may well have formed as a part of an interrelated karyotypic alteration simultaneously affecting all the chromosomes in some brief series of cytogenetic events.

The study of the nucleolus organizer regions (NOR) provides evidence for further cytogenetic distinction. Techniques for localizing NORs and evaluating their function reveal that these regions vary in number and location in different species. In the great apes and humans there are 4 to 6 pairs of chromosomes containing a NOR region, while in the old world monkeys only one pair of chromosomes has a NOR region (Chiarelli, 1971; Tantravahi et al., 1976). These regions are composed largely of repetitive copies of the DNA sequences for encoding ribosomal RNAs. The silver staining NOR procedure of Howell et al. (1975) positively stains these regions when they are transcriptionally active (Howell, 1977). _In situ_ DNA hybridization techniques demonstrate that the 18s and 28s ribosomal DNA is, in fact, localized in the gibbon and siamang negative staining marker region of both the siamang pair 20 and the gibbon pair 15 marker chromosomes as well as the same respective regions in the siabon hybrid marker chromosomes (Shafer et al., 1980; Henderson et al., 1976). An intriguing finding was that NOR silver staining showed that these same respective regions stained positively in either the gibbon or siamang parent chromosomes, but in the siabon hybrid, only the gibbon-type marker chromosome showed positive NOR staining while the siamang-type marker chromosome did not silver stain (Shafer and Howell, 1978). The implication of this finding is that the siamang ribosomal genes are specifically suppressed while the gibbon ribosomal genes are active (Figure 3). This situation is also comparable to that which occurs in cellular hybrids. For example, when human and mouse cells are fused, the human ribosomal genes are specifically suppressed both in actual gene function and in NOR silver staining (Miller et al., 1976). While the physiological and evolutionary significance of this difference between the gibbon and siamang is unclear, it is apparent that some functional genetic difference must at least occur in this particular specialized gene region.

DISCUSSION

Evolutionary implications

The viability of the siabon hybrids gives empirical evidence that the structural genes of lar gibbons and siamang have remained nearly identical during some 5 to 15 million years since these species diverged. The chemical building blocks and developmental sequences provided by both the gibbon and siamang parental DNA must therefore be considered essentially equivalent. This conclusion is supported by direct biochemical evidence that lar gibbons and siamang have retained nearly identical sequences of amino acids in various structural proteins (Goodman, 1975).

The lack of significant evolutionary divergence at the level of specific genes challenges the traditional neo-Darwinian viewpoint of the evolutionary process particularly since such distinct differences are now evident at the chromosomal level. Rather than individual genes gradually evolving over millions of years in response to the adaptive pressures of natural selection, it seems that a brief series of dramatic cytogenetic alterations may cause sharp speciation without any significant evolution of genes. This emerging macroevolutionary hypothesis would support the proposal that the species divergence of lar gibbon and siamang resulted not from successive gene mutations and subsequent natural selection, but rather, from the rearrangement of genes and the alteration of their expression produced by such chromosome reorganization (Shafer, 1979; Shafer and Myers, 1976; Bush et al., 1977; Wilson et al., 1975). A similar process of cytogenetic speciation may also be evident in the case of H. concolor. The concolor, unlike most of the other gibbons, has 52 chromosomes with unique banding patterns (Dutrillaux et al., 1975) - 40% being similar to the siamang - as well as large terminal C-bands like the

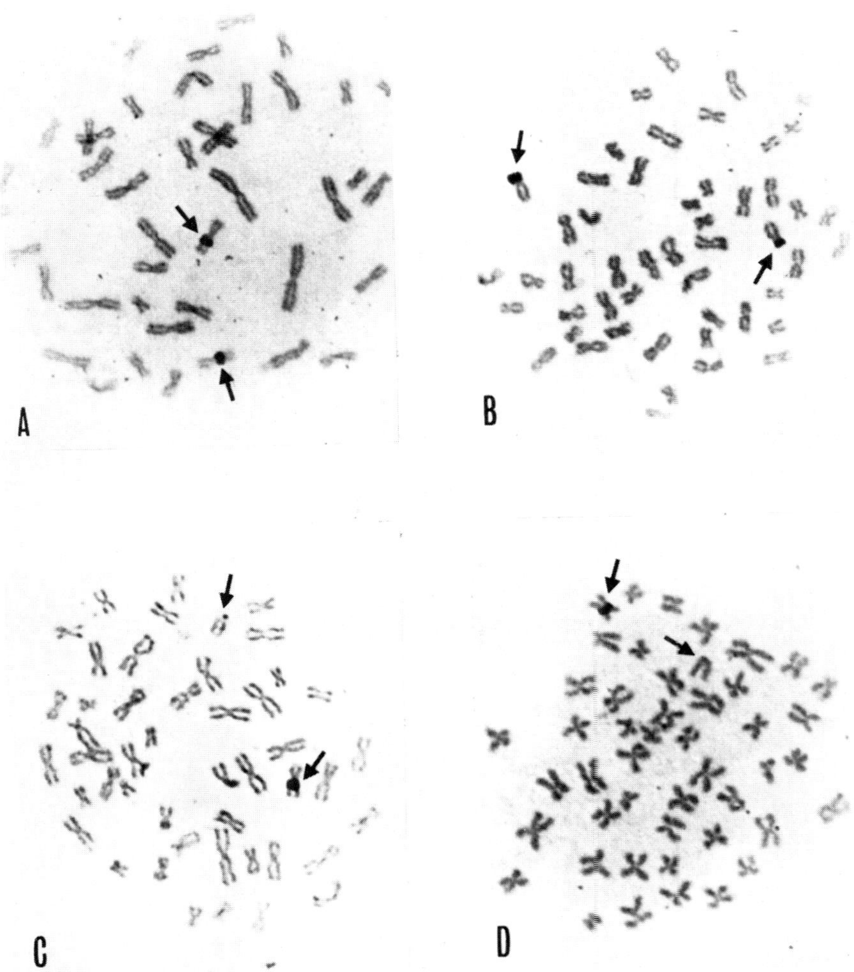

Figure 3. Silver nucleolus organizer (NOR) staining of gibbon parent (A), siamang parent (B), and siabon hybrid (C and D) chromosomes. Both gibbon marker 15 chromosomes (A, arrows) show dark silver NOR stain at the central, constricted, ribosomal gene region while the same NOR stain effect occurs in the siamang marker 20 chromosomes (B, arrows) at the terminal, constricted, ribosomal gene region. In the siabon, only the gibbon-type marker chromosome shows the silver NOR staining indicative of ribosomal gene activity (C and D, arrows).

siamang (vanTuinen and Ledbetter, 1982). In light of these cytogenetic differences, it is likely that siamang and concolor are derived from a common ancestor which became cytogenetically rather than geographically isolated from the other lesser apes. At the same time, they would have become fully emergent as separate species on morphological and behavioral grounds as a direct product of their chromosomal transformation.

The evidence suggests, furthermore, that this transformation could take place as a consequence of the rearrangement of structural gene regions (the largely euchromatic regions identified by G-, R-, or Q-banding) as well as by the striking C-band modifications. While it is uncertain that the formation of the larger and more numerous terminal C-bands in the siamang and concolor preceded significant G-band rearrangements, these unusual C-bands may nonetheless be important to their distant evolutionary status among lesser apes. Studies of C-band variation betwen closely related species show that the C-band, heterochromatic, repetitive DNA regions will decrease recombination in adjacent euchromatic, structural gene regions (Miklos and Nankivell, 1976). Therefore, these species may have become inhibited in further small scale species divergence by their large, terminal C-bands. (Indeed, it may be argued that with such large terminal C-bands, any further evolution may have largely required major rearrangements.) Recently it has also been determined that among the gibbons, _H. hoolock_ is cytogenetically distinct as it has only 38 chromosomes (Prouty et al., 1982). While the banding comparisons are incomplete, present indications suggest that this species must, at the least, be distinguished by a number of banding rearrangements.

In summary, the cytogenetic evidence suggests that the lesser apes are not simply a multibranched, long-term evolutionary radiation of ancestral species responding to the adaptive pressures of variant ecological niches. Rather, it appears that there were a few significant cytogenetic events - an initial major C-band alteration in the common ancestor of siamang and concolor and subsequent major G-band rearrangements occurring separately within this line and in other lesser apes such as the ancestors of _H. hoolock_ (Figure 4). These chromosomal processes could radically transform these lines phylogenetically into separate species as siamang, concolor, and hoolock without requiring bit by bit accumulation of adaptively selected gene mutations. On the other hand, gradual natural selection probably operated in establishing the present differences in coloration, vocalization, and behavior evident in the gibbons with 44 chromosomes. Finally, the evidence discussed here suggests ancestral cytogenetic formation of four primary species or subgenera of lesser ape - the siamang, the concolor, the hoolock, and the _lar_ gibbons. While further banding studies of the _lar_ gibbon subgroups may clarify their evolutionary status, from a genetic

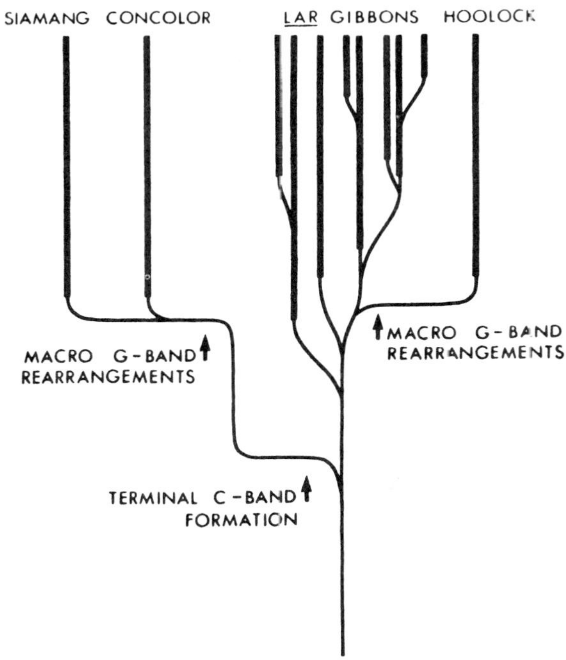

Figure 4. Hypothetical phylogenetic tree of the lesser apes based upon cytogenetic criteria. Siamang and concolor are presumed to have separated from the gibbon line by the formation of terminal C-band material in their common ancestor and then to have separated from each other, subsequently, by considerable G-band rearrangement. Since the C-band status of the hoolock is unknown, its cytogenetic relation to the other lesser apes is uncertain. Its cytogenetic isolation by major G-band rearrangement from the gibbon line may have been recent, ancient, or concurrent with the divergence of siamang and concolor. The remaining lar gibbon forms are presumed to have minor genetic and/or cytogenetic distinctions. However, banding studies have yet to be made on most of these species or subspecies.

standpoint it is uncertain whether the morpho-behavioral distinctions between these forms are sufficient to reflect species, subspecies, or racial differences.

REFERENCES

Arrighi, F. E., Saunders, P. P., Saunders, G. F., and Hsu, T. C.: Distribution of repetitious DNA in human chromosomes. Experientia 27:964-966. (1971).

Baldwin, L. A., and Teleki, G.: Patterns of gibbon

behavior on Hall's Island, Bermuda: a preliminary ethogram for Hylobates lar. In D. M. Rumbaugh. The Gibbon and Siamang. Vol. 4., Basel, Karger. (1976).

Brumback, R. A.: Fluorescent-banded karyotype of a hybrid macaque. J. Human Evol. 4:383-384. (1975).

Bush, G. L., Case, S. M., Wilson, A. C., and Patton, J. L.: Rapid speciation and chromosome evolution in mammals. Proc. Natl. Acad. Sci. USA 74:3942-3946. (1977).

Cambefort, Y., Mounie, Ch., and Colombies, P.: Topograhies des bandes chromosomiques chez Papio papio. Annales de Genetique. 19:5-9. (1976a).

Cambefort, Y., Mounie, Ch., and Colombies, P.: Topographies des bandes chromosomiques chez Papio hamadryas. Annales de Genetique. 19:269-272. (1976b).

Carpenter, C. R.: A field study in Siam of the behavior and social relations of the gibbon (Hylobates lar). Comp. Psychol. Monogr. 16:1-212. (1940).

Chiarelli, B. A.: Comparative cytogenetics in primates and its relevance for human cytogenetics, In B. A. Chiarelli. Comparative Genetics in Monkeys, Apes and Man. Academic Press, New York. (1971).

Chivers, D. J.: The siamang and the gibbon in the Malay peninsula. In D. M. Rumbaugh. The Gibbon and Siamang. Vol. 1. pp. 103-135. Basel, Karger. (1972).

Donehower, L., and Gillespie, D.: Restriction site periodicities in highly repetitive DNA of primates. J. Mol. Biol. 134:805-834. (1979).

Dutrillaux, B.: Very large analogy of chromosomes banding between Cebus capucinus (Platyrrhini) and man. Cytogenet. Cell Genet. 24:84-94. (1979).

Dutrillaux, B., Rethore, M. O., Aurias, A., and Goustard, M.: Analyse due caryotype de deux expeces de gibbons (Hylobates lar et H. concolor) par differentes techniques de marguage. Cytogenet. Cell Genet. 15:81-91. (1975).

Dutrillaux, B., Rethore, M. O., and Lejeune, J.: Analyse du caryotype de Pan paniscus. Comparison avec les autres pongidae et l'homme. Humangenetik. 28:113-119. (1975).

Dutrillaux, B., Rethore, M. O., and Lejeune, J. A.: Comparison du caryotype de l'orang-outang (Pongo pygmaeus) a celui de l'homme, du chimpanze et du gorille. Ann. Genet. 18:153-161. (1975).

Dutrillaux, B., and Rumpler, Y.: Chromosome banding analogies between a prosimian (Microcebus murinus), a platyrrhini (Cebus capucinus) and man. Amer. J. Phys. Anthropol. 52:133-137. (1980).

Ellefson, J. O.: A natural history of gibbons in the Malay penisula. Diss. Abstr. 28B:1322-1323. (1967).

Ellefson, J. O.: A natural history of white-handed gibbons in the Malayan peninsula, In D. M. Rumbaugh. The Gibbon and Siamang. Vol. 3. pp. 1-136. Basel, Karger. (1974).

Goodman, M.: Protein sequences and immunological

specificity: their role in phylogenetic studies of primates. In W. O. Luckett and F. S. Szalay. Phylogeny of the Primates. pp. 227-248. Plenum Press, New York. (1975).

deGrouchy, J., Turleau, C., Roubin, M., and Chavin-Colin, F.: Chromosomal evolution of man and the primates (*Pan troglodytes*, *Gorilla gorilla*, *Pongo pygmaeus*). In T. Caspersson and L. Zech. Chromosome Identification. pp. 124-131. Academic Press, New York. (1973).

Harlow, H. F., and Novak, M. A.: Psychopathological Perspectives. Perspect. Biol. Med. 16:461-478. (1973).

Henderson, A. S., Atwood, K. C., and Warburton, D.: Chromosomal distribution of DNA in *Pan paniscus*, *Gorilla gorilla beringei*, and *Symphalangus syndactylus*: comparison to related primates. Chromosoma. 56:29-32. (1976).

Howell, W. M., Denton, T. E., and Diamond, J. R.: Differential staining of the satellite regions of human acrocentric chromosomes. Experientia. 31:260-262. (1975).

Howell, W. M.: Visualization of ribosomal gene activity: silver stains proteins associated with rRNA transcribed from oocyte chromosomes. Chromosoma. 62:361-367. (1977).

Lin, C. C., Chiarelli, A. B., deBoer, L. E. M., and Cohen, M. M.: A comparison of the fluorescent karyotypes of the chimpanzee (*Pan troglodytes*) and man. J. Human Evol. 2:311-321. (1973).

Marshall, J. T., and Marshall, E.: Letter on Hylobatid taxonomy and distribution. Mamm. Chromosome Newsl. 16:30-33. (1975).

Marshall, J. T., and Marshall, D. R.: Gibbons and their territorial songs. Science. 193:235-237. (1976).

Miklos, G. L. G., and Nankivell, R. N.: Telomeric satellite DNA functions in regulating recombination. Chromosoma. 56:143-167. (1976).

Miller, D. A.: Evolution of primate chromosomes. Science. 198:1116-1124. (1977).

Miller, D. A., Dev, V. G., Tantravahi, R., and Miller, O. J.: Suppression of human nucleolus organizer activity in mouse-human somatic hybrid cells. Exp. Cell Res. 101:235-243. (1976).

Miller, O. J., Miller, D. A., Dev., V. G., Tantravahi, R., and Croce, C. M.: Expression of human and suppression of mouse nucleolus organizer activity in mouse-human somatic cell hybrids. Proc. Natl. Acad. Sci. USA 73:4531-4535. (1976).

Myers, R. H., and Shafer, D. A.: Hybrid ape offspring of a mating of gibbon and siamang. Science. 205:308-310. (1979).

Paris Conference Supplement Report: Standardization in human cytogenetics. Birth Defects: Original Article Series. Vol. 11, No. 9. The National Foundation, New York. (1975).

Prouty, L., Buchanan, P., Pollitzer, W. S., and Mootnick, A.: A new hylobatid chromosomal subgenus. The IXth Congress of the International Primatological Society. Atlanta, Georgia. (1982).

Rumbaugh, D. M., Wolkin, J. R., Wilkerson, B. J., and Myers, R. H.: A hybrid ape (Hylobates lar moloch X Symphalangus syndactylus). Lab Primate Newsl. 15:32. (1976).

Schultz, A. H.: The skeleton of the Hylobatidae and other observations of their morphology. In D. M. Rumbaugh. The Gibbon and Siamang. Vol. 2. pp. 1-54. Basel, Karger. (1973).

Seuanez, H., Fletcher, J., Evans, H. J., and Martin, D. E.: A chromosome rearrangement in orangutan studied with Q-, C- and G-banding techniques. Cytogenet. Cell Genet. 17:26-34. (1976).

Seuanez, H.: The Phylogeny of Human Chromosomes. Springer-Verlag, New York. (1979).

Shafer, D. A., and Howell, W. M.: unpublished findings. (1977).

Shafer, D. A., and Myers, R. H.: Cytogenetic studies of an intergeneric hybrid gibbon-siamang, Mamm. Chromosome Newsl. 18:48. (1977).

Shafer, D. A., Warburton, D., and Henderson, A. S.: unpublished findings. (1980).

Stock, A. D., and Hsu, T. C.: Evolutionary conservatism in the arrangement of genetic material. A comparative analysis of chromosome banding between the rhesus macaque and the African green monkey. Chromosoma. 43:211-224. (1973).

Tantravahi, R., Dev., V. G., Firschein, I. L., Miller, D. A., and Miller, O. J.: Karyotype of the gibbons Hylobates lar and H. moloch. Inversion in chromosome 7. Cytogenet. Cell Genet. 15:92-102. (1975).

Tantravahi, R., Miller, D. A., Dev, V. G., and Miller, O. J.: Detection of nucleolus organizer regions in chromosomes of human, chimpanzee, gorilla, orangutan and gibbon. Chromosoma. 56:15-27. (1976).

vanTuinen, P., and Ledbetter, D. H.: Cytogenetic comparison of Hylobates (Symphalangus) Syndactylus and Hylobates (Nomascus) Concolor (Hominoidea; Primates). Mamm. Chromosome Newsl. 23:49. (1982).

Turleau, C., deGrouchy, J., and Chavin-Colin, F.: Inversion pericentrique du 3, homozygote et heterozygote, et translation centromerique du 12 dans une famille d'orangs-outangs. Annales de Genetique. 18:227-233. (1975).

Turleau, C., deGrouchy, J., and Klein: Phylogenie chromosomique de l'homme et des primates hominiens, (Pan troglodytes, Gorilla gorilla et Pongo pygmaeus). Annales de Genetique. 15:225-240. (1972).

deVries, G. F., deFrance, H. F., and Schevers, J. A.: Identical Giemsa banding patterns of two macaca species: Macaca mulatta and M. fasicularis. A densitometric study. Cytogenet. Cell Genet. 14:26-33. (1975).

Warburton, D., Firschein, I. L., Miller, D. A., and Warburton, F. E.: Karyotype of the Chimpanzee, _Pan troglodytes_, based on measurements and banding patterns: comparison to the human karyotype. Cytogenet. Cell Genet. 12:453-461. (1973).

Warburton, D., Henderson, A. S., and Atwood, K. C.: Localization of rDNA and Giemsa banded chromosome complement of the white-handed gibbon. _Hylobates lar_. Chromosoma. 51:35-40. (1975).

Wilson, A. C., Bush, G. L., Case, S. M., and King, M. C.: Social structuring of mammalian populations and rate of chromosomal evolution. Proc. Natl. Acad. Sci. USA 72:5061-5065. (1975).

Wolkin, J. R.: Structural and behavioral characteristics of a young hybrid ape (_Hylobates lar moloch_ x _Symphalangus syndactylus_). M. A. thesis, Georgia State Univ., Atlanta, Georgia. (1977).

Wolkin, J. R., and Myers, R. H.: Characteristics of a gibbon - siamang hybrid ape. Internat. J. Primatol. 1:203-221. (1980).

PART III

Biomedicine

21

Studies on Lymphocyte Subpopulations in Primates

K. Ben, B. Song, W. Dai, and R. Zhou

Lymphocyte subpopulations in peripheral blood of *Macaca mulatta*, *M. assamensis*, *M. nemestrina*, *M. arctoides*, and *Tupaia belangeri* and in lymphoid tissues of *M. mulatta* and *T. belangeri* were determined. The following markers were used: 2-S-amino-ethyl isothiouronium bromide or neuraminidase-treated sheep red blood cells (E_{AET} or E_{VCN}); zymosan-yeast-complement complex (ZYC); stump-tailed monkey red blood cells (SMRBC) and alpha-naphthyl acetate esterase (ANAE) staining combined with the above markers. The range of mean percentage of E_{AET} or E_{VCN} rosette forming cell (RFC) and of ZYC rosette cells in peripheral blood of macaques was 76-82% and 9.0-15.0%, respectively. The percent of E_{AET} RFC in thymus, spleen, and lymph node of *M. mulatta* was 95.2±2.9, 28.8±10.3 and 44.8±6.2, respectively. SMRBC rosette cells in *M. mulatta* were not B cells. Most T cells of macaques were ANAE positive; however, B cells and null cells might also be ANAE positive. ANAE staining is therefore not specific for determining the lymphocyte subpopulations in primates.

INTRODUCTION

Lymphocyte isolation and classification of subsets from peripheral blood and lymph nodes in primates are basic techniques which are fundamental in the study of primate medicine and immunobiology. The determination of E and EAC lymphocyte receptors in some primates has been reported (Ganguly et al., 1977; Taylor et al., 1980; Terrell et al., 1977).

ZYC is a simple and satisfactory marker of B lymphocytes in humans and other mammalian species (Ross, 1980). Pellegrino et al. (1975) have also reported that stump-tailed monkey red blood cells formed rosettes with human B lymphocytes. Mueller et al. (1975) reported that ANAE staining was a method for the determination of T cells, but recent studies have shown that this technique is not specific for T cells (Yang et al., 1979).

In order to determine if these methods are satisfactory for the classification of nonhuman primate lymphocyte subpopulations and to study lymphocytes further, using several markers simultaneously or alone, the lymphocyte subpopulations were determined in humans and five species of nonhuman primates.

METHODS AND MATERIALS

Animals

Specimens for examination were collected from rhesus monkeys (*Macaca mulatta*, 12 males and 11 females, 2-8 years old), pig-tailed monkeys (*M. nemestrina*, 1 male and 2 females, 3-6 years old), assamese monkeys (*M. assamensis*, 4 males, 3-9 years old), a stump-tailed monkey (*M. arctoides*, 1 male, 3 years old), and tree shrews (*Tupaia belangeri*, 4 males and 5 females, all adult). These animals were all clinically normal at the time specimens were collected.

Isolation of peripheral blood lymphocyte (PBL)

Blood was collected in preservative-free heparin (20-40 u/ml) and diluted with 3 volumes of PBS (pH 7.2). Lymphocytes were isolated by Ficoll-metrizoate (d = 1.077 gm/cm^3) density gradient centrifugation for 20 minutes at 1,200g. The cells were washed three times with PBC and resuspended in TC 199 medium. During the last wash, they were supplemented with 20% heat-inactivated calf serum. Viability was tested by ethidium bromide-acridine orange staining (Ford, 1978) and was greater than 99% in all cases.

Preparation of cell suspension of lymphoid tissues

Thymus, spleen and inguinal lymph nodes were collected within half an hour of humane sacrifice of rhesus monkeys for polio vaccine production and from anaesthetized tree shrews, then placed separately into cold Dulbecco's solution. The tissues were minced with scissors and homogenized with small glass homogenizers. Cell suspensions were immediately washed three times with TC 199 medium. Cell concentrations were adjusted to 2 times 10^6/ml. Viability was greater than 95% in all cases.

Erythrocyte preparation

Sheep red blood cells (E) and stump-tailed monkey red blood cells (SMRBC) were stored at 4°C in Alsever's solution and used within one week after collection. E_{AET} and E_{VCN} preparation were made by the methods of Kaplan (1976) and Galili and Schlesinger (1974). AET was synthesized in our laboratory (Doherty et al., 1957). Vibrio comma neuraminidase (VCN) was purchased from British Drug House. The concentrations of various red blood cells were diluted with TC 199 medium to a concentration of 2 times 10^8/ml just before rosette assays (Kaplan et al., 1976).

Zymosan-yeast-complement complex preparation

The yeast was made by Kunming Food Incorporation. Serum complement was taken from mice. ZYC was prepared by the slightly modified method of Guan and Yu (1980), suspended at 4-5 times 10^8/ml, stored at 4°C, and used within one week after preparation.

Erythrocyte rosette assays

Four volumes of red blood cells, 2 volumes of heat-inactivated calf serum, and 4 volumes of lymphocyte suspension were added to a round-bottom test tube, mixed thoroughly, centrifuged at 200g for 5 minutes and stored at 4°C for 1-2 hours. One drop of 1% glutaraldehyde solution was then added to 1 ml of rosette suspension, resuspended gently by hand, and placed at room temperature for 20 minutes.

ZYC rosette procedure

One volume of lymphocyte suspension was added to one volume of ZYC suspension, mixed and incubated at 37°C for 5 minutes, centrifuged at 200g for 5 minutes and stored at 4°C for 1-2 hours. A 1% glutaraldehyde solution was then added in the same way as for the erythrocyte rosette assays (Guan and Yu, 1980).

Mixed rosette assays

One volume of E_{AET} and one volume of ZYC suspension or one volume of ZYC and one volume of SMRBC suspension were mixed with one volume of lymphocyte suspension and one volume of calf serum. This suspension was then processed as described above for ZYC rosette assay.

Cytocentrifuged slide preparation

One drop of rosette suspension and 9 drops of PBS were mixed, and 1-3 drops of the diluted rosette suspension were added to one plastic cup in Cytospin (Shandon, England). These were centrifuged at 90g for 5 minutes and the slide preparations were air-dried.

Giemsa staining

Air-dried slide preparations were stained with diluted Giemsa stain for 5-10 minutes and cover slipped with Canada balsam for permanent storage.

ANAE staining

Cytocentrifuged slide preparations were fixed in formalin-acetone solution for one minute, washed with tapwater for 3 minutes, and then incubated in fresh hexazotized pararosaniline-alpha-naphthyl acetate solution for 3 hours. The slides were then counterstained in 0.5% methyl green solution for 1-2 minutes, rinsed in distilled water, and cover slipped with Canada balsam after air-dry.

Scoring and counting of rosette and esterase activity

In the erythrocyte rosette assays, adhesion of three or more erythrocytes to a lymphocyte was scored as a rosette. In the ZYC rosette assay, adhesion of five or more yeast cells to a lymphocyte was scored as a rosette. A mixed rosette was scored when each kind of marker adhered to a lymphocyte and each lymphocyte had 3 or more erythrocytes and 5 or more yeast cells. The ANAE stained slides were graded as ANAE pluses if the cell contained one or more red spots; ANAE plus if the cell contained dispersed red granules; and ANAE minus if there was no red staining in the lymphocyte. At least 200 lymphocytes were counted in each slide.

RESULTS

Rosette-forming cells (RFC) in PBL

In PBL of rhesus monkeys the mean percentage of RFC with E, E_{AET}, E_{VCN}, ZYC, and SMRBC was 52.1, 76.1, 75.5, 11.4, and 10.2, respectively. The absolute numbers of RFC were 3555, 5158, 4830, 749, and 727 per microliter, respectively. The mean percentage of mixed rosette-forming cells with E_{AET} plus ZYC and SMRBC plus ZYC was 1.0 and 0.7, respectively. This indicates that

TABLE 1. The percent and absolute numbers of various rosette forming cells in peripheral blood lymphocyte of nonhuman primates.*

Species	Sheep Erythrocytes: E %	E Conc.	E_{AET} %	E_{AET} Conc.	E_{VCN} %	E_{VCN} Conc.	E_{ZYC} (A E T + Z Y C): %	E_{ZYC} Conc.	$E_{AET+ZYC}$ %	$E_{AET+ZYC}$ Conc.
Macaca mulatta	52.2 ±8.7 (3)	3555 ±453	76.1 ±6.4 (23)	5158 ±218	75.5 ±2.8 (4)	4830 ±1670	11.4 ±4.2 (23)	749 ±456	1.0 ±0.9	97 ±121
Macaca nemestrina	49 (1)	2964	76.0 ±6.4 (3)	5540 ±2505	-	-	12.8 ±2.2 (3)	880 ±290	1.0 ±0.8	168 ±168
Macaca assamensis	61 (2)	6450	77.7 ±4.2 (4)	6670 ±1787	-	-	9.0 ±3.8	706 ±180	-	-
Macaca arctoides	-	-	82 (1)	6546	-	-	15 (1)	1432	-	-
Tupaia belangeri	-	-	24.9 ±7.2 (8)	194 ±185	22.1 (2)	108	15.7 ±7.4 (7)	170 ±93	-	-

*The numbers in parentheses stand for the number of animals examined.
Conc. = concentration of RFC per microliter. - = not done.

rhesus lymphocytes which form rosettes with SMRBC are probably not B lymphocytes in contrast to that reported for humans.

The mean percentage of RFC with E_{AET} in PBL of the pig-tailed monkey, assamese monkey, stump-tailed monkey, and tree shrew was 76, 77.7, 82 and 24.9, respectively. ZYC-RFC in these species were 11.4, 12.8, 9.0, 15, and 15.6, respectively (Table 1).

RFC in lymphoid tissues

The mean percentage of RFC with E_{AET} and ZYC in lymphoid tissues of the rhesus monkey and tree shrew are summarized in Table 2. With the exception of non-viable cells, almost all thymocytes in the rhesus monkey formed rosettes with E_{AET}, thus indicating that E_{AET} RFC in the rhesus monkey does represent T cells. In contrast, the nature of E_{AET} RFC in the tree shrew is uncertain, since only about 30% of the lymphocytes in the thymus, spleen, and PBL were E_{AET} RFC.

TABLE 2. The percent of E_{AET} and ZYC rosette forming cells in lymphoid tissues of *Macaca mulatta* and *Tupaia belangeri*.*

	Thymus			Spleen		
Species	E_{AET}	ZYC	E_{AET} + ZYC	E_{AET}	ZYC	E_{AET} + ZYC
Macaca mulatta	95.2 ±2.9 (9)	-	-	28.8 ±10.3 (7)	7.0 ±3.5 (3)	0.33 ±0.23 (3)
Tupaia belangeri	38.5 (2)	1.25	0	28.5 (2)	7.3 (2)	0

	Lymph Nodes		
Species	E_{AET}	ZYC	E_{AET} + ZYC
Macaca mulatta	44.8 ±6.2 (6)	4.1 ±2.0 (3)	0 (3)
Tupaia belangeri	-	-	-

*The numbers in parentheses stand for the number of animals examined.
- = not done.

TABLE 3. ANAE activity in peripheral blood lymphocyte from 4 species of macaques.*

Species	No. of Animals	ANAE++	ANAE+	ANAE-
Macaca mulatta	20	55.3±6.3	29.3±5.1	15.4±4.2
Macaca nemestrina	3	50.0±4.6	33.8±5.1	16.2±0.8
Macaca assamensis	4	47.5±8.1	36.7±6.7	15.8±3.2
Macaca arctoides	1	64.3	25.6	10.1

*ANAE++ = 1-3 red spots in lymphocyte;
ANAE+ = dispersed red micrograules in lymphocyte;
ANAE- = no red substance in lymphocyte.

ANAE activity of peripheral blood lymphocyte

ANAE activity of PBL in 4 species of macaques are shown in Table 3. The findings are comparable for each of the species. The percentage of ANAE pluses and ANAE plus lymphocytes was over 80% and similar to the E_{AET} RFC.

After mixed rosette assay with both E_{AET} and ZYC, ANAE activity of PBL was determined. The results are shown in Table 4. ANAE positive cells in the T lymphocyte population of rhesus monkeys averaged 95.5%; in the B lymphocyte population there was an average of 52.8% ANAE positive cells; and 78.8% of the null cell population (9 rhesus examined) was ANAE positive. Therefore, the T, B, and null cell populations of the rhesus monkey all show ANAE positivity.

DISCUSSION

Taylor et al. (1980) found marked differences in the percentage of ERFC of PBL from 5 species of macaques. However, our results showed that in 4 species of macaques the percent of E_{AET} RFC was quite similar and higher than the results obtained by others. The percent of ZYC-RFC in our studies was lower than that for EAC-RFC as determined by Taylor and others (Table 5; Taylor et al., 1980; Terrell et al., 1977; and Ganguly et al., 1977).

We do not know whether E_{AET} RFC in the pig-tailed monkey, assamese monkey, and stump-tailed monkey represent T lymphocytes since thymocytes from these species for study were not available.

ANAE activity has been reported in human and bovine T and B lymphocytes (Yang et al., 1979). Similar ANAE activity was found in nonhuman primates in the present study. We found ANAE activity in T, B, and null cells and concluded that ANAE staining is not a specific method for determining the lymphocyte subpopulations in nonhuman primates.

TABLE 4. ANAE activity of E_{AET} RFC and ZYC-RFC in PBL from 4 species of macaques.*

Species	ANAE activity	R+	R-	Z+	Z-
Macaca mulatta	ANAE++,+	68.5±4.9	16.6±4.9	7.6±2.7	76.5±5.8
(12)	ANAE-	3.2±1.4	11.7±2.2	6.7±9.2	9.2±5.6
Macaca nemestrina	ANAE++,+	62.1±6.8	22.9±5.5	11.8±1.9	77.8±3.5
(3)	ANAE-	3.9±0.6	11.2±1.0	5.5±1.2	8.9±0.5
Macaca assamensis	ANAE++,+	62.8±1.5	22.7±4.7	6.0±2.3	77.1±5.8
(4)	ANAE-	4.1±2.9	10.5±3.3	6.7±3.0	10.2±1.8
Macaca arctoides	ANAE++,+	75.6	14.7	8.1	77.7
(1)	ANAE-	5.3	4.6	7.9	6.3

*The numbers in parentheses stand for the number of animals examined.

ANAE++,+ = including ANAE++ and ANAE+. Values are mean percentage ± standard deviation. R+ = E_{AET} RFC; R- = not E_{AET} RFC; Z+ = ZYC-RFC; and Z- = not ZYC-RFC.

TABLE 5. A comparison between results of various RFC in PBL from some macaques determined by different authors.*

	Present Paper			Taylor et al. (1980)	Terrell et al. (1977)	Ganguly et al. (1977)				
Marker	E	E_{AET}	E_{VCN}	E	E	E	ZYC	EAC	EAC	EAC
Macaca mulatta	52 (3)	76.1 (23)	75.5 (4)	63 (20)	63.5 (37)	57.7 (30)	11.4 (23)	24	14.9 (20)	22.8 (30)
Macaca nemestrina		76.0 (3)	-	23 (1)	-	-	12.8 (3)	26 (1)	-	-
Macaca assamensis		77.7 (4)	-	-	-	-	9.0 (4)	-	-	-
Macaca arctoides		82 (1)	-	19 (1)	-	-	15 (1)	30 (1)	-	-
Homo sapiens		82 (1)		58 ± 7 (6)	-	-	13 (1)	30 (6)	-	27.1

*The numbers in parentheses stand for the number of animals examined. - = not done.

REFERENCES

Doherty, D. G., Shapira, G., and Burnett, W. T.: Synthesis of aminoalkylisothiuronium salts and their conversion to mercaptoalkyguanidines and thiazolines. J. Amer. Chem. Soc. 97:5667-5671. (1957).

Ford, W. L.: The preparation and labelling of lymphocytes. In Weir. Handbook of Experimental Immunology. 23.10. Blackwell Scientific Publication, Oxford. (1978).

Galili, U., and Schlesinger, M.: The formation of stable rosettes after neuraminidase treatment of either human peripheral blood lymphocytes or sheep red blood cells. J. Immunol. 112:1628-1634. (1974).

Ganguly, N. K., Mohan, C., Sapru, R. P., and Kumar, M.: T and B cell populations in the peripheral blood of rhesus monkeys. Int. Archs. Allergy Appl. Immun. 53:290-292. (1977).

Guan, Y., and Yu, S.: A simple method of determination of B lymphocytes. Chinese J. Hematol. 1:243-244. (1980).

Kaplan, M. E., Woodson, M., and Clark, C.: Detection of human T lymphocytes by rosette formation with AET-treated sheep red cells. In Bloom. In vitro Methods in Cell-mediated and Tumor Immunity. pp. 83-88. Academic Press, New York. (1976).

Mueller, J., Brun del Re, G., Buerki, H., Keller, H. U., Hess, M. W., and Cottier, H.: Nonspecific acid esterase activity: a criterion for differentiation of T and B lymphocytes in mouse lymph nodes. Eur J. Immunol. 5:270. (1975).

Pellegrino, M. A., Ferrone, S., and Theofilopoulos, A. N.: Rosette formation of human lymphoid cells with monkey red blood cells. J. Immunol. 115:1065-1071. (1975).

Ross, G. D.: Analysis of the different types of leucocyte membrane complement systems. J. Immunol. Methods. 37:197-211. (1980).

Taylor, D. W., Perri, S. F., Sorensen, C. A., and Palumbo, N. E.: A comparative study of rosette formation in 12 species of nonhuman primates. J. Med. Primatol. 9:76-82. (1980).

Terrell, T. G., Holmberg, C. A., and Osburn, B. I.: Immunologic surface markers on nonhuman primate lymphocytes. Am. J. Vet. Res. 38:505-517. (1977).

Yang, T. J., Jantzen, P. A., and Williams, L. F.: Acid alpha-naphthyl acetate esterase: presence of activity in bovine and human T and B lymphocytes. Immunol. 38:85-93. (1979).

22

Color Vision Tests of Japanese and Rhesus Monkeys

T. Oyama, T. Furusaka, and T. Kito

Two Japanese monkeys (*M. fuscata*) and two rhesus monkeys (*M. mulatta*) learned to discriminate yellow from either red or green monochromatic light, and to discriminate blue-green from either blue or green monochromatic light (color discrimination learning). The monkeys were then tested for discrimination between yellow and various mixtures of red and green monochromatic lights, and between blue-green and various mixtures of blue and green monochromatic lights (color mixture tests). Human subjects were tested for color-similarity ratings with the same apparatus. The results obtained from the monkeys and color-normal humans showed inverted U-shape curves with peaks at almost the same mixing ratios, while tests on color-anomalous human subjects showed peaks at different ratios. The two Japanese monkeys were also tested for discrimination between white light and 23 monochromatic lights chosen from a spectral range of 417 to 638 nm in about 10 nm steps (neutral point tests). They showed no confusion (no neutral points). All results indicated that these macaques had color vision very similar to normal human vision.

INTRODUCTION

De Valois and his collaborators (De Valois, 1965; De Valois and Jacobs, 1971; De Valois et al., 1974) compared macaque monkeys (*Macaca fascicularis*, *M. nemestrina*, and *M. speciosa*) with color-normal and color-defective humans in scotopic and photopic luminosity functions, wavelength and saturation discrimination, and anomaloscopic and neutral point tests. They obtained similar results from monkeys and color-normal humans.

Anomaloscopic and neutral point tests are very

important for detection and classification of color deficiencies in humans (Hurvich, 1972; Jacobs, 1981; Pokorny et al., 1979). An anomaloscopic test is a color-mixture test in which, usually, yellow monochromatic light is presented with a mixture of red and green monochromatic lights. For normal (trichromatic) humans, the mixture can be matched with yellow monochromatic light in a special mixing ratio, by adjusting the luminance of the yellow light. For anomalous trichromatic humans (protanomalous and deuteranomalous), the mixture is matched with standard light in mixing ratios which differ from the normal and from each other. For dichromatic humans (protanopes and deuteranopes), the mixture can be matched with the standard over the whole range of the mixing ratio, by adjusting luminance only. The results of De Valois et al., indicated that the mixture was matched with the standard at the same ratio by both monkeys and normal humans. Anomaloscopic tests were also given to carp by Oyama and Jitsumori (1973), and to pigeons by Jitsumori (1976), to study their color visions.

The neutral point is the spectral region at which discrimination from white light fails. Normal human trichromats do not have any neutral point at all. Neither do anomalous trichromats, but they do have some spectral regions at which discrimination from white light becomes relatively difficult. Dichromats have neutral points at around 495 nm for protanopes, 500 nm for deuteranopes, and 570 nm for tritanopes (Pokorny et al., 1979). The results of De Valois et al. (1974) indicated that the macaques did not have any neutral points. This is very important because it also denies another possibility; that the macaques have dichromatic vision different from human dichromats.

In the present study, anomaloscopic tests (color mixture tests in the red-green range), and neutral point test were made to re-confirm the findings of De Valois et al. (1974) in Japanese and rhesus monkeys. A special attempt was made to perform another color mixture test in the blue-green region where human tritanopes and tritanomalous have deficiencies.

METHODS AND MATERIALS

Subjects

Two Japanese monkeys (*Macaca fuscata*) a male, J1, and a female, J2; and two male rhesus monkeys (*Macaca mulatta*) RA and RB, were used. All were 6 years old at the start of the experiments. They had previously experienced experiments on discrimination learning with colored objects.

For comparison, nine 19 to 27 year-old university students served as subjects. Six had normal color vision, one was a protanomalous trichromat and two were deuteranomalous trichromats. This was determined by Ishihara color vision and anomaloscopic tests.

Apparatus

The experimental apparatus was a black wooden box 86 cm wide, 86 cm deep, and 190 cm high (see Figure 1). During experiments with the monkeys, this apparatus was attached to the front of the subject's individual home cage. At the front of the apparatus at the level of the subject's eyes, two 12 by 12 cm windows were located, with a 25 cm center-to-center horizontal separation. Two black plastic stimulus panels, hinged at the top and equipped with microswitches were inside the windows. A translucent white plastic plate, 2.5 by 2.5 cm, was in the center of each panel. Stimulus lights were projected onto each plate from one or two of three projectors located behind each stimulus panel. Between the panel and each projector were filter-holders. Interference filters, neutral density filters, and pairs of polarizing filters were inserted into the holders to control the color and

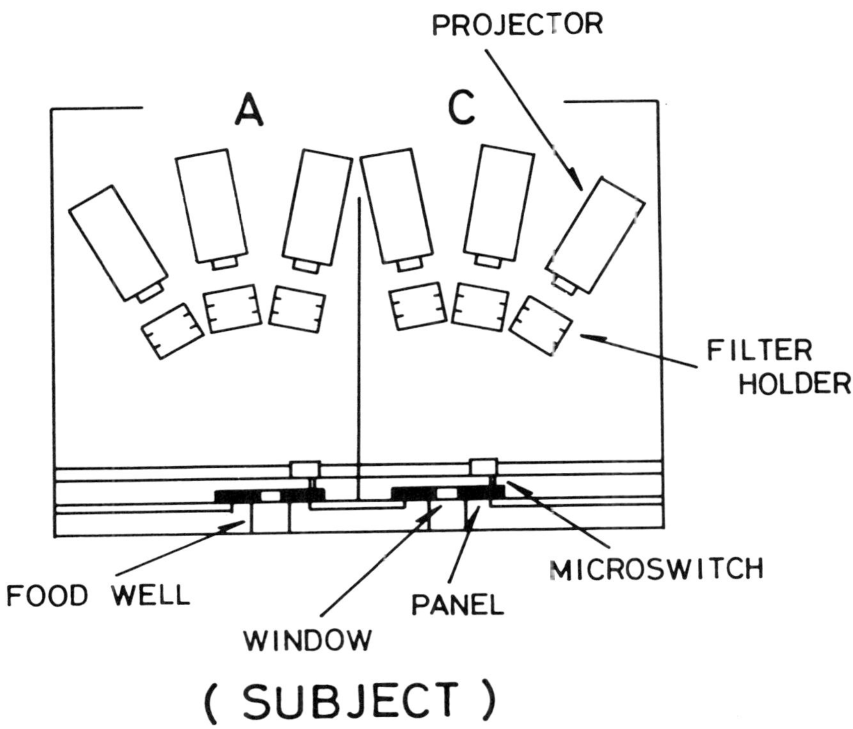

Figure 1. Plan of the apparatus used in this study.

intensity of the stimulus lights according to the conditions of each trial. Light sources for the projectors were 150 W tungsten filament incandescent lamps. The apparatus was symmetrical on the left and right sides of the subject, which will be called A and C sides in the following. When pressed, the stimulus panel in each side operated the microswitch which turned on a pilot lamp indicating the subject's response, and turned off the stimulus lights. Thirteen cm under each stimulus panel was a white food well into which two raisins were delivered every time the monkey pushed the correct panel in both the learning and test sessions. A large black guillotine door dropped covering the whole front of the apparatus immediately after the subject responded.

General procedure for monkeys

The monkeys were fed once a day after the experiments. Experiments were run in a dark room after a 15 minute dark adaptation. The non-correction method was used.

Experiment 1: Color mixture test in the red-to-green range

In this experiment, the color vision of monkeys was compared with that of normal and anomalous human subjects in discrimination (or matching) between yellow monochromatic light and various mixtures of red and green light.

The monkeys were trained to discriminate yellow monochromatic light from either red or green monochromatic light. After this discrimination learning was completed, the monkeys were tested with various mixtures of red and green monochromatic lights.

Stimulus conditions

In discrimination learning, two sets of interference filters (red, green and yellow) and neutral density filters were used at the A and C sides of the apparatus. The wavelengths, half-band widths of the interference filters, and the luminance on the stimulus panels measured with a Spectra Prichard Photometer, are shown in Table 1.

Slight differences in wavelengths existed between the corresponding pairs in the A and C sets, because it was very difficult to obtain interference filters of exactly the same quality. Three different levels of luminance were used for each color to avoid the possibility that brightness rather than hue was used as a cue for discrimination. The levels were obtained by combining different neutral density filters with the same

TABLE 1. Wavelengths of maximum transmittance, half-band widths, and luminance of stimulus lights used in Experiment 1

		Yellow			Red			Green	
		A	C		A	C		A	C
Wavelength of maximum transmittance (nm)		574	578		644	647		530	531
Half-band width (nm)		10	10		11	18		10	10
Luminance (cd/m^2	(Y)	9.9	7.5	(R)	4.0	3.3	(G)	9.0	13.0
	(Y')	5.3	4.1	(R')	1.9	1.7	(G')	5.1	7.2
	(Y")	2.6	2.0	(R")	1.0	1.0	(G")	2.7	3.5

interference filter. For instance, Y, Y', and Y" represent three different luminance levels of yellow light.

In the color mixture test, red and green lights were simultaneously projected onto the same stimulus panel. Various combinations of angles of polarizing filters inserted between the projectors and the stimulus panels produced nine different combinations of red-green lights on both the A and C sides. The luminance of the red and green components, and the R% (the percentage of the red light luminance in the total luminance of the mixed light) of these mixed lights are shown in Table 2.

The luminance unit (cd/m^2) was standardized for the humans, and was also used for the monkey subjects because the spectral luminosity function of macaque monkeys is known to be similar to that of humans (De Valois et al., 1974; Schrier and Blough, 1966).

In the wavelength discrimination test of the monkeys, five additional interference filters were used as negative stimuli. The wavelengths of their maximum transmittances were 557, 569, 577, 589, and 598 nm, with half-band widths of 7 to 9 nm. The luminance of the stimulus lights using these interference filters was 1.0 to 1.8 cd/m^2.

Procedure

Discrimination learning of monkeys

After four days of preliminary training to shape up panel-pushing behavior, discrimination learning began. It

TABLE 2. Luminance of red and green components and R% of nine mixed lights in A and C sets (Experiment 1)

	1	2	3	4	5	6	7	8	9
					A				
Red component (cd/m^2)	0.86	0.93	1.5	1.5	2.2	1.5	2.5	2.2	2.6
Green component (cd/m^2)	2.68	2.78	2.8	2.4	2.7	1.8	2.3	1.7	1.6
R%	24	25	34	38	45	46	52	57	62
					C				
Red component (cd/m^2)	0.92	1.0	1.4	1.4	2.0	1.5	2.3	2.0	2.5
Green component (cd/m^2)	3.01	3.2	3.1	2.6	2.9	1.9	2.5	1.8	1.8
R%	23	24	31	36	41	43	48	54	58

was divided into three phases. In Phase 1, six pairs of stimulus lights, (Y-R, Y-G, Y'-R, Y'-G, Y-R', and Y-G'), were presented twice on each of the two spatial arrangements (left-right and right-left) in random order for 24 trials in each daily session. Yellow lights were always positive stimuli and red and green were negative. A response to the positive stimulus was always reinforced with two raisins. Such training was repeated daily until each monkey reached a learning criterion of more than 90% correct responses for two successive days. At the beginning of every session, four warm-up trials were held with Y-R and Y-G combinations.

Phase 2 began the day after Phase 1 was completed. Procedure was the same as in Phase 1 except that Y-R, Y-G, Y"-R, Y"-G, Y-R", and Y-G" were used as stimulus pairs. The same criterion was applied again.

Phase 3 began 1 to 12 days after the end of Phase 2. The interval differed among subjects. The purpose of Phase 3 was to refresh for monkeys' training at the beginning of the color mixture test. Procedure and criterion were exactly the same as in Phase 2.

Color mixture test on monkeys

The color mixture test on each monkey started the day after the end of Phase 3 of the discrimination learning.

In the color mixture test, each of the yellow (positive) stimuli of three intensities (Y, Y', and Y") was paired with each negative stimuli (R and G as shown in Table 1 and nine red-green mixtures as shown in Table 2). All combinations of positive and negative stimuli and the two spatial arrangements, (3 times 11 times 2 = 66), were presented in every two successive days; i.e., 33 test trials per day. In ten days, 330 test trials were given. Each combination was presented five times in each spatial position during the test period. Four warm-up trials were given at the beginning of each session. During the test trials, responses to positive stimuli (Y, Y', and Y") were always reinforced as in the learning sessions.

Wavelength discrimination test on monkeys

To determine the wavelength (or hue) discrimination sensitivity of monkeys, discrimination tests using monochromatic lights with small wavelength differences were given three weeks after the end of the color mixture test. In this test, the positive stimuli (Y, Y', Y" in Table 1) were paired with seven negative stimuli; five monochromatic lights of 557 to 598 nm shown above, and the red and green lights used in the discrimination learning. Each combination was presented five times in each of the two spatial arrangements in ten days. Four warm-up trials were given on each day before the test. All responses to positive stimuli were reinforced as before.

Before the test, discrimination learning sessions were held using the same procedures as in Phases 2 and 3 of the discrimination learning to reconfirm their learning.

Color mixture test on humans

The color mixture test on humans was performed using the same apparatus and the same stimulus lights as for monkeys. The human subjects sat in a chair in front of the apparatus and, after a 15 minute dark adaptation, compared the stimulus lights on the two stimulus panels, and assigned a number from 1 to 10 to indicate apparent color similarity of the two stimulus lights. "Ten" meant "the same", and "1" indicated "quite different". Decimal fractions were allowed. All 66 combinations of stimulus lights were presented once each session. Each human subject participated in one session and assigned a number to each combination of stimuli.

Results and Discussion

The numbers of days needed to reach the learning criterion in Phases 1, 2, and 3 of the discrimination learning were as follows: 9, 3, and 4 days for J1; 5, 2, and 2 days for J2; 14, 4, and 2 days for RA; and 9, 4, and 2 days for RB. These results clearly indicate that both the Japanese and rhesus monkeys discriminated between yellow light and either red or green light, because the luminance of both the positive and negative stimulus lights was varied randomly at three levels. This means that the monkeys did not have such color-deficiencies as human protanopes and deuteranopes.

The results of the color mixture tests on the A and C sides of the apparatus are shown separately in Figures 2 and 3, because the wavelengths and other characteristics of the stimulus lights differed slightly on the A and C sides as shown in Tables 1 and 2. These results show just the same things. The inversed U-shaped curves from the normal subjects indicate that mixed red and green monochromatic lights in a suitable ratio (approximately 40% R, an arbitrary unit in this study) result in high ratings, indicating a strong similarity to the yellow monochromatic light. Mixed lights with a higher red or green component gave lower scores, indicating less similarity to the yellow. The curves from the anomalous human subjects suggested that they would have maximum points of similarity to the yellow lights

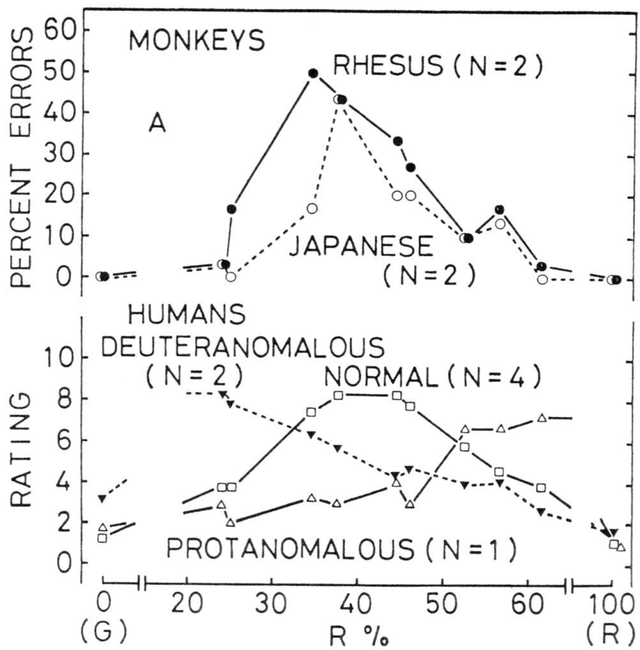

Figure 2. Results obtained from the A side of the apparatus in Experiment 1. The abscissae show the mixing ratio, R% and the ordinates, the percent errors for monkeys, or the mean similarity ratings for humans.

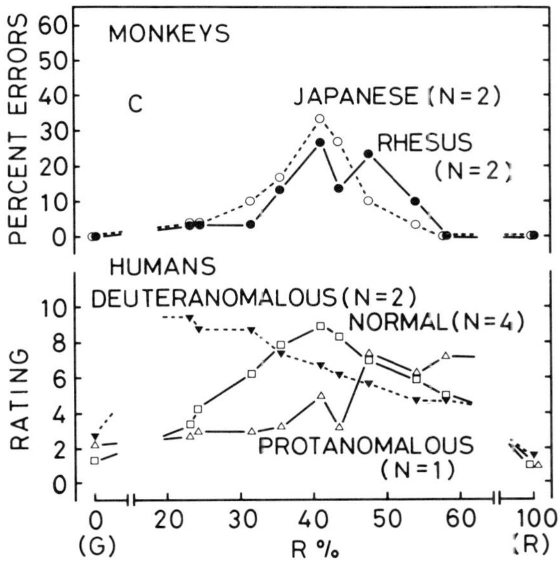

Figure 3. Results obtained from the C side of the apparatus in Experiment 1.

outside the R% range used. The protanomalous subjects needed more red, and the deuteranomalous subjects needed more green light than normal subjects to match the mixed lights with the yellow light. These results reconfirm well-known facts (Hurvich, 1972; Pokorny et al., 1979).

The percentage error results of the Japanese and rhesus monkeys also give inverted U-shaped curves with the maximum at approximately 40% R. Errors in this experiment meant confusions of red and green mixed lights with the yellow monochromatic lights (positive stimuli). The peaks of the curves approach the chance level (50%). In sharp contrast to the differences between normal and anomalous humans, monkeys and normal humans produced very similar curves, although there were some slight irregularities. This indicates that the monkeys had neither protanomalous nor deuteranomalous deficiency, and strongly suggests that the color vision of these monkeys is very similar to normal human color vision as far as red and green systems are concerned.

The relearning of discrimination before the wavelength discrimination was completed in two days for all subjects. In the total of 210 trials of wavelength discrimination tests for each, errors were 38, 32, 32, and 34 for J1, J2, RA and RB respectively. These errors occurred mostly with a negative stimulus of 577 nm whose wavelength is just between the two positive stimuli on the A and C sides (574 and 578 nm). The responses were practically correct. Only six errors were made for the other stimuli by all subjects; three for 589 nm (J1, J2, and RA), two for 567 nm (J1), and one for 557 nm (J1). These results show high sensitivity of monkeys in wavelength discrimination. Consequently, errors in the color mixture test could not be attributed to a low sensitivity of the monkeys in hue discrimination, but to the color mixture, as for the human subjects.

Experiment 2: Color mixture test in the blue-to-green range

In this experiment, the color vision of monkeys was compared with that of normal human subjects in discrimination between blue-green monochromatic light and various mixtures of blue and green lights.

Stimulus conditions

In discrimination learning, two sets of three interference filters (blue, blue-green, and green), as well as neutral density filters were used on the A and C sides of the apparatus as shown in Table 3. Three levels of luminance for each color were used to make the brightness irrelevant to discrimination.

In the color mixture test, blue and green lights were simultaneously projected onto the same stimulus panel. Various combinations of polarization filters produced eleven blue-green mixtures, as shown in Table 4.

In the wavelength discrimination test on monkeys, five additional interference filters were used as negative stimuli. Their wavelengths were 478, 488, 499, 508, and 518 nm, and their half-band widths were 7 to 9

TABLE 3. Wavelengths of maximum transmittance, half-band widths, and luminance of stimulus lights used in Experiment 2

		Blue-Green			Blue			Green	
		A	C		A	C		A	C
Wavelength of maximum transmittance (nm)		494	494		469	469		530	531
Half-band width (nm)		7	8		8	9		10	10
Luminance (cd/m^2)	(BG)	2.2	2.5	(B)	1.6	1.6	(G)	4.9	5.0
	(BG')	1.1	1.3	(B')	0.93	0.86	(G')	2.6	2.7
	(BG")	0.65	0.75	(B")	0.51	0.48	(G")	1.4	1.5

TABLE 4. Luminances of blue and green components and B% of eleven mixed lights in A and C sets (Experiment 2)

	1	2	3	4	5	6	7	8	9	10	11
					A						
Blue component (cd/m^2)	0.32	0.33	0.36	0.35	0.39	0.42	0.35	0.42	0.44	0.42	0.44
Green component (cd/m^2)	3.46	3.22	3.19	2.84	2.98	2.57	1.88	2.02	1.92	1.27	1.10
B%	8.5	9.3	10.1	11.0	11.6	14.0	15.7	17.2	18.6	25.0	28.6
					C						
Blue component (cd/m^2)	0.25	0.27	0.30	0.31	0.35	0.38	0.32	0.36	0.40	0.39	0.41
Green component (cd/m^2)	3.49	3.29	3.56	2.98	3.12	2.78	2.09	1.99	2.02	1.30	1.16
B%	6.9	7.6	7.8	9.4	10.1	12.0	13.3	15.3	16.5	23.1	26.1

nm. The luminance of the stimulus lights obtained with these interference filters was 2.1 to 6.0 cd/m^2.

Procedure

The procedure was generally the same as in Experiment 1. Only the different points will be described in the following.

Discrimination learning of monkeys

The same four monkeys were used as subjects about 10 months after Experiment 1.

No preliminary training was needed. Stimulus pairs used were BG-B, BG-G, BG'-B, BG'-G, BG-B', and BG-G' in Phase 1; and BG-B, BG-G, BG"-B, BG"-G, BG-B", and BG-G" in Phase 2. Blue-green lights were always positive stimuli and blue and green lights were negative. When the monkeys reached the learning criterion, the color mixture test began in the following day without Phase 3.

Color mixture test for monkeys

The positive stimuli were blue-green lights of three luminances (BG, BG', and BG"), and the negative stimuli were B, G and the eleven blue-green mixtures shown in Table 4. Thirty-nine test trials per day were given for ten days.

Wavelength discrimination test on monkeys

This test was made immediately after the color mixture test. The positive stimuli (BG, BG' and BG" in Table 3 but with luminance multiplied by 3) were paired with seven negative stimuli; the five monochromatic lights of 478 to 518 nm shown above and the same blue and green lights as in the learning (luminance was changed to 2.1 and 2.6 cd/m^2 on the A side, and to 2.5 and 2.7 cd/m^2 on the C side).

Color mixture test on humans

Only four students with normal color vision, two of whom experienced Experiment 1, served as subjects, because tritanomalous subjects were not available.

Results and Discussion

The number of days needed for monkeys to reach the learning criterion in Phases 1 and 2 of the discrimination learning was 18 and 2 for J1; 5 and 2 for J2; 12 and 2 for RA; and 9 and 4 for RB, respectively. This indicates that the monkeys could discriminate blue-green monochromatic light from either blue or green monochromatic light in situations where brightness could not be used as a cue for discrimination. It means that the monkeys did not have tritanopic deficiency.

Figures 4 and 5 show the results of the color mixture test with blue and green lights on human and monkey subjects, on the A and C sides of the apparatus, respectively. The abscissae indicate the blue percentage in the luminance of the mixed lights. The ordinates represent the mean rating of color similarity for human subjects and the percentage errors for the monkey subjects. All curves show inverted U-shapes with peaks at a blue percentage of 10 to 15. This indicates that mixtures of blue and green monochromatic lights in certain ratios were apt to be confused with blue-green monochromatic light by monkeys, as by color-normal humans. The closeness of peak points between them suggests that the monkeys did not have tritanomalous deficiency.

In the total of 210 trials of the wavelength discrimination tests, on each monkey, errors were 22, 24, 26 and 18 for J1, J2, RA, and RB, respectively. These errors were made mostly with the negative stimuli of 488 and 499 nm, which were very close in wavelength to the

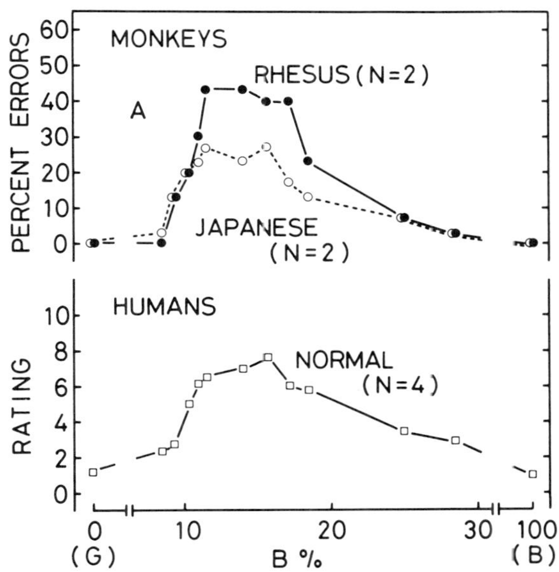

Figure 4. Results obtained from the A side of the apparatus in Experiment 2. The abscissae show the mixing ratio, B%.

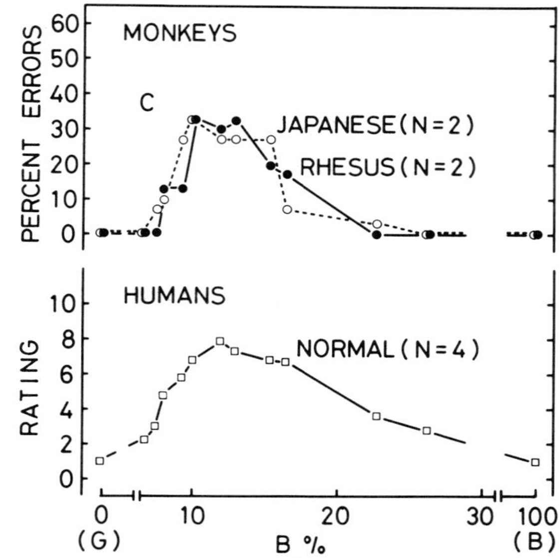

Figure 5. Results obtained from the C side of the apparatus in Experiment 2.

positive stimuli (494 nm). Only RA made 4 errors with the other stimuli; 2 for 478 nm, and 2 for 518 nm. These results show a high sensitivity of monkeys in wavelength discrimination in the blue-green range. The errors in the color mixture test can not be attributed to the monkeys' insensitivity in color discrimination.

All results of Experiment 2 showed that the monkeys' color vision was very similar to normal human vision in the blue-green range.

Experiment 3: Neutral point test

The results of Experiments 1 and 2 indicate that the color vision of the two Japanese monkeys and the two rhesus monkeys was closer to normal human vision than to human protanopic, deuteranopic, tritanopic, protanomalous, deuteranomalous, or tritanomalous vision. However, there remains the possibility that the monkeys had some deficiencies which were quite different from any human color vision deficiency. If the monkeys were dichromats rather than trichromats, there should be a part or parts (neutral points) of the spectrum which they can not discriminate from white light. If the monkeys were anomalous trichromats, a part or parts of the spectrum would be desaturated and not easily discriminated from white light (De Valois, 1965).

The following experiment was performed to check

whether the monkeys had any neutral point or desaturated range in the visible spectrum.

Only the two Japanese monkeys were used as subjects in this experiment, 6 months after the end of Experiment 2.

Stimulus conditions

White light from projectors with incandescent lamps was used as the positive stimuli. The color temperatures were approximately 3300 K on the A side and 3200 K on the C side. The negative stimuli were 23 monochromatic lights from 417 to 638 nm, obtained by inserting interference filters between the projectors and the stimulus panels. The wavelengths of maximum transmittance were varied by approximately 10 nm as follows: 417, 429, 439, 450, 458, 468, 478, 488, 499, 508, 518, 529, 539, 548, 557, 569, 577, 589, 598, 610, 619, 628, and 638. The half-band widths were 7 to 11 nm. Luminance was varied from 0.64 to 39 cd/m^2. The luminance of the white light paired with each monochromatic light was varied at three levels, by inserting suitable neutral density filters. The levels were approximately the same luminance as the monochromatic light used, three to five times higher, and one-third to one-fifth lower.

Procedure

Monochromatic lights were presented one by one from the shortest wavelength for J1, and from the longest wavelength for J2. Each monochromatic light was presented 12 times in random order, 6 times on the A side and 6 times on the C side. White lights at the three luminance levels were presented twice on each side. The white light and the monochromatic light were always paired in the presentation. Responses to the white lights were always reinforced with two raisins. Twenty-four to 36 trials were given each day for nine days.

Results and Discussion

The subjects made very few errors. In a total of 276 trials, J1 made errors in only four trials, one for each of 417, 478, 488, and 577 nm; and J2 made no errors at all. The errors at 478 and 488 nm (around the neutral points of the protanopes and deuteranopes) suggests the possibility of J1 being protanomalous or deuteranomalous (Pokorny et al., 1979). However the results of Experiment 1 deny this possibility.

Another possibility is that the 10 nm steps were too great and missed neutral points or desaturated ranges in the spectrum. However, Grether (1939) and De Valois (1965) used 10 nm steps, and were successful in finding the desaturated ranges for capuchin (Cebus) and squirrel (Saimiri) monkeys. Their results would eliminate this possibility.

Therefore we can safely conclude from our results that the Japanese monkeys were neither dichromats nor anomalous trichromats of any type (including possible unknown types).

CONCLUSIONS

All results obtained in Experiments 1, 2, and 3 indicate that the color vision of Japanese and rhesus monkeys, at least of the individuals used, is very similar to normal human color vision, and reconfirmed the conclusions of De Valois et al. (1974), on different kinds of macaques.

ACKNOWLEDGMENTS

This research was supported in part by a grant to Professor Ben Yagi, Aoyama Gakuin University, from the Ministry of Education, Japan. The authors are indebted to Professor J. M. Pette, Tokyo University of Agriculture and Technology, for his helpful suggestions concerning English expression.

REFERENCES

De Valois, R. L.: Behavioral and electrophysiological studies of primate vision. In W. D. Neff (ed.). Contributions to Sensory Physiology, Vol. 1. Academic Press, New York. (1965).

De Valois, R. L., Jacobs, G. H.: Vision. In A. M. Schrier, H. F. Harlow and F. Stollnitz (eds.). Behavior of Nonhuman Primates, Vol. 3. Academic Press, New York. (1971).

De Valois, R. L., Morgan, H. C., Polson, M. C., Mead, W. R., and Hull, E. M.: Psychophysical studies of monkey vision-I: macaque luminosity and color vision tests. Vision Res. 14:53-67. (1974).

Grether, W. F.: Color vision and color blindness in monkeys. Comp. Psychol. Monogr. 15(4):1-38. (1939).

Hurvich, L. M.: Color vision deficiencies. In D. Jameson, and L. M. Hurvich (eds.). Handbook of Sensory Physiology, Vol. VII/4. Springer-Verlag, Berlin. (1972).

Jacobs, G. H.: Comparative color vision. Academic Press, New York. (1981).

Jitsumori, M.: Anomaloscope experiment for a study of color mixture in the pigeon. Jap. Psychol. Res. 18:126-135. (1976).

Oyama, T., and Jitsumori, M.: A behavioral study of color

mixture in the carp. Vision Res. 13:2299-2308. (1973).

Pokorny, J., Smith, V. C., Verriest, G., and Pinckers, A. J. L. G.: Congenital and Acquired Color Vision Defects. Grune and Stratton, New York. (1979).

Schrier, A. M., and Blough, D. S.: Photopic spectral sensitivity of macaque monkeys. J. Comp. Physiol. Psychol. 62:457-458. (1966).

23

Handedness and Cerebral Asymmetry in Nonhuman Primates

B. Preilowski, M. Reger, and H. C. Engele

In a task requiring the production of a specific pressure between the fingertips of one hand, rhesus monkeys showed extreme degrees of hand preference. With that same task, no intermanual transfer of training was observed, suggesting that the skill remained lateralized within one hemisphere. The results of a time series and cross correlation analysis, applied to the asymptotic performance data of left and right hand, support the idea that in rhesus monkeys there is a potential for separate and independent development of sensory-motor skill in each of the two hands and their respective cerebral control areas. It is suggested that handedness, as other forms of functional asymmetry as well, are adaptive specializations, which may have developed independently in man and nonhuman primates. In each case, persistence of asymmetric motor habits and lack of interhemispheric transfer may be the decisive factors.

INTRODUCTION

In recent years, the study of handedness in man and in animals has received renewed attention because it is hoped that it might contribute to an understanding of the functional asymmetries of the human brain. In order to decide whether handedness as a genuine expression of cerebral asymmetry exists in nonhuman primates three criteria have predominantly been applied: (1) intraindividual consistency of strong hand preference, which is task independent and resistant to learning and experience; (2) interindividual consistency of preference represented by a distribution significantly shifted to one side; and (3) differential effects of unilateral brain lesions depending on the direction of hand preference. After reviewing the existing literature and his own work of many years, Warren (1981), applying these criteria, concluded that handedness in monkeys is not homologous to handedness in man and does not reflect any functional asymmetry between the cerebral hemispheres.

One can agree with Warrens's statement and still consider the issue of whether there is handedness and cerebral asymmetry in nonhuman primates very much alive. To begin with, one can show that our present knowledge of handedness and its relationship to cerebral asymmetry in humans is rather sketchy and equivocal and certainly not sufficient to serve as a standard with which results from studies of other primates are to be compared (Preilowski, 1978, 1981). More important still, the study of laterality in animals only makes sense if we consider functional asymmetries to be general biological phenomena: evolutionary products which may appear in different species in different forms due to different selective pressures (Preilowski, 1972, 1981). With regard to our problem this would mean, that it may be of evolutionary advantage for a monkey to be handed, but not necessarily to be right-handed rather than left-handed. Thus it would seem to be important that one should look not only at distributions of direction of preference, but also at what determines the degree of hand preference. We have done that and found that the degree of hand preference in rhesus monkeys is independent of the direction of preference. It is also independent of the level of success achieved while performing with a particular hand and independent of previous test experience. However, it varies systematically with the age of the animal and the requirements of the task: Older animals show stronger preferences within a particular task than younger animals; the preference is stronger the more sensory-motor effort and cerebral-cortical control a task requires (Preilowski, 1981). The latter is of particular interest with regard to the relationship between handedness and cerebral asymmetry. It points to another consequence of accepting handedness as a possible form of adaptive specialization, namely, the importance of performance handedness in comparison with preference handedness. And indeed, if tasks are applied which maximize lateralization of cerebral sensory-motor involvement, performance differences between hands are found in rhesus monkeys. Also, for such tasks no immediate intermanual transfer of learning is apparent (Preilowski, 1977, 1979).

In the following we would like to present further data from experiments investigating performance differences between right and left hands as well as intermanual transfer in rhesus monkeys.

METHODS AND RESULTS

Method/Experiment 1

Two rhesus monkeys (1 male-3kg, 1 female-4kg) were trained to produce a specific continuous pressure between two fingers of one hand on a "pressure pill" transducer similar to the one described by Wiesendanger and Hepp-Reymond (1970). The difficulty of the task could be varied by increasing the duration of the required continuous pressure and by changing the distance between an upper and a lower limit in-between which the pressure had to lie in order to be correct.

Each day (Monday through Friday) four blocks of 36 trials maximum were run. Each trial was 10 seconds long and signalled by a 350 Hz tone. This trial-on-tone was replaced by a 500 Hz tone as long as the pressure produced by the monkey fell within the tolerance range. If the correct pressure was held continuously for the required duration a 1000 Hz signal was presented, the trial terminated and the reward in the form of a 190 mg banana flavored pellet was delivered. A block was terminated after 18 trials were successfully concluded. Figure 1 gives a schematic representation of the experimental set-up.

For each block of trials a performance score X was computed: $X = S/T \times (2-A/T) \times 100$, where S represents the number of successfully completed trials, T the number of trials presented and A the number of trials during which at least one attempt was made by the animal. This measure of performance X, varying between 0 and 100, represents an attempt to correct for extreme variations in motivational level and thus allow for reliable determination of performance. This becomes a problem especially as the task is made more difficult and the success rate drops.

During the first phase of the experiment the animals were given free choice of hand and trained at progressively higher levels of difficulty ranging from level 1 (200 p tolerance, .25 second duration) up to level 7 (20 p tolerance, 1.5 second duration). Progression from one level of difficulty to the next was decided on the basis of a performance criterion ($X \geq 90$ on three successive blocks). This first phase was terminated if criterion was not reached within 24 days.

In a subsequent second phase the animal was forced to use one hand during the first two blocks and the other hand during the remaining two blocks of trials. The sequence of hands was alternated from day to day. Starting again at the first level of difficulty this phase was run under otherwise identical conditions to those of the first phase.

Results

From the start of phase 1 on, one animal only worked with its right hand, the other only with its left hand. During several thousand trials no switch of hand use or attempt to use the other hand was observed. During phase 2 no transfer of training between hand was noted: The originally trained hand rapidly progressed from one level of difficulty to the next, reaching its previous maximum

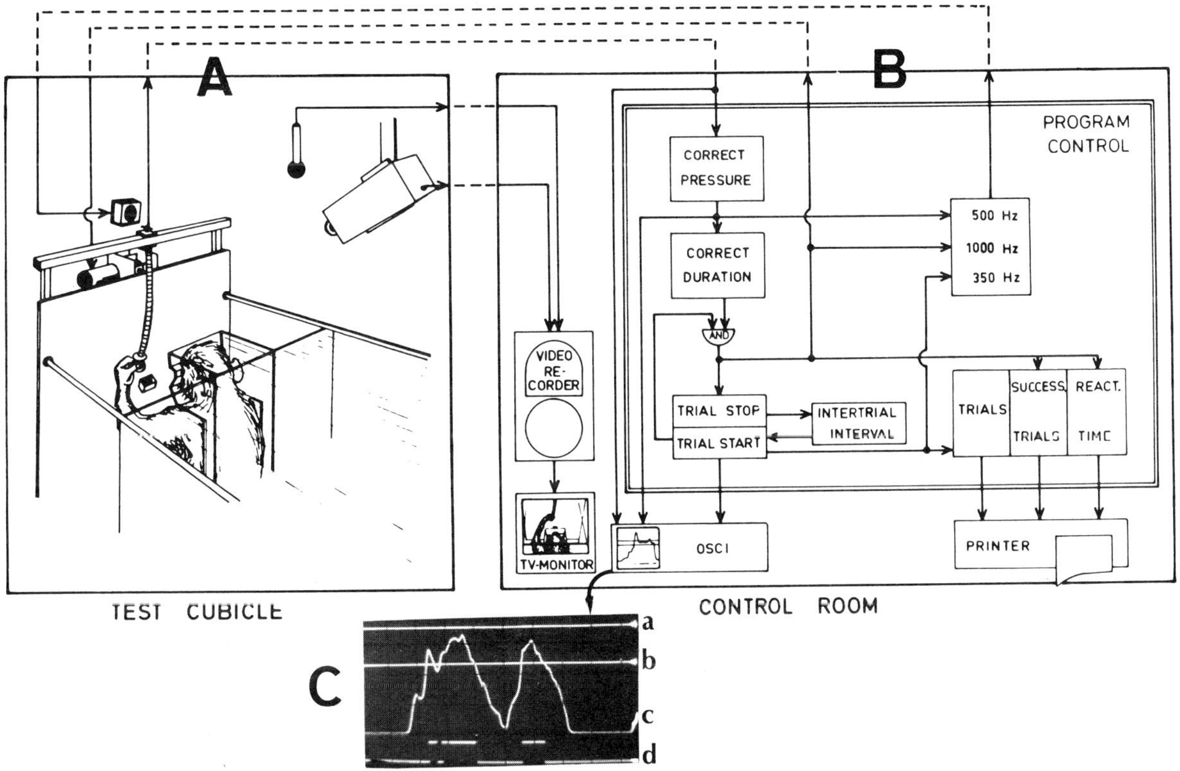

Figure 1. Schematic representation of the experimental set-up. A-Test cubicle; B-Control; C-Oscilloscope recording of a 10-second trial: a-upper limit, b-lower limit, c-pressure produced by the monkey, d-control trace indicating correct performance.

level of performance as rapidly as the number of required criterion trials would allow. At the same time the other hand had to be retrained from the beginning. Performance with that hand resembled that of a naive animal. Figure 2 shows the performance of one of the animals. The similarity between performance curves of the original training of one hand during phase 1 and the subsequent training of the other hand in phase 2 is quite striking. The result is basically the same for the second animal. In addition this monkey used a strikingly different finger posture when performing with right and left hand during phase 2.

A comparison between the performance of both hands at their highest level of difficulty showed a difference in favor of the right hand of both animals.

Method/Experiment 2

In this experiment six rhesus monkeys (3 males, 3 females; 5-6kg) were used as subjects. In order to eliminate a possible experimenter effect in retraining the naive hand during the transfer phase, a different manipulandum was used, which required much less shaping than the "pill transducer" used in Experiment 1. With the new manipulandum (Figure 3), the animal reached into a slot of a loosely mounted cylinder and squeezed two metal tongues. A maximum isometric pressure of 800 p could be measured this way. The basic procedure was identical to that of Experiment 1. However, this time six different experimental phases were conducted.

In the first phase three animals were trained with only one hand; for the other three animals hand use was

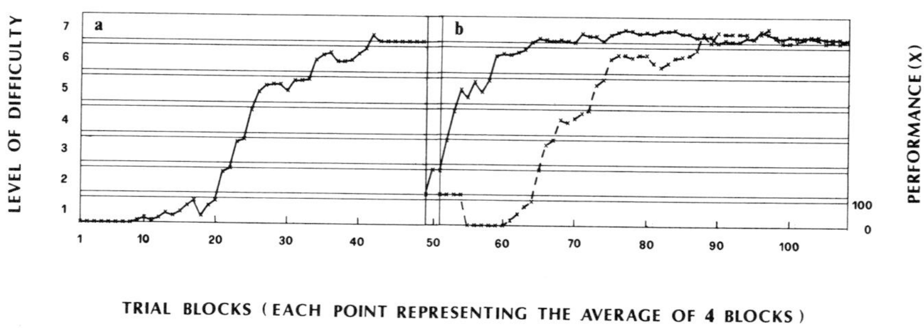

Figure 2. Performance curves of a single animal over seven levels of difficulty. a- Right hand performance (original learning). b- Retraining of right hand (uninterrupted curve), transfer learning with left hand (interrupted curve). Each point on the abscissa represents the average of four blocks of trials, with the exception of points number 49 and 50 which represent a single block each.

Figure 3. Manipulandum used in Experiments 2 and 3.

alternated from day to day. Training was continued until the first five levels of difficulty were mastered (level 1 - 800 p tolerance, 0.1 second duration; level 5 - 800 p tolerance, 1.5 second duration). In the second phase two days of testing with free choice of hands at level 5 were conducted. In phase three all animals were trained at progressively more difficult levels with only one hand, beginning at level 5 up until level 9 (80 p tolerance, 1.5 second duration), which however was not reached by all animals. For two animals of the original "one-hand-group" the hand was now different from the one used during phase 1. One of these animals now used an inadequate finger grip between third and fourth fingers. Since even prolonged attempts to shape a grip between thumb and forefingers were unsuccessful, the animal had to be removed from the experiment. In phase four a test of transfer to the opposite hand was run. During phase five on each day two blocks of trials were given for each hand, with levels of difficulty adjusted to the performance of each hand (similar to phase 2 of Experiment 1). The order of testing of right and left hand was changed from day to day. A last phase (six) consisted of one day of preference tests and two more days of testing with free choice of hands two months later.

Results

The "two-hand-group" of the first phase was meant to control for transfer of general experimental experience. However, even in this phase we found some indication of a lack of intermanual transfer, since the animals which were trained with both hands needed, on the average, as many trials to reach criterion with each hand as those trained with only one hand. In other words training with the left hand did not profit from training of the right hand on the previous day and vice versa. Since there is no sensible way to test statistically the difference between the curves shown in Figure 4, a possible interpretation of these data in terms of a negative transfer between hands has to remain speculative.

During the second phase the animals of the one-hand-group showed a 100 percent preference for the trained hand. The animals of the two-hand-group showed preferences of 65, 81, and 100 percent on the first day; on the second day the preference scores were 93, 98, and 100 percent respectively. The choice of hand appeared to be independent of the level of performance during the first phase.

The two animals which in the third phase were forced to change hands provide some information concerning an intermanual transfer at a low level of difficulty. Only one of the two animals changed its finger posture when changing hands, however none showed any sign of immediate transfer of training. Both needed some time to reach their original level of performance after using the new hand.

A more striking lack of transfer was found after several months of training when the five remaining animals again switched hands in phase four. This time no retraining was done; rather, the transfer test was conducted at the previously achieved level of difficulty. However, in three animals we had to regress to a lower level of difficulty after continuous failure at that level. Again with the change of hands a change in finger posture was observed for all five animals. In the

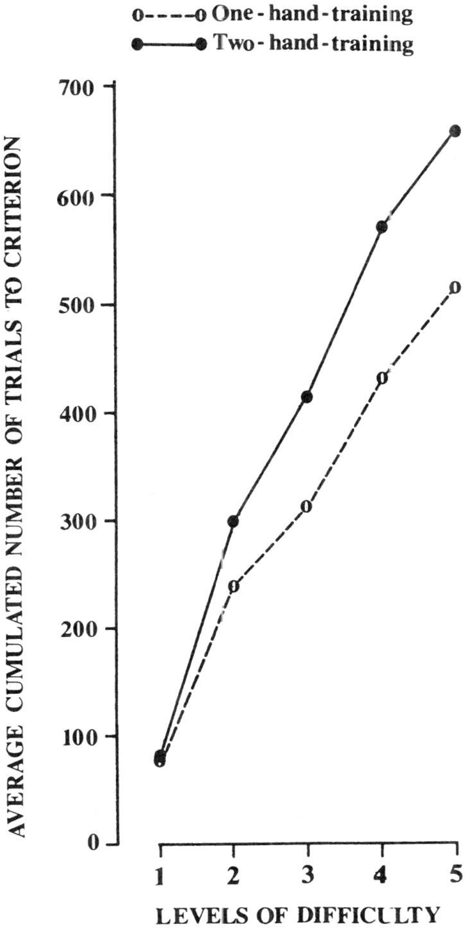

Figure 4. Average cumulated number of trials needed to reach criterion at the first five levels of difficulty for three animals trained with one hand only (interrupted line), and for three animals trained with both hands changing hands from day to day (uninterrupted line). For the latter group the data include the averages of right and left hand performance.

two-hand-group of phase 1 it was the finger grip used during that phase.

During the performance test of phase five, four of the five animals showed persistently better performance with their right hand. One animal worked better with its left hand.

The preferences tested during the last phase were 100 percent for one particular hand in all animals. However, no connection was seen between the level of performance achieved in phase five and the direction of preference.

Method/Experiment 3

This experimental test was run with the five animals which completed Experiment 2. After a rest of four months the animals were retrained with both hands at increasing levels of difficulty adjusted individually to each hand. During the comparative testing of right and left hand performance that level of difficulty was chosen which had been successfully achieved with both hands. During each day of testing (Mondays through Fridays) two blocks of trials were run, one with each hand. The hand tested first was alternated from day to day.

For the intended data analysis an asymptotic level of performance was mandatory. Since the two female monkeys continued to show improvements in performance even at this stage, only the data of the three male animals are considered here. The six time series, one for each hand of the three animals are presented in Figure 5. They comprise data from 67, 68, and 88 days of testing.

A time series analysis (TSA) according to Box and Jenkins (1976) was performed with these data in order to remove their serial dependence and to construct linear stochastic models describing this dependence. It should be mentioned here, that the application of TSA to performance data demands some caveats: for example, possible ceiling effects, nonlinearity of the performance measure used, the relative arbitrariness in model identification, or the interpretation problems presented by non-equidistant samples. On the other hand, TSA is more conservative with respect to statistical adequacy and more sensitive than the traditional approach of comparing means and sums of squares calculated directly from the raw data.

Results

Performance differences between hands

Only one animal (C in Figure 5) exhibited between-hands differences in performance levels that were significant at the 10% level.

Differences in the structure of serial dependence

In order to compare the structure of serial dependence between the hands of an animal, we estimated a common ("overfit") model for both hands and found significant differences in this dependence for two of the three animals (see Table 1). In all cases a larger AR

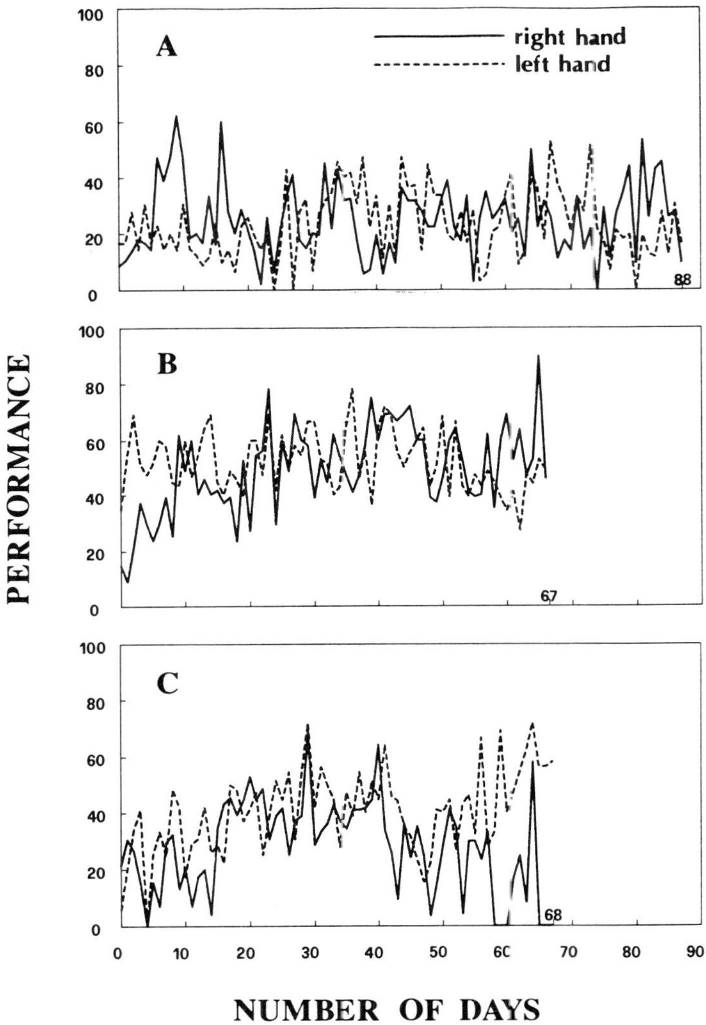

Figure 5. Asymptotic performance of three male animals (A, B, C).

parameter for the right hand indicated a higher "moment of inertia" (higher dependence on history) for this hand; in view of the trend of a somewhat lower level of performance of the right hand it is tempting to see a causal relationship here.

While a difference between first (and perhaps even second) order parameters (Phi 1, Phi 2 in Table 1) admits an interpretation in terms of different temporal coherence ("memory") in the two motor systems of left and

TABLE 1. Intermanual comparison of asymptotic performance (results of a time series analysis)

Animal	Hand	Identified model	Differences in structure of serial dependence*	Differences in level*	Significant cross correlation coefficients ** lag=0	lag≠0
A	R	AR(1)	$\phi 1(R) > \phi 1(L)$***			
	L	AR(6)		none	none	-3
B	R	AR(7)	$\phi 2(R) > \phi 2(L)$****			
	L	Random	$\phi 7(R) > \phi 7(L)$***	none	none	none
C	R	AR(1)	none	c(L)>c(R) ***		
	L	AR(1)			none	none

ϕj = j'th order autoregressive parameter
c = level of process
* = specifications refer to overfitted models (see text)
** = specifications refer to "best" (most parsimonious) models
*** = $p < 0.10$
**** = $p < 0.05$

right hand, a seventh order parameter (Phi 7 in Table 1) can hardly be given a meaningful interpretation due to the fact that we have skipped weekends during testing.

Intermanual dependence

As a measure of intermanual dependence (intermanual transfer being a special case of such a dependence), we decided to use the cross correlation of prewhitened data. It should be insignificant if there is no dependence at all, and it could be significant at a single lag only, if there is some special type of interdependence. Intermanual "transfer" in this sense was found only in animal A. However, it is difficult to conceive a mechanism of transfer between left hand performance on one day and right hand performance three days later.

DISCUSSION

In summarizing, we would like to stress the following results: (1) The animals show extreme degrees of hand preference with this task; none of them switched from the use of one hand to the other spontaneously over 18 months of testing and several thousand trials. (2) We found no indication of immediate intermanual transfer regardless of the level of difficulty. In most animals even basic finger posture did not transfer. (3) A time series analysis indicated structural differences within these animals concerning the way and degree in which the performance of each hand could be predicted from its previous performance. (4) Synchronous cross correlations of prewhitened performance data gave no indication of a direct coupling between hands.

Subtle forms of interaction between both hands cannot be excluded with the present form of analysis. Since it also cannot yet be excluded that some form of state dependent learning may be responsible for the observed lack of transfer, none of the above mentioned results taken by itself, is totally convincing. However, taken together the results do form a coherent picture suggesting separate, independent development of skill in each hand.

For lack of space it is not possible here to discuss the existing literature on handedness in man and other primates but this has been done elsewhere (Preilowski and Engele, 1981). As far as a comparison with human handedness is concerned we have already pointed to the problems which force us to consider handedness, and other forms of cerebral asymmetry as well, to be a general biological phenomenon. From this point of view it appears possible that under appropriate selection pressure, the capacity for independent development of skill within each brain half, together with a lack of spontaneous interhemispheric transfer, may have resulted in handedness in man as well as in nonhuman primates. Additional selection pressure could have been responsible for further developments in man, such as the preponderance of right-handers.

As far as the cognitive and language related asymmetries are concerned, which so far have been convincingly shown only in man, we presently have no way of deciding whether they have evolved by way of a similar process, whether they are based on the development of handedness, or whether they are the product of a different and independent evolutionary process.

ACKNOWLEDGMENTS

This work was supported by the Deutsche Forschungsgemeinschaft through grants Pr 117/5 and Pr 117/5z and the University of Konstanz.

REFERENCES

Box, G. E. P., and Jenkins, G. M.: Time Series Analysis. Holden-Day, San Francisco. (1976).

Preilowski, B.: Is there a specialization of the cerebral hemispheres in nonhuman primates? Paper presented at the NATO Advanced Study Institute on Comparative Biology in Primates. Torino, Italy. June 7-19. (1972).

Preilowski, B.: Lack of intermanual transfer of a fine sensory-motor skill in normal rhesus monkeys. Poster presented at the European Brain and Behaviour Society Workshop on Callosal Functions. Rotterdam, The Netherlands. March 30-April 2. (1977).

Preilowski, B.: Problems with "grand schemes" and handedness. The Behavioural and Brain Sciences. 1:312-313. (1978).

Preilowski, B.: Performance differences between hands and lack of transfer of finger posture and sensory-motor skill in intact rhesus monkeys: Possible model for the origin of cerebral asymmetry. Neuroscience Letters. 3:89. (1979).

Preilowski, B.: The study of laterality in nonhuman primates: Problems and promises for an explanation of lateral cerebral asymmetries. In Preilowski and Engele. Is there a cerebral hemispheric specialization in nonhuman primates? pp. 51-72. University of Tubingen, Tubingen. (1981).

Preilowski, B., and Engele, H. C.: Is there a cerebral hemispheric asymmetry in nonhuman primates? University of Tubingen, Tubingen. (1981).

Warren, J. M.: Handedness and laterality in monkeys and humans. In Preilowski and Engele. Is there a cerebral hemispheric specialization in nonhuman primates? pp. 27-49. University of Tubingen, Tubingen. (1981).

Wiesendanger, M., and Hepp-Reymond, M. C.: Cerebrale Organisation der Motorik und Fertigkeitsbewegungen. In Battig. Psychologische Experimente. Huber, Bern. (1970).

24

Retention and Pathologic Effects After Plutonium Injection into Tupaias

A. Seidel, G. Darai, P. Moller, W. Sontag, D. M. Taylor, and W. G. Thiess

The retention of intramuscularly injected plutonium-239 citrate in the *Tupaia belangeri*, its macroscopic and microscopic distribution within the skeleton, and pathological observations are described. Resorption of 239 Pu from the injection site is rapid; the main depository organs are the skeleton and liver. About 90% of the liver burden is removed with a half-life of 130 days, whereas the half-life in the skeleton is in the range of several years. The relative gross distribution of 239 Pu within the different parts of the skeleton hardly changes with time. The nuclide is deposited on the bone surfaces with only a slow rate of redistribution. Biochemical evidence suggests that 239 Pu is bound to lysosomes in the liver. The data are compared with observations in other animal species.

INTRODUCTION

There is widespread concern regarding potential harmful effects of very small amounts of plutonium currently present in our environment and that which may be released from nuclear power plants. Detailed study of the toxicology of plutonium in humans is an important area of radiotoxicological investigation, irrespective of the decisions which may be made about the future development of nuclear-fission power programs.

Most of our toxicologic information for plutonium was derived from studies in rats and dogs. Considerable variation exists between these species, especially in the liver, one of the main organs of deposition. There are few experimental studies of the retention of systematically deposited actinides in primates (Cohen et al., 1983; Durbin et al., 1983; Metivier et al., 1976; references also in Seidel et al., 1982). The *Tupaia* belangeri, possibly the most primitive form of primate, has a reasonably long life span of 10-14 years. This and its small size make it a potentially useful species in which to study the radiotoxicology of plutonium.

These pilot studies were undertaken to provide information on the metabolism and toxicity of plutonium in Tupaias and to compare the data with those from other species.

METHODS AND MATERIALS

The studies were carried out in 20 Tupaia belangeri, 8 to 18 months of age, which were obtained from a breeding group at the University of Bayreuth.

Each animal received a single intramuscular injection of monomeric plutonium-239 citrate, plus a molar excess of citrate, injected at a constant depth into the muscles of the right thigh. For most studies the dose administered was 23 KBq/kg body weight (0.6 muCi/kg), but for subcellular binding studies approximately 30 times less activity was administered. Details concerning the determination of the radioactivity content in the organs by liquid scintillation counting, the subcellular distribution studies, and the analysis of the microdistribution in bone are given elsewhere (Seidel and Volf, 1972; Winter and Seidel, 1982; Sontag et al., 1984; Sutterlin, 1982).

For the determination of plutonium in density gradient fractions, the samples were dried and wet washed and the plutonium purified by ion exchange (on Dowex 1 by 4, 50-100 mesh) according to a slightly modified procedure described by Frindik (1980). After electroplating, the radioactivity was determined by an Ortec 576 alpha-spectrometer.

RESULTS AND DISCUSSION

Clearance from the injection site was relatively rapid and only a small percent of the dose was not resorbed from the injection site (Figure 1). As previously observed in other species, the principle organs of deposition were the skeleton and liver (Figure 2). The initial deposition was approximately 35% in the skeleton and 20% in liver. Skeletal retention was prolonged, with a biological half-life of at least 2,000 days (6 years).

Liver retention was biphasic with about 90% of the liver burden being removed with a half-life of about 130 days and the remainder being retained with an apparently infinite half-life.

Kidney retention showed a similar biphasic pattern (Figure 3). However, the rate of loss from the major compartment, 95% of the total kidney activity, was more rapid than from the liver, with a half-life of about 40

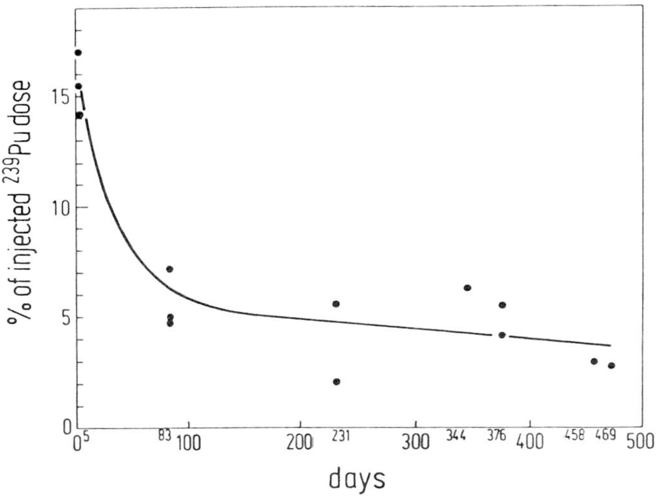

Figure 1. Retention of 239 Pu at the injection site (thigh muscle of Tupaias).

days. Uptake in spleen and lungs was generally less than 0.2% of the injected activity.

The biphasic pattern of liver retention found in the Tupaias is also found in the rat and mouse but not in dogs or Syrian or Chinese hamsters (Figure 4). The size of the initial compartment in the Tupaias is similar to that in the rat and mouse, but the half-life of retention is about 20 times longer than that in the rat. Little information is available for other primate species, but studies in baboons (Metivier et al., 1976) and macaque monkeys (Durbin, 1983) suggest that the rate of loss of plutonium from the liver may be similar to that observed in Tupaias. These observations indicate that mice, rats, Tupaias, and some higher primates show a pattern of liver retention of plutonium which is markedly different from that seen in dogs and hamsters. Human data derived from a few patients deliberately injected with plutonium in the 1940s show a wide scatter. An analysis of data on the content of fall-out plutonium in human liver may suggest that the nuclide is slowly eliminated from this organ (Taylor, 1984).

In the liver of Tupaias most plutonium is found in the mitochondria-lysosome (ML) fraction, from day 7 after injection onward. This is in good agreement with findings from four rodent species (Winter and Seidel, 1982) and from the beagle dog liver (Bruenger et al., 1976). The behavior of subcellular organelles from the liver of Tupaias after centrifugation of the ML fraction in a linear sucrose density gradient is shown in Figure 5. The profiles and their median densities are in close agreement with those from four rodent species (Winter and

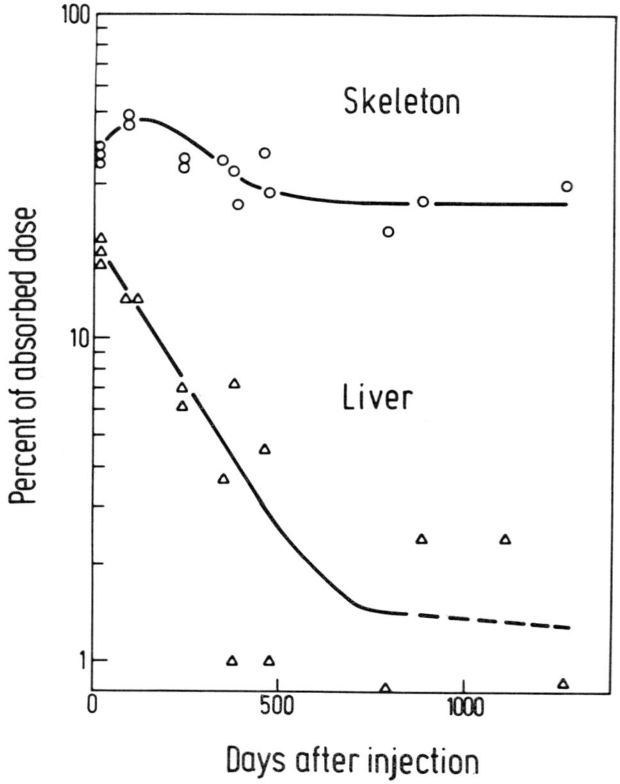

Figure 2. Retention of intramuscularly injected 239 Pu in skeleton and liver of Tupaias. Each point represents one animal. The data are expressed as percentage of that 239 Pu fraction, which had been absorbed from the injection site until the day of sacrifice.

Seidel, 1982). As with rodents, the plutonium profile at day 7 after injection overlaps with the curves of the marker enzymes, except for alkaline phosphodiesterase which was used for pericellular membranes.

Centrifugation of the ML fraction in metrizamide gradients results in considerable damage to lysosomes, as indicated by the high percentage of lysosomal enzymes in the top fractions of the gradient (Figure 6); this higher fragility of lysosomes is in contrast to the results with rodents. Nevertheless, it can be concluded that plutonium is most probably not associated with endoplasmic reticulum membranes (G-6-P) and mitochondria (GDH). These results must, of course, be corroborated by further experiments, but an association of plutonium with lysosomes has also been demonstrated in rodent liver (Winter and Seidel, 1982; Sutterlin, 1982).

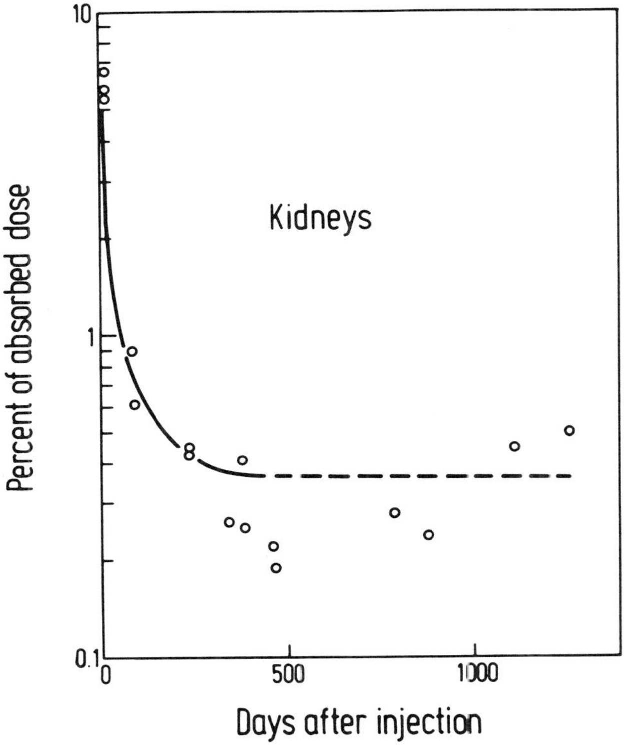

Figure 3. Retention of intramuscularly injected 239 Pu in kidneys from Tupaias. (See Figure 2).

The skeletal retention of plutonium in the Tupaias appears to be uniformly long in all bones (Figure 7) when total retention is considered. However, as has been found in other species, the distribution of plutonium within the bone is very non-uniform with a marked tendency to deposit on bone surfaces, especially on the endosteum. This is illustrated in Figure 8 which shows computer scans of neutron-induced autoradiographs of femur, tibia, humerus, and lumbar vertebrae from animals killed at 5 or 462 days post-injection. The association with the surfaces is clearly seen, as are the changes which occur due to bone remodeling and the deposition of some of the original surface deposits under newly-formed bone. Due to the very short range of the alpha -particles from 239 Pu, the change in the plutonium distribution within the bone would be expected to lead to marked changes in the microdistribution of radiation dose in the region of the radiosensitive osteoprogenitor cells which lie in close proximity to the bone surfaces. The distribution of

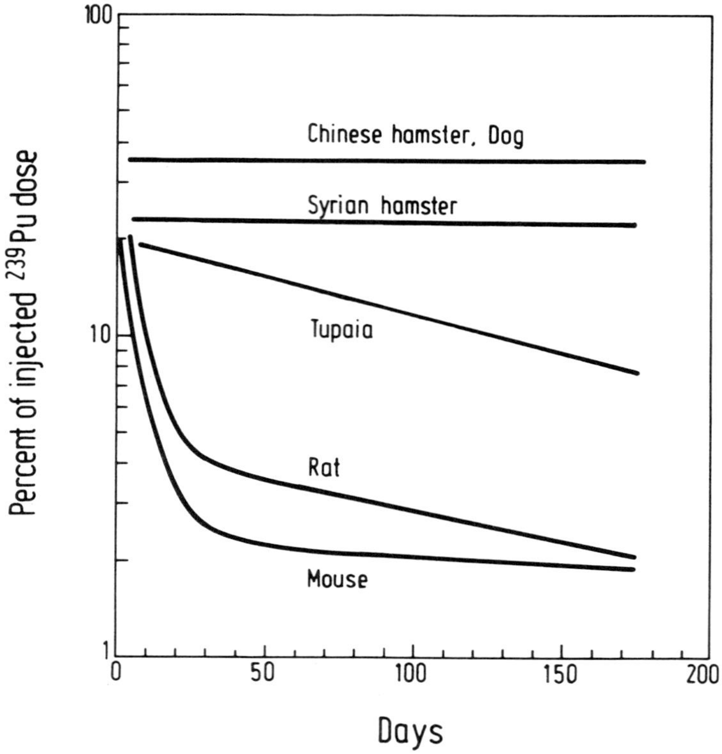

Figure 4. Comparison of the retention of 239 Pu in liver of different animal species. Schematic representation of the data is derived from the references given in Seidel et al., 1982.

radiation dose within bone can be estimated from autoradiographs using a computer coupled microscopic scanning system, which was developed at our Institute. Table 1 shows a comparison of several morphological and radiological parameters of bone from various animal species after 239 Pu injection. Comparisons of this type are of great importance for the interpretation of toxicological data, such as tumor incidence rates, and for extrapolation to humans.

Due to the small number of animals studied it is not possible to draw firm conclusions concerning the observations made in these animals. The major pathological observations are summarized in Table 2. Liver damage was seen in four animals, and one showed a hepatic carcinoma and a uterine adenocarcinoma. One osteosarcoma and a histologically unidentified cranial tumor were found in the 9 Tupaias examined 350 days or later after plutonium injection.

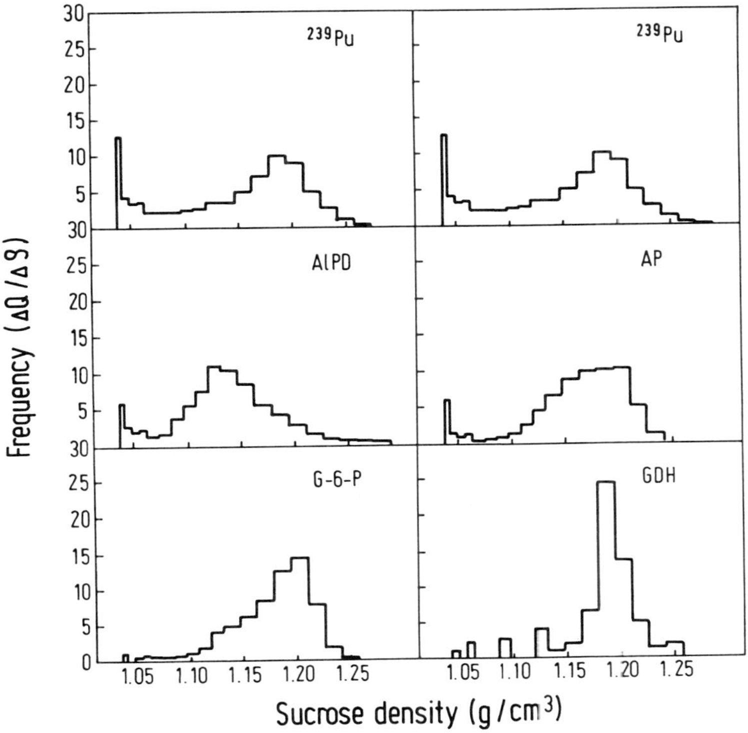

Figure 5. Profiles of 239 Pu (7 days after injection) and the marker enzymes for lysosomes (AP = acid phosphatase), pericellular membranes (AlPD = alkaline phosphodiesterase), mitochondria (GDH = glutamate-dehydrogenase), and endoplasmic reticulum membranes (G-6-P) from Tupaia liver after 4 hours centrifugation of the mitochondria-lysosome fraction at 88,000 gav in a linear sucrose density gradient. Enzyme profiles representative for five experiments, plutonium from one Tupaia. Ordinate = frequency, where ΔQ = fractional amount of constitutent in the fraction and Δp = density increment from one fraction to the other (Beaufay and Amar-Costesec, 1976).

The incidence of tumors in Tupaias maintained under laboratory conditions has been published (Hofman et al., 1981). Although spontaneous osteosarcoma was not observed, it is difficult to determine whether the osteosarcoma of the rib in our series may have been radiation-induced. However, a second osteosarcoma was recently observed during an ongoing experiment with a high plutonium dose. Two hepatocellular carcinomas were seen in the group of spontaneous tumors, which raises

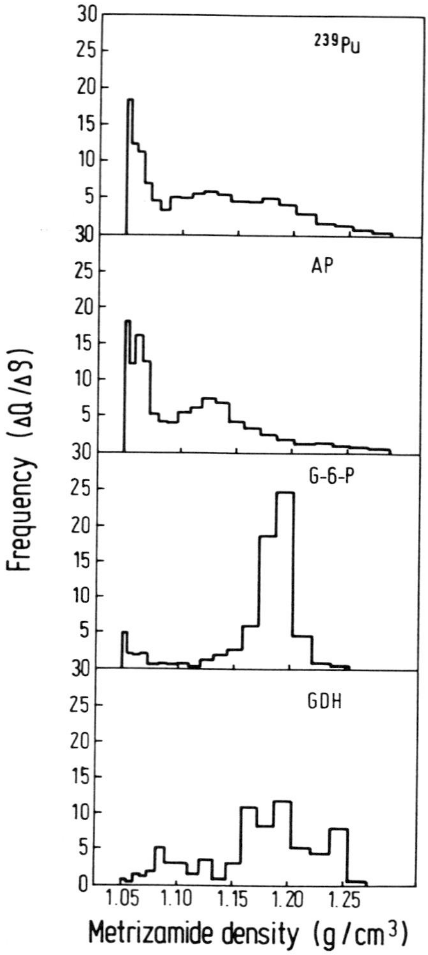

Figure 6. Profiles of 239 Pu (7 days after injection) and marker enzymes from Tupaia liver after 6-hour centrifugation of the mitochondria-lysosome fraction at 88,000 gav in a linear metrizamide gradient (for explanation, see Figure 5).

considerable doubt regarding the relationship between the hepatic malignancy observed in our series and the plutonium exposure.

In conclusion, these studies have confirmed that the metabolism of plutonium in the Tupaias has the same general pattern observed in other species. Skeletal uptake is high and prolonged, with a half-life of at least 6 years. Within the bone there are species differences in the redistribution of plutonium, but the

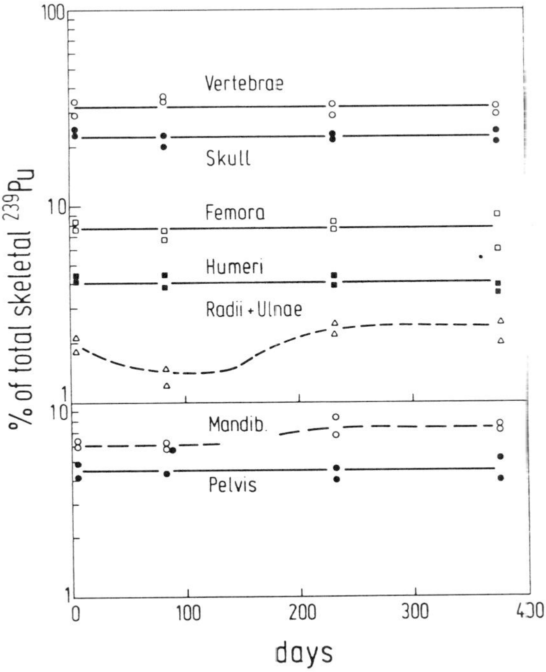

Figure 7. Retention of 239 Pu in different bones of Tupaias. Each point represents one animal.

TABLE 1. Major pathological observations in Tupaia after plutonium injection

Animal No.	Day of Sacrifice	Remarks*
13} 2 } 11}	~350	Liver weight 3 times normal, increased GPT, GOT, and g-GT concentration in blood serum
17} 1 }	~460	Extensive damage of liver parenchyma with fatty degeneration and single cell necrosis
14	790	Osteosarcoma (origin rib or sterum)
5	878	Skull tumor (type not characterized)
6	1,104	Primary polycentric malignant hepatoma plus adenocarcinoma of the uterus
12	1,265	Undifferentiated nasal carcinoma

* GPT = glutamate-pyruvate-transaminase;
GOT = glutamate-oxalate-transaminase;
g-GT = gamma-glutamyl-transferase.

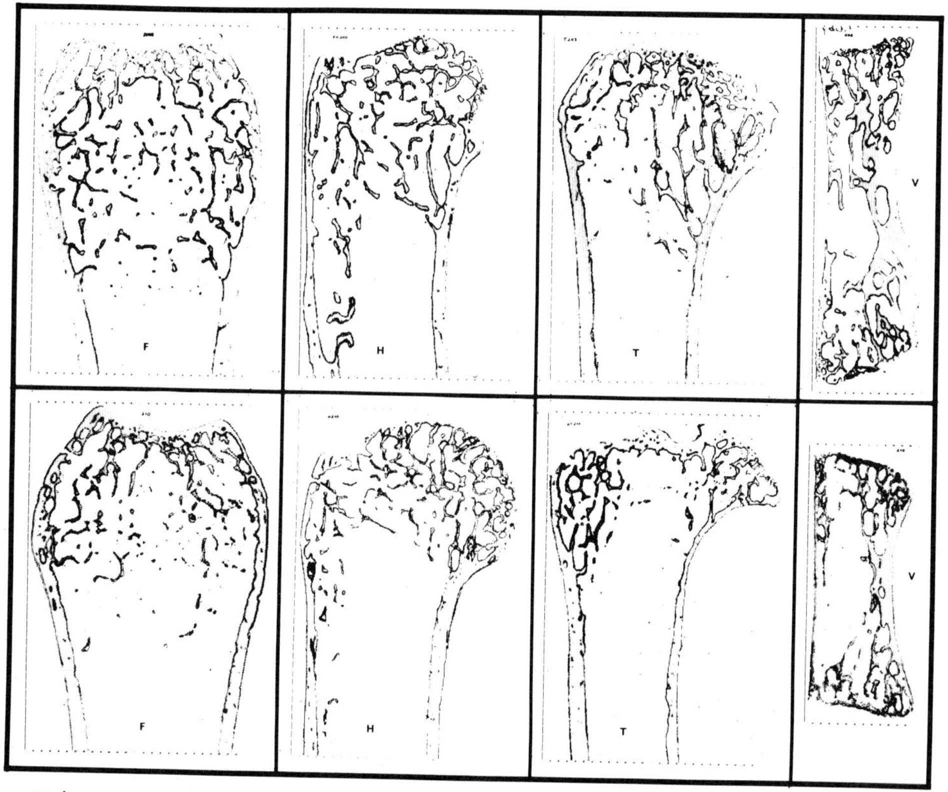

Figure 8. Computer scan of neutron-induced autoradiographs of several Tupaia bones from animals killed 5 (upper row) or 482 (lower row) days after 239 Pu injection. The black dots represents 239 Pu activity. F = femur; H = humerus; T = tibia; and V = vertebra.

TABLE 2. Morphological and radiological parameters of bone from various animal species

Animal Species	Retention Half-life (Days)	Bones	Bone Surface/ Volume-Ratio (cm^2/cm^3)	Bone Formation Rate (%/yr)
Rat	~ 400a	Femur	~160a	~ 100d
		L.vertebrae	~140a	~ 100d
Tupaia	~2000	Femur	180	≥ 83
		Humerus	180	≥ 54
		L.vertebrae	180	≥ 62
Beagle	~3000b	Humerus	200c	139c
		L.vertebrae	200c	183c

Animal Species	Endosteal surface Activity 5 Days After Injection* (Bq/cm^2)	Endosteal Surface Activity 365 Days After Injection* (Bq/cm^2)	Ratio $\frac{SA\ 365\ Days}{SA\ 5\ Days}$ (%)
Rat	~ 13.1a	~ 10.5a	~ 80
	~ 13.2a	~ 9.2a	~ 70
Tupaia	15.3	5.8	38
	17.2	15.5	90
	10.0	4.3	43
Beagle	18.5ef	6.3f	34
	18.5f	4.4f	24

* Values are normalized to 1 muCi/kg bodyweight and 50% initial deposition of skeleton. SA = surface activity.
a Unpublished data (preliminary results).
b Data from Stover et al. (1969).
c Data from Kimmel et al. (1980b).
d Data from Polig and Sontag (1981).
e Data from Kimmel et al. (1980a).
f Data from Wronski et al. (1980).

general picture in bone from Tupaias is similar to that found in adult individuals of other species. The presently available data indicate that the biological half-life of 239 Pu in the liver of Tupaias is much shorter than in the human liver. Liver retention appears to be closer to that seen in rats or mice than that found in hamsters and dogs. More detailed radiotoxicological studies would be of interest and might further establish Tupaias as appropriate models for such investigations.

ACKNOWLEDGMENTS

The authors are very grateful to Mrs. G. Kampmann for the preparation of the samples for plutonium low-level counting and to Mrs. E. Kruger for general technical assistance.

REFERENCES

Beaufay, H., and Amar-Costesec, A.: Cell fractionation techniques. In Methods in Membrane Biology. Plenum Press, New York. pp. 1-100. (1976).

Bruenger, F. W., Stevens, W., and Atherton, D. R.: The subcellular distribution of some actinide elements in the beagle liver. In Health Effects of Plutonium and Radium. J. W. Press, Salt Lake City. pp. 211-221. (1976).

Cohen, N., Burkhart, W., Lo Sasso, Th., Parkinson, W. W., and Henley, L. C.: Curium excretion studies in man and baboon: A predictive animal model. Health Phys. Suppl. 1., 44:403-409. (1983).

Durbin, P. W., Jeung, N., and Schmidt, Ch. T.: Distribution and retention of 238 Pu in Macaque monkeys. In Proceedings of the 7th International Congress of Radiation Research, Amsterdam. E5-03. (1983).

Frindik, O.: Alphaspectrometrische Methode zur Bestimmung von Plutonium und Uran in Lebensmitteln, biologischem Material und Boden. Report of Bundesforschungsanstalt fur Ernahrung, Karlsruhe, BFE-1980/6. (1980).

Hofman, W., Moller, P., Schwaier, A., Flugel, R. M., Zoller, L., and Darai, G.: Malignant tumors in Tupaia (Tree Shrew). J. Med. Primatol. 10:155-163. (1981).

Kimmel, D., Shafer, J., and Jee, W. S. S.: A semi automated method for analyzing neutron-induced autoradiographs of bone. In Annual Report of the Radiobiology Division of Utah, College of Medicine, Salt Lake City, C00-119-256. pp. 55-73. (1980a).

Kimmel, D., Smith, C., and Jee, W. S. S.: Comparison of bone turnover in various regions of trabecular bone in the young adult beagle. In Annual Report of the Radiobiology Division of Utah, College of Medicine, Salt Lake City, C00-119-256. pp. 74-89. (1980b).

Metivier, H., Masse, R., Nenot, J. C., Nolibe, D., and Lafuma, J.: Etude experimentale de la contamination par un melange issu de la combustion du sodium en presence d'oxyde de plutonium. In Diagnosis and Treatment of Incorporated Radionuclides IAEA, Vienna . pp. 107-117. (1976).

Polig, E., and Sontag, W.: Alpha-dosimetry and fluorescence label analysis in bone. In Bone and Bone Seeking Radionuclides: Physiology, Dosimetry and Effects. Harwood Academic Publisher, Brussels. pp. 73-90. (1981).

Seidel, A., and Volf, V.: Rapid determination of some transuranium elements in biological material by liquid scintillation counting. Int. J. Appl. Radiat. Isotopes. 23:1-4. (1972).

Seidel, A., Darai, G., Flugel, R., Hofmann, W., and Sontag, W.: Distribution and retention of 239 Pu in the tree shrew (<u>Tupaia belangeri</u>) following intramuscular injection. Health Phys. 43:239-246. (1982).

Sontag, W., Seidel, A., and Moller, P.: Distribution of 239-Pu in the skeleton of a tree shew (Tupaia belangeri) during the first 15 months after injection. Int. J. Radiat. Biol. 45(1):51-64. (1984).

Stover, B. S., Atherton, D. R., Bruenger, F. W., and Buster, D. S.: 239 Pu (IV): its distribution in the beagle. In Delayed Effects of Bone-seeking Radionuclides. University of Utah Press, Salt Lake City. pp. 109-123. (1969).

Sutterlin, U.: Der Einfluss der Zeit auf die subzellulare Verteilung von Transuranen und 59 Fe in der Leber von Ratte und Chinesischem Hamster. Report of Kernforschungszentrum Karlsruhe. KFK-3385. (1982).

Taylor, D. M.: The retention of plutonium and americium in liver: An interspecies comparison. In Proceedings of the 6th International Congress of the International Radiation Protection Association, Berlin, 1984.

Winter, R., and Seidel, A.: Comparison of the subcellular distribution of monomeric 239 Pu and 59 Fe in the liver of rat, mouse and Syrian and Chinese hamster. Radiat. Res. 89:113-123. (1982).

Wronski, T. J., Smith, J. M., and Jee, W. S. S.: The microdistribution and retention of injected 239 Pu on trabecular bone surfaces of beagle: Implications for the induction of osteosarcoma. Radiat. Res. 83:74-89. (1980).

25

Natural Trypanosome Infection in Neotropical Monkeys With Special Reference to *Saimiri sciureus*

B. L. Travi, O. J. Colillas, and E. L. Segura

Serological screening for *Trypanosoma cruzi* infection was carried out in 127 *Saimiri sciureus*, 40 *Cebus apella*, and 30 *Alouatta caraya*. *S. sciureus* had the highest prevalence of infection (43%), followed by *C. apella* (20%) and *A. caraya* (13%) according to the rapid flagellar agglutination test. Xenodiagnosis was the most sensitive diagnostic method (65% positive) when compared with complement fixation (51%), indirect hemagglutination (45%), and indirect immunofluorescence (21%) in a randomly selected group of *S. sciureus*. Nine of the 11 *T. cruzi*-like strains isolated by xenodiagnosis from squirrel monkeys grew in culture media with patterns of development similar to those of human *T. cruzi* isolates. Statistical analysis of the reproductive efficiency in a randomly selected group of female *S. sciureus* showed that the *T. cruzi* infected group had significantly higher neonatal mortality ($p < 0.005$) and lower total reproductive efficiency ($p < 0.005$) by the chi-square method.

INTRODUCTION

In Central and South America, nonhuman primates are parasitized by several species of trypanosomes of uncertain classification and their biological properties have not been studied completely (Dunn et al., 1963). Since the early description by Chagas (1909) of *Trypanosoma minasense* infecting a marmoset, subsequent reports have shown that nearly all species of neotropical primates can be naturally infected with these protozoan flagellates (Dunn et al., 1963; Marinkelle, 1966; Hoare, 1972).

The fact that monkeys share *T. rangeli* and *T. cruzi* with humans (Marinkelle, 1966; Hoare, 1972; Sousa et al.,

1974) has medical and epidemiological significance. Despite considerable literature on the subject, it is still difficult to answer most of the basic questions pertaining to this subject. For this reason, the present work seeks to contribute information on some aspects related to natural trypanosome infections in nonhuman primates.

We surveyed different species of monkeys kept at Centro Argentino de Primates, Corrientes Province, Argentina, (CAPRIM) and free ranging _Alouatta caraya_ in the surrounding area for the prevalence of natural infections. In addition, diagnostic studies were conducted on a group of squirrel monkeys. The results of these studies are reported herein.

METHODS AND MATERIALS

Species of monkeys

Saimiri sciureus and _Cebus apella_ were wild-caught monkeys from Bolivia and Paraguay, respectively, that were housed at CAPRIM while the _Alouatta caraya_ were free-ranging monkeys inhabiting Corrientes province and nearby islands in the Parana river.

Serology

The prevalence of _T. cruzi_ infection in the different species was determined by the rapid flagellar agglutination (RFA) method and by the indirect hemagglutination (IH) technique, both developed at Instituto Nacional de Diagnostico e Investigacion de la Enfermedad de Chagas, Buenos Aires, Argentina, (INDIECH), (Lansetti, 1980). RFA antigen is an enriched flagellar fraction obtained by rupturing _T. cruzi_ parasites in a tissue microhomogenizer, followed by differential centrifugation. In the IH test, type O human red blood cells are used as an antigen support, and the antigenic fraction is derived from a supernatant of homogenized _T. cruzi_ epimastigotes centrifuged at 30,000g.

In the comparison of different diagnostic methods, the IH technique was tested simultaneously in 37 _S. sciureus_ with the complement fixation (CF) and indirect immunofluorescence (IIF) tests according to methodologies employed in human diagnosis (Cerisola, 1958; Alvarez, 1968). The xenodiagnosis was accomplished using 10 laboratory-reared, third-instar _Triatoma infestans_ per monkey. The reduviid bugs were allowed to feed on the abdomen of anesthetized monkeys for 20 minutes. Thirty and 60 days later the feces of the _T. infestans_ were examined for the presence of the metacyclic stage of the trypanosomes.

Trypanosome culture

Eleven *T. cruzi*-like strains were obtained by xenodiagnosis on *S. sciureus*. The trypanosomes present in the feces of the *T. infestans* were seeded in a diphasic culture medium composed of nutrient agar (Oxode Lab) with 10% defibrinated sheep blood, covered with Locke's solution. Incubation was carried out at 30°C for a 6-week period.

Experimental inoculation

Five *T. cruzi*-like strains were inoculated intraperitoneally into Swiss mice, 6 animals per group, using fecal suspensions obtained from infected bugs. Every 3 days samples of tail blood were examined microscopically for the presence of parasites and to estimate the parasitemia level. Between 40 and 110 days postinfection the mice were sacrificed, and tissue sections stained with hematoxylin and eosin were prepared from the heart, spleen and liver.

Reproductive efficiency

Forty-five multiparous females were randomly selected and separated into groups according to whether they were positive or negative by xenodiagnosis. Subsequently, the reproductive records of these females (Table 1) were statistically analyzed by the chi-square method.

TABLE 1. Reproductive record of 45 multiparous *Saimiri sciureus* females

	Negative by Xenodiagnosis	Positive by Xenodiagnosis	Total
Females	18	27	45
Opportunities*	50	80	130
Pregnancies**	32 (64%)	54 (67%)	86
Abortions***	11 (34%)	30 (55%)	41
Live births	21 (66%)	24 (44%)	45
Neonatal mortality#	3 (9%)	14 (26%)	17
Weaning	14 (44%)	5 (9%)	19

* Breeding opportunities for the whole group.
** Diagnosed by transabdominal palpation.
*** Absence of pregnancy with further palpations or finding of placental or fetal remains.
\# Between 0 and 10 days postpartum.

TABLE 2. Serological screening for Trypanosoma cruzi infection in New World monkeys

	Indirect Hemagglutination % (Pos./Total)	Rapid Flagellar Agglutination % (Pos./Total)
Saimiri sciureus	31 (39/127)	43 (46/107)
Cebus apella	5 (2/40)	20 (8/40)
Alouatta caraya	3 (1/30)	13 (4/30)

RESULTS

Serological screening showed that S. sciureus was the species with the highest prevalence of trypanosome infection (Table 2).

Among the different diagnostic methods, xenodiagnosis was the most sensitive technique followed by CF, IH, and IIF tests (Table 3). Although the sensitivity of the IIF test is poor, it had the highest concordance with xenodiagnosis (Table 3). Nine of the 11 strains isolated by xenodiagnosis readily grew in the culture media with patterns of development and morphological characteristics similar to those of human T. cruzi isolates. The appearance of typical rosettes and active epimastigotes were evident around day 7 (Figure 1).

All 5 strains inoculated into mice were infective, four of them becoming patent in 11-13 days and the fifth on day 63. In all cases the parasitemia was low and there was no animal mortality (Table 4).

Histological sections of different organs did not reveal amastigote nests, but in one mouse sacrificed 110 days postinfection a clear leukocytic infiltration of the myocardium was observed (Figure 2).

The statistical analysis of the reproductive potential among 45 multiparous females showed there was no difference in pregnancies and abortions between the uninfected and parasitized monkeys. On the other hand, neonatal mortality was significantly higher ($p < 0.005$)

TABLE 3. Comparison of diagnostic methods for Trypanosoma cruzi infection in Saimiri sciureus (n = 37)

Method	% Positive Monkeys	% Concordance with Xenodiagnosis
Xenodiagnosis	65	-
Complement fixation	51	79
Indirect hemagglutination	45	70
Indirect immunofluorescence	21	88

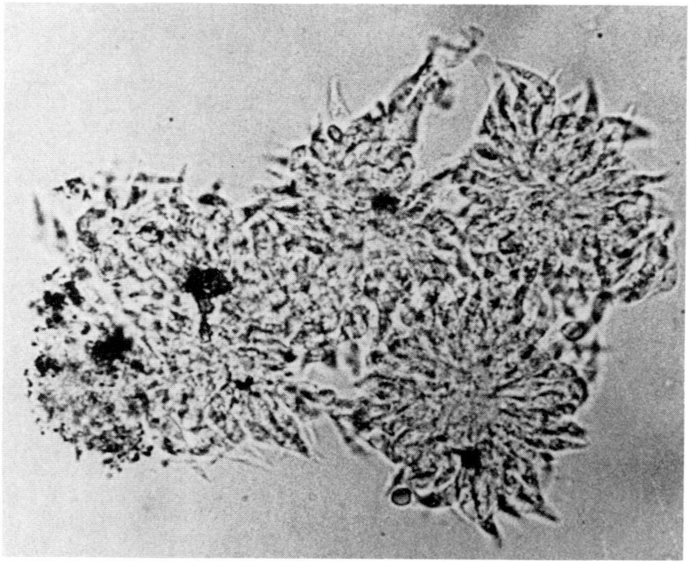

Figure 1. *T. cruzi* rosette formation in a culture isolated from *S. sciureus* (magnification x 750).

in the *T. cruzi* infected group. Furthermore, when all the reproductive data were tested together, they showed that the total reproductive efficiency was significantly lower (p < 0.005) in the *T. cruzi* infected females.

DISCUSSION

The serological results showed that squirrel monkeys of Bolivian origin from the Sta. Cruz de la Sierra region are frequently exposed to trypanosomes in their natural habitat. In this area 90.4% of the human habitations are infested with reduviid bugs, 70.5% of them with *T. cruzi* (Jauregui and Valdivia, 1972). These epidemiological data suggest a close relationship between the sylvatic and domiciliary cycle of *T. cruzi*.

Xenodiagnosis was the most effective method in detecting infected animals when compared with the serological techniques. It also has the advantage of isolating the different strains for further taxonomic or biological studies. From a practical point of view, it has the disadvantage of being an uncommon diagnostic method in those countries where Chagas' disease is not a medical problem.

It is believed that the low sensitivity of the IIF test was due to the utilization of human anti-gammaglobulin, and that future tests with monkey antisera could greatly improve this method for screening purposes.

TABLE 4. Comparison of different _Trypanosoma cruzi_-like strains from neotropical monkeys experimentally inoculated into mice

	Age of Mice (Days)	Infectivity %	Mortality %	Prepatent Period (Days)	Tissue Sections	Natural Host
Albuquerque and Barretto (1969)	20	100	63	8.4	Amastigotes in heart and other organs	_Callithria penicillata jordani_
Albuquerque and Barretto (1969)	20	100	26.7	11.3	Amastigotes in heart	_Callicebus nigrifrons_
de Melo Rocha and Barretto (1969)	?	100	6.7	7.3	Few amastigotes in several organs	_Callithrix geoffroyi_
Deane (1962)	20	pos	?	24	Amastigotes in heart and other organs	_Callithrix jacchus_
Funayama, et al. (1969)	20	100	63.3	5.3	Amastigotes in heart and other organs	_Alouatta caraya_
Dunn, et al. (1963)	1-2	100	100		Amastigotes in heart and other organs	_Tamarinus nigricollis_
	14-21	62	50	?		
Present Study	10	100	0	13	negative	_Saimiri sciureus_ M832
	10	100	0	13	negative	F769
	10-28	100	0	11	negative	M773
	14	50	0	63	leukocytic infiltration	F751
	7	100	0	11	negative	M733

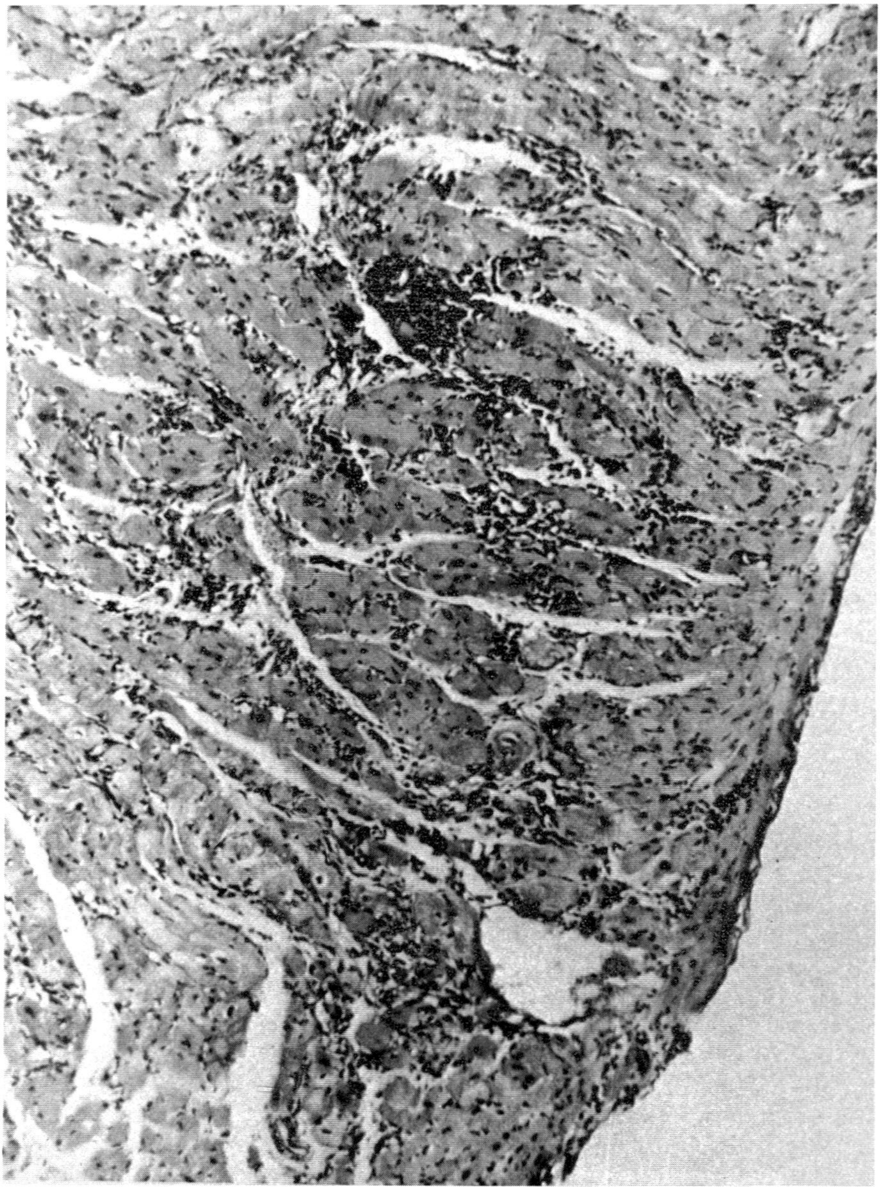

Figure 2. Leukocytic infiltration of the myocardium in a Swiss mouse inoculated with a *T. cruzi* strain isolated from *S. sciureus* (magnification x 139).

The low parasitemia of the mice observed in this and other studies (Marinkelle, 1982) stresses the necessity of a previous parasite adaptation to the new host through serial passages before evaluating its virulence. This has been done with T. cruzi isolates from chagasic human patients (Gonzalez Cappa et al., 1980). The irregular behavior of nonadapted T. cruzi isolates is exemplified by the strain isolated from monkey F751 (Table 4). When inoculated into mice it exhibited a lower infectivity and longer prepatent period than the other strains, but it did produce a leukocytic infiltration of the myocardium.

Clinical evaluation in a randomly selected group of S. sciureus is still in progress at CAPRIM. Partial left bundle branch block was found in 11.6% (5/43) of the specimens by means of electrocardiography. Although no extensive studies were done, the present data suggest that the clinicopathological significance of T. cruzi infection may be similar to that reported for humans (Laranja et al., 1956).

It is unknown whether the prevalence of cardiac alterations is a reflection of animal susceptibility or the pathogenicity of the T. cruzi strain. In the last decade interest in the biological behavior of different T. cruzi strains has increased, particularly those that produce a specific chagasic pathology such as megaviscera or heart disease (Melo and Brener, 1978). Efforts to characterize trypanosomes by means of their isoenzymic profiles (Miles et al., 1980, 1981) and k-DNA restriction endonuclease fingerprints (Morel et al., 1980) are in progress, and it is likely that, in the future, animal susceptibility will be tested with well standardized T. cruzi strains.

The statistical analyses showed that T. cruzi infections resulted in an impairment of the reproductive potential in squirrel monkey females. Although the risk of abortion seems to be higher in the positive xenodiagnosis group (55% versus 34% for the uninfected group), this difference is not significant by the chi-square method. Early pregnancy is similar in both groups, but there is an increased neonatal mortality in the offspring delivered by infected females. This implies that T. cruzi does not interfere with the early placentation, but it is speculated that it may produce a progressive placental insufficiency causing premature delivery and low birth weight. A case of congenital infection resulting in late-term abortion in a female marmoset (Lushbaugh et al., 1969) supports this hypothesis and is in agreement with observations made in humans that interruption of pregnancy occurs at a well advanced stage (Bittencourt and Barbosa, 1972). Histologic studies of the placenta and fetus of S. sciureus at different periods of pregnancy are planned to investigate this particular feature of chagasic pathology. There are multiple reasons for detecting T. cruzi infections in monkeys, many of which have been previously stated (Cunn et al., 1963; Marinkelle, 1966;

Sousa et al., 1974; Travi et al., 1982; Eberhard and D'Alessandro, 1982). The most important points are those related to zoonotic risks and potential discrepancies of results in biological studies. These problems cannot be avoided simply by working with laboratory-born monkeys because, although they may appear healthy, there is the possibility of congenital infection (Eberhard and D'Alessandro, 1982).

Like any other major parasitic infection, T. cruzi-infected squirrel monkeys should be treated or removed from breeding colonies in order to improve reproductive efficiency. These measures will diminish the number of feral monkeys needed for research.

The present work shows that S. sciureus is a suitable animal model for Chagas' disease that encompasses, in natural infections, several important features of American trypanosomiasis.

ACKNOWLEDGMENTS

The authors wish to express their gratitude to Dr. J. Lazzari for the interpretation of the electrocardiograms and to Dr. R. C. Lowrie, Jr., and Dr. M. L. Eberhard for reviewing the manuscript. This study was supported by funds provided by CONICET (Consejo Nacional de Investigaciones Cientificas y Tecnicas) and SECYT (Secretaria de Estado de Ciencia y Technologia), Argentina.

REFERENCES

Alvarez, M., Cerisola, J. A., and Rohweder, R. W.: Test de immunofluorescencia para el diagnostico de la enfermedad de Chagas. Bol. Chileno de Parasitol. 23:4-7. (1968).

Bittencourt, A. L., and Barbosa, H. S.: Incidencia da transmissao congenita da doenca de Chagas em abortos. Rev. Inst. Med. Trop. Sao Paulo. 14:257-259. (1972).

Bullock, B. C., Wolf, R. H., and Clarkson, T.B.: Myocarditis associated with trypanosomiasis in a cebus monkey (Cebus albifrons). J. Am. Vet. Med. Assoc. 151:920-922. (1967).

Cerisola, J. A., and Rosenbaum, M. B.: La reaccion de fijacion de complemento para el diagnostico de la enfermedad de Chagus I. Tecnica Prensa Med. Arg. 45:1551-1553. (1958).

Chagas, C.: Neue Trypanosomen. Arch. Schiffs Tropen. Hyg. 13:120-122. (1909).

Dunn, F. L., Lambrecht, E. L., and du Plessis, R.: Trypanosomes of South American monkeys and marmosets. Am. J. Trop. Med. Hyg. 12:524-534. (1963).

Eberhard, M, and D'Alessandro, A.: Congenital Trypanosoma

cruzi infection in a laboratory-born squirrel monkey, *Saimiri sciureus*. Am. J. Trop. Med. Hyg. 31:931-933. (1982).

Gonzalez Cappa, S. M., Chiale, P., Del Prado, G. E., Katzin, A. M., De Martini, G. W., De Isola, E. D., Abramo Orrego, L., and Segura, E. L.: Aislamiento de una cepa de *Trypanosoma cruzi* de un paciente con miocardiopatia chagasica cronica y su caracterizacion biologica. Medicina (Buenos Aires) 40 (Supl. 1): 63-68. (1980).

Hoare, C. A.: The Trypanosomes of Mammals. A Zoological Monograph. Blackwell, Oxford. (1972).

Jauregui, L., and Valdivia, C. J.: Epidemiologia de la enfermedad de Chagas en Bolivia. Simposio Internacional de la Enfermedad de Chagas, Bs. As. 171-177. (1972).

Lansetti, J. C., Giordano, A. D., Subias, E., and Segura, E. L.: Reacciones serologicas de descarte de la infeccion chagasica. Medicina (Bs. As.) 40 (Supl. 1):258-259. (1980).

Laranja, F. S., Dias, E., Nobrega, G., and Miranda, A.: A clinical, epidemiologic and pathologic study. Circulation. 14:1035-1060. (1956).

Lushbaugh, C. C., Humason, G., and Gengozian, N.: Intrauterine death from congenital Chagas' disease in laboratory-bred marmosets (*Saguinus fuscicollis lagonotus*). Am. J. Trop. Med. Hyg. 18:662-665. (1969).

Marinkelle, C. J.: Observations on human, monkey and bat trypanosomes and their vectors in Colombia. Trans. Roy. Soc. Trop. Med. Hyg. 60:109-116. (1966).

Marinkelle, C. J.: The prevalence of *Trypanosoma (Schizotrypanum) cruzi cruzi* infection in Colombian monkeys and marmosets. Ann. Trop. Med. Parasitol. 76:121-124. (1982).

Melo, R. C., and Brener, Z.: Tissue tropison of different *Trypanosoma cruzi* strains. J. Parasitol. 64:475-482. (1978).

Miles, M. A., Lanham, S. M., de Souza, A. A., and Povoa, M.: Further enzymic characters of *Trypanosoma cruzi* and their evaluation for strain identification. Trans. Roy. Soc. Trop. Med. Hyg. 74:221-237. (1980).

Miles, M. A., Povoa, M. M., de Souza, A. A., Lainson, R., Shaw, J. J., and Ketteridge, D. S.: Chagas' disease in the Amazon Basin: II. The distribution of *Trypanosoma cruzi* zymodemes 1 and 3 in Para State, north Brazil. Trans. Roy. Soc. Trop. Med. Hyg. 75:667-674. (1981).

Morel, C., Plessmann, C. E., Camargo, E., Mattei, D. M., Romanha, A. D., and Simpson, L.: Strains and clones of *Trypanosoma cruzi* can be characterized by pattern of restriction endonuclease products of kinetoplast DNA minicircles. Proc. Nat. Acad. Sci. USA 77:6810-6814. (1980).

Sousa, O. E., Rossan, R. N., and Baerg, D. C.: The prevalence of trypanosomes and microfilariae in

Panamian monkeys. Am. J. Trop. Med. Hyg. 23:862-868. (1974).

Travi, B. L., Ruiz, A. M., Colillas, O. J., and Segura, E. L.: Trypanosomiasis en monos neotropicales - I: prevalencia de la infeccion en colonias de cria y vida salvaje. Medicina (Bs. As.) 42:55-60. (1982).

26

Procurement and Standardization of Nonhuman Primates for Biomedical Research in South Africa

D. J. Goosen, M. W. Odendaal,
H. Van Der Laarse, and E. Alberts

The close relationship of nonhuman primates to man makes them the ideal research animal for much biomedical research. Because South Africa has large, natural populations of primates including the chacma baboon (*Papio ursinus*) and the vervet monkey (*Cercopithecus pygerythrus*), there is a ready supply of these species for use in essential biomedical researches, and indeed, these two species (which are pest animals in most areas of the country) have been used extensively in the laboratory. While there seems to be little correlation between availability and demand, 2078 vervets and 2262 baboons were used in medical research in South Africa from 1979 to 1980. During the period reported here, these primates have been used in a wide variety of researches, including transplantations, orthopedics, urology, endocrinology, virology, vaccine production, tropical disease, nutrition, dental and pharmacology.

METHODS AND MATERIALS

Two methods of capturing these nonhuman primates have been used with success: trapping with cages and immobilization through darting (Goosen et al., 1980). Trapping is the most effective method for capturing animals which are wild and not used to the presence of humans. Many types of cages with a guillotine door, used with favored food items as bait, work equally well. The effectiveness of this method can be enhanced, if first the troop of primates is familiarized with the cages and bait items before setting the trap doors. The major disadvantage in this method is that it allows little control over which animals are trapped.

Darting, on the other hand, allows one to obtain precisely the class of animal desired, but works only in situations where the animals are habituated to human presence sufficiently to allow the close-range use (within 20 meters) of a gas operated pistol or blow pipe. Our experience with this method is that a dose of Ketamine of 20-25 mg/kg will immobilize both baboons and vervets for 5 to 10 minutes.

DISCUSSION

After capture, animals are quarantined to produce an acclimatized, healthy and to the extent possible, standardized research animal. Careful attention is given to disease entities during this period, such as tuberculosis, herpes viruses, Marburg virus, hepatitis, etc. (See Whitney et al., 1973).

As the animals are usually obtained from free-ranging troops, they are allowed to stabilize to the laboratory environment for one week, during which time they are subjected to a daily sanitation program, a new diet and human contact. After this time, they are anesthetized and given a thorough examination. This clinical examination includes: oral exam, identification, TB testing (palpebral method) and the taking of a number of samples for laboratory analysis. These analyses include: hematology profiles, biochemistry (liver, kidney electrolites, proteins, etc.), serology for viruses such as polio, rubella, influenza, herpes, hepatitis, rubeola and chickungunya, bacteriology of rectal and throat swabs, especially for salmonella, escherischia coli, klebsiella, proteus mirabilis, and shigella.

After this first sampling, the animals are dewormed with a broad spectrum antihelminitic, given a vitamin B-complex injection and sprayed with a fungicide and organic phosphate dip. Six weeks later these sampling procedures are repeated, and the final laboratory and clinical results evaluated. In hopes that some of the data we have obtained from freshly captured, wild baboons and vervets may be of value to researchers elsewhere, we provide a compendium of these values in Tables 1 through 5.

TABLE 1. Blood Biochemistry

TEST	VERVET MONKEY n=21		CHACMA BABOON n=21		HUMAN Range	UNIT
	MEAN	SD	MEAN	SD		
SODIUM	153,23	5,37	146,4	5,2	146-139	mmol/1
POTASSIUM	3,66	0,84	4,26	1,58	5,20-3,70	mmol/1
CHLORIDE	110,19	10,06	106,7	3,5	107,00-99	mmol/1
CO_2	20,39	7,0	20,5	7,55	30-23	mmol/1
UREA	8,12	2,11	5,4	1,3	7,40-3,40	mmol/1
URATE	0,02	0,0068	0,01	0,0078	0,52-0,25	mmol/1
CREATININE	138,05	209,92	111,6	99,5	125-60	mmol/1
PHOSPHATE	1,91	0,27	1,7	0,51	1,40-0,80	mmol/1
PHOSPHATASE (ALK)	170,70	116,03	402,72	338,4	132-20	IE/L
PROTEIEN (TOT)	68,7	5,42	70,3	3,42	79-66	g/1
ALBUMINE	45,1	4,85	42,18	4,15	54-43	g/1
BILIRUBINE (TOT)	4,26	1,71	3,39	1,2	32-7	u mol/1
CALSIUM	2,06	0,62	2,34	0,117	2,50-2,20	mmol/1
MAGNESIUM	0,7	0,06	1,22	0,85	0,95-0,80	mmol/1
GLUCOSE	6,04	1,18	8,94	4,56	6,20-3,50	mmol/1
CHOLESTROL	4,24	0,95	2,54	0,64	7,20-3,50	mmol/1
TRIGLISIRIDE	0,91	0,43	0,77	0,54	2,00-0,60	mmol/1
CPK	164,57	98,48	389,42	255,5	165-45	IE/L
LDH	699.57	505,01	138,57	75,8	260-150	IE/L
GMT	40,14	12,01	53,16	10,34	40-8	IE/L
AMILASE	292	101,35	387,85	280,51	121-23	E/L

TABLE 2. Hematology

TEST	VERVET MONKEY n=21		CHACMA BABOON n=21		HUMAN Range	UNIT
	MEAN	SD	MEAN	SD		
RBT	5,96	0,99	5,698	0,31	6,2-4,2	x 10^{12}/1
WBC	7,91	2,2	8,69	2,26	10,8-4,0	x 10^{9}/1
HT	49,81	6,19	41,3	3,33	52-37	%
HB	14,91	2,07	14,73	1,05	18-12	g/dl
NEUTROPHILS	41,58	14,18	39,1	10,96		%
LYMPHOCYTES	57	14,76	49,9	14,76		%
MONOCYTES	0,5	0,79	8	11,42		%
EOSINOPHILS	0,41	1,16	5,5	3,62		%
BASOPHILS	0,08	0,28	1,1	1,44		%

TABLE 3. Viral Serology (July 1981 - July 1982)

VIRUSSE	VERVET MONKEY	CHACMA BABOON	POSITIVE	NEGATIVE	ANTI COMPLEMENTORY
HEPATITISA	5	97	31	71	0
HERPES 1	3	138	99	42	0
HERPES 2	0	120	67*	53	0
POLIO 1	5	73	38	40	0
POLIO 2	5	73	47	31	0
POLIO 3	5	73	55	23	0
CHIKUNGUNYA	5	96	27**	74	0
RUBELLA	3	140	31	12	0
HbSAG	0	43	0	43	0
MEASLES	8	132	16***	43	81
MYCOP. PNEUM	8	118	6***	37	83
MUMPS	8	77	11***	23	51
INFLUEN 2	0	71	1	21	49
P-INFLUEN	0	71	1	21	49
ADENO.	0	71	1	21	49
PSITTAK.	0	71	1	21	49
Q-FEVER	0	71	0	22	49
RESP. SYW.	0	71	2	20	49
	55	1606	434	718	509

* Titre of 1/10; ** Titre of 1/20; *** Titre of 1/16

TABLE 4. Number of organisms isolated from throat and rectal swabs from P. ursinus and C. pygerythrus (July 1981 - July 1982)

	P. ursinus	C. pygerythrus
Throat	158	62
Rectal	131	65
Other	26	-
TOTAL	329	127

TABLE 5. Organisms

Gram Positive Cocci	Throat	Rectal	Total
Staphylococcus aureus	53	11	64
Staphylococcus epidermis	7	1	8
Streptococcus Lancefield Group A	1		1
Streptococcus Lancefield Group D	7	2	9
Streptococcus Lancefield Group F	0	3	3
Streptococcus Lancefield Group G	6	8	14
Streptococcus dysgalactiae	3	1	4
Streptococcus equisimilis	1	0	1
Streptococcus fecalis	2	1	3
Streptococcus salivarius	4	0	4
Streptococcus bovis	6	7	13
Streptococcus sangius	1		1

REFERENCES

Goosen, D. J., and Alberts, E.: Annual Report. H. A. Grove Animal Research Center. p. 17. (1980).

Whitney, R. A., Johnson, D. J., and Cole, W. C.: Laboratory Primate Handbook, First Edition. pp. 25-27. Academic Press, New York. (1973).

INDEX